Phycology-Based Approaches for Wastewater Treatment and Resource Recovery

Phycology-Based Approaches for Wastewater Treatment and Resource Recovery

Edited by

Pradeep Verma and Maulin P. Shah

CRC Press
Taylor & Francis Group
Boca Raton London New York

CRC Press is an imprint of the
Taylor & Francis Group, an **informa** business

First edition published 2022
by CRC Press
6000 Broken Sound Parkway NW, Suite 300, Boca Raton, FL 33487-2742

and by CRC Press
2 Park Square, Milton Park, Abingdon, Oxon, OX14 4RN

CRC Press is an imprint of Taylor & Francis Group, LLC

ISBN: 9780367726447 (hbk)
ISBN: 9780367726454 (pbk)
ISBN: 9781003155713 (ebk)

DOI: 10.1201/9781003155713

Typeset in Times LT Std
by KnowledgeWorks Global Ltd.

Contents

Preface

Phycology, also known as "algology", is defined as the study of a large heterogeneous group of aquatic organisms having characteristics of plants. It ranges from prokaryotic (cyanobacteria) to eukaryotic (phytoplankton and microalgae) to multicellular and large organisms (seaweeds). Algae are a direct source of food material for several organisms. They are also used as a secondary source of food supplements for humans. Algae are also a great source of commercially viable products such as agar, iodine, etc. Algae are capable of growing in freshwater, salt water and even wastewater where it derives its nutrients for its growth by autotrophic as well as heterotrophic mode. Algae are capable of growing in normal conditions as well as harsh conditions; thus, they are often suggested as a great source for wastewater treatment.

Water is one of the essential components for survival of life on earth. However, anthropogenic activities result in utilization of freshwater and generate a huge amount of effluents and wastewater. Earlier, considering the enormous availability of water, people cared less about water preservation, leading to its scarcity, which was further exacerbated by increasing population. Wastewater is considered a rich source of essential chemicals and energy. Thus, it's been suggested to treat wastewater for its utilization in various anthropogenic activities.

Several chemical and physical approaches have been suggested for wastewater treatment. Recently, biological treatment and algal/phycology-based approaches have been gaining the interest of the scientific community. This book is an attempt to highlight some of the advanced algal-based technologies developed or under consideration for wastewater treatment. It will also highlight the opportunities that existing technologies can provide at an industrial scale. This book integrates information in the field of algal-based wastewater treatment and resource recovery approaches from academicians, researchers, and industrialists around the globe for the translation of ideas into practice. The book consists of chapters highlighting major research findings on the application of wastewater as a growth medium for microalgae, seaweeds and cyanobacteria. Several chapters highlight the critical evaluation of phycology-based approaches for the removal of chemicals, heavy metals and organic wastes from domestic, pharmaceutical wastewater and industrial effluents along with simultaneous resource recovery. Several chapters emphasize the commercialization potential of phycology-based wastewater treatment and resource recovery via microalgal-based biorefinery. Chapters contributed by several scientists working in industrial venues highlight limitations associated with existing and recently developed technologies during large-scale applications. The book was developed with an objective to be useful inside and outside the laboratory, i.e., for students, researchers (especially biochemists, environmentalists, and biotechnologists) as well as engineers and policy makers. In addition, the book has the potential to garner a huge readership among environmentalists, biotechnologists, and academicians.

Pradeep Verma
Maulin P. Shah

Acknowledgments

First of all, we would like to convey our gratitude toward CRC Press/Taylor & Francis Group, for accepting our proposal to publish this book. We convey our heartfelt gratitude toward all the researchers and academicians who have contributed chapters to the book. We are thankful for all the technical support provided by Mr. Bikash Kumar (PhD Scholar, Department of Microbiology, Central University of Rajasthan, Ajmer) during the compilation and formatting of the book chapters. Prof. Pradeep Verma is thankful to Central University of Rajasthan (CURAJ), Ajmer, India, for providing infrastructural support, and a suitable teaching and research environment. CURAJ provided him with the necessary milieu conducive to the needs of academicians, students, and researchers and was greatly helpful during the development of the book. Prof. Pradeep Verma is also thankful to the Department of Biotechnology (BT/304/NE/TBP/2012 and BT/PR7333/PBD/26/373/2012) for providing the funds through sponsored projects, for setting up the Bioprocess and Bioenergy Laboratory at CURAJ.

We are always thankful to God and our parents for their blessings. We also express our deep sense of gratitude to our families for their support during the development of this book and in life.

About the Editors

Prof. Pradeep Verma, PhD, earned his PhD from Sardar Patel University, Gujarat, India, in 2002. In the same year, he was selected as UNESCO fellow and joined Czech Academy of Sciences, Prague, Czech Republic. He later moved to Charles University, Prague, to work as a Post-Doctoral Fellow. In 2004, he joined as a visiting scientist at UFZ Centre for Environmental Research, Halle, Germany. He was awarded a DFG fellowship which provided him another opportunity to work as a Post-Doctoral Fellow at Gottingen University, Germany. He returned to India in 2007 where he joined Reliance Life Sciences, Mumbai, and worked extensively on biobutanol production which attributed a few patents to his name. Later he was awarded with JSPS Post-Doctoral Fellowship Programme and joined Laboratory of Biomass Conversion, Research Institute of Sustainable Humanosphere (RISH), Kyoto University, Japan. He is also a recipient of various prestigious awards such as Ron-Cockcroft award by Swedish society, and UNESCO Fellow ASCR Prague.

Prof. Verma began his independent academic career in 2009 as a Reader and Founder Head of the Department of Microbiology at Assam University. In 2011, he moved to Department of Biotechnology at Guru Ghasidas Vishwavidyalaya (a Central University), Bilaspur, and served as an Associate Professor untill 2013. He is currently working as Professor (former Head and Dean, School of Life Sciences) at Department of Microbiology, CURAJ. He is a member of various national and international societies/academies. He has completed two collaborated projects worth 150 million INR in the area of microbial diversity and bioenergy.

Prof. Verma is a Group leader of Bioprocess and Bioenergy laboratory at Department of Microbiology, School of Life Sciences, CURAJ. His areas of expertise involve Microbial Diversity, Bioremediation, Bioprocess Development, Lignocellulosic and Algal Biomass-based Biorefinery. He also holds 12 international patents in the field of microwave-assisted biomass pretreatment and bio-butanol production. He has authored more than 62+ research articles in peer-reviewed international journals and contributed to several book chapters (28 published; 15 in press) in various edited books. He has also edited 3 books for international publishers such as Springer and Elsevier. He is a guest editor to several journals such as *Biomass Conversion* and *Biorefinery* (Springer), *Frontiers in Nanotechnology* (Frontiers), and *International Journal of Environmental Research and Public Health* (MDPI). He is also an editorial board member for the journal *Current Nanomedicine* (Bentham Sciences). He is serving as a reviewer for more than 40 journals from various publishers such as Springer, Elsevier, RSC, ACS, Nature, Frontiers, MDPI, etc.

Dr. Maulin P. Shah is an active researcher and scientific writer in his field for more than 20 years. He earned a B.Sc. degree (1999) in Microbiology from Gujarat University, Godhra (Gujarat), India. He also earned his PhD (2005) in Environmental Microbiology from Sardar Patel University, Vallabh Vidyanagar (Gujarat), India. He is Chief Scientist and Head of the Industrial Waste Water Research Lab, Division of Applied and Environmental Microbiology Lab at Enviro Technology Ltd., Ankleshwar, Gujarat, India. His work focuses on the impact of industrial pollution on the microbial diversity of wastewater, and genetically engineering high-impact microbes for the degradation of hazardous materials. His research interests include Biological Wastewater Treatment, Environmental Microbiology, Biodegradation, Bioremediation, and Phytoremediation of Environmental Pollutants from Industrial Wastewaters. He has published more than 250 research papers in national and international journals of repute on various aspects of microbial biodegradation and bioremediation of environmental pollutants. He is the editor of more than 50 books of international repute (Elsevier, Springer, RSC and CRC Press). He is an active editorial board member on top-rated journals. He is on the Advisory Board of *CLEAN—Soil, Air, Water* (Wiley); editor of *Current Pollution Reports* (Springer Nature), *Environmental Technology & Innovation* (Elsevier), *Current Microbiology* (Springer Nature), *Journal of Biotechnology & Biotechnological Equipment* (Taylor & Francis), *Ecotoxicology* (Microbial Ecotoxicology) (Springer Nature), and *Current Microbiology* (Springer Nature); and associate editor of *GeoMicrobiology* (Taylor & Francis) and *Applied Water Science* (Springer Nature).

Contributors

Komal Agrawal
Bioprocess and Bioenergy Laboratory
Department of Microbiology
Central University of Rajasthan
Ajmer, Rajasthan, India

A. S. Ahluwalia
Eternal University
Baru Sahib, Himachal Pradesh, India

Glácio Souza Araújo
Federal Institute of Education, Science and Technology of Ceará – IFCE
Aracati, Ceará, Brazil

Muthu Arumugam
Microbial Processes and Technology Division
Council of Scientific and Industrial Research – National Institute for Interdisciplinary Science and Technology
Thiruvananthapuram, India
and Academy of Scientific and Innovative Research
Ghaziabad, India

Sneha Athalye
Reliance Industries Ltd
Reliance Corporate Park
Ghansoli, Navi Mumbai, India

Trisha Bagchi
Department of Botany
West Bengal State University
Barasat, West Bengal, India

Prithu Baruah
Environmental Biotechnology Laboratory
Department of Biotechnology and Bioinformatics
North-Eastern Hill University
Shillong, India

Nitika Bhardwaj
Department of Botany
Panjab University
Chandigarh, India

Ashok Kumar Bishoyi
Department of Microbiology
Marwadi University
Rajkot, Gujarat, India

Ayushi Bisht
Amity University
Noida, India

K. Chandrasekhar
School of Civil and Environmental Engineering
Yonsei University
Seoul, Republic of Korea

Venkatesh Chaturvedi
School of Biotechnology
Institute of Science
Banaras Hindu University
Varanasi, India

Neha Chaurasia
Environmental Biotechnology Laboratory
Department of Biotechnology and Bioinformatics
North-Eastern Hill University
Shillong, India

Murthy Chavali
Office of the Dean (Research) & Department of Chemistry
Alliance College of Engineering and Design

Faculty of Science & Technology
Alliance University (Central Campus) Bengaluru & NTRC-MCETRC and Aarshanano Composite Technologies Pvt. Ltd.
Guntur, Andhra Pradesh, India

João Cotas
Department of Life Sciences
Marine and Environmental Sciences Centre (MARE)
University of Coimbra
Coimbra, Portugal

Santanu Dasgupta
Reliance Industries Ltd
Reliance Corporate Park
Ghansoli, Navi Mumbai, India

Dishant Desai
Reliance Industries Ltd
Jamnagar, Gujarat, India

Joana Maia Dias
Laboratory for Process Engineering, Environment, Biotechnology and Energy (LEPABE)
Department of Metallurgical and Materials Engineering
Faculty of Engineering of University of Porto
Porto, Portugal

Vinay Dwivedi
Reliance Industries Ltd
Jamnagar, Gujarat, India

Ana M. M. Gonçalves
Department of Life Sciences
Marine and Environmental Sciences Centre (MARE)
University of Coimbra
Coimbra, Portugal
and Department of Biology and CESAM
University of Aveiro
Aveiro, Portugal

Prabuddha Gupta
Department of Microbiology
Marwadi University
Rajkot, Gujarat, India

Santosh Kodgire
Reliance Industries Ltd
Jamnagar, Gujarat, India

Sunil Kumar
Department of Botany
Panjab University
Chandigarh, India

Silvia Lomartire
University of Coimbra
Marine and Environmental Sciences Centre (MARE)
Department of Life Sciences
Coimbra, Portugal

Navonil Mal
Department of Botany
University of Calcutta
Kolkata, India

S. K. Mandotra
Department of Botany
Panjab University
Chandigarh, India

João C. Marques
University of Coimbra
Marine and Environmental Sciences Centre (MARE)
Department of Life Sciences
Coimbra, Portugal

Reecha Mohapatra
Department of Life Science
NIT
Rourkela, Odisha, India

Diana Pacheco
University of Coimbra

Marine and Environmental Sciences Centre (MARE)
Department of Life Sciences
Coimbra, Portugal

Sara Pardilhó
Laboratory for Process Engineering, Environment, Biotechnology and Energy (LEPABE)
Department of Metallurgical and Materials Engineering
Faculty of Engineering of University of Porto
Porto, Portugal

Leonel Pereira
Department of Life Sciences
Marine and Environmental Sciences Centre (MARE)
University of Coimbra
Coimbra, Portugal

Chandra Prakash
School of Biology
Institute of Science
Banaras Hindu University
Varanasi, India

Shyam Prasad
Reliance Industries Ltd
Jamnagar, Gujarat, India

Chandrashekaraiah P.S.
Reliance Industries Ltd
Jamnagar, Gujarat, India

Mahendrapal Singh Rajput
Department of Microbiology
Marwadi University
Rajkot, Gujarat, India

Gaurav Sanghvi
Department of Microbiology
Marwadi University
Rajkot, Gujarat, India

Debanjan Sanyal
Reliance Industries Ltd
Jamnagar, Gujarat, India

Yagya Sharma
Department of Biotechnology
St. Joseph's College
Bengaluru, India

Meenakshi Singh
Department of Botany
Faculty of Science
Maharaja Sayajirao University of Baroda
Vadodara, Gujarat, India

Sweta Singh
Department of Botany
Utkal University
Bhubaneswar, India

Ujwal Trivedi
Department of Microbiology
Marwadi University
Rajkot, Gujarat, India

Pradeep Verma
Bioprocess and Bioenergy Laboratory
Department of Microbiology
Central University of Rajasthan
Ajmer, Rajasthan, India

Durairaj Vijayan
Microbial Processes and Technology Division
Council of Scientific and Industrial Research – National Institute for Interdisciplinary Science and Technology
Thiruvananthapuram, India

1 Biotechnological Advances for Utilization of Algae, Microalgae, and Cyanobacteria for Wastewater Treatment and Resource Recovery

Prabuddha Gupta, Ashok Kumar Bishoyi, Mahendrapal Singh Rajput, Ujwal Trivedi, and Gaurav Sanghvi

CONTENTS

DOI: 10.1201/9781003155713-1

1.1 INTRODUCTION

Since ancient times, wastewater treatment processes (WWTPs) have been known to be effective in water recycling and reclamation. The older cities of the Indus Valley Civilization (~3000 BC), Europe, and the Mediterranean region made use of traditional construction and engineering techniques to carry out wastewater management for houses, irrigation channels, and drainage systems (Angelakis et al. 2005; Fardin et al. 2013; Jansen 1989). As urbanization and development accelerated, several modifications in wastewater management practices were adapted gradually. Conventional WWTPs further evolved by the integration of physical, chemical, and biological principles and processes (Hendricks 2010). The prime function of conventional wastewater treatment is to eliminate solids (suspended/dissolved) and organic matter and increase the dissolved oxygen by means of primary, secondary, and tertiary treatment processes (Davis 2017). Although conventional wastewater processes are efficient and robust, there are major problems associated, such as economic feasibility, environmental impact, energy consumption, and greenhouse gas emissions (Liang et al. 2021; Molinos-Senante et al. 2010; Sala-Garrido et al. 2012). For example, it is estimated that 0.6–3% of the total electricity generated in developed nations is spent on WWTPs. In the European Union (EU) region, the energy demand for WWTPs itself is estimated to be more than 1% of the expenditure of electricity in Europe (Gurung et al. 2018). Moreover, depending on the source of energy used, the carbon emissions associated with WWTPs are also significant. WWTPs emit carbon dioxide (CO_2), methane (CH_4), and nitrous oxide (N_2O) directly during the biological treatment processes, while indirect CO_2 emission is contributed at various energy-intensive steps such as WWTP operation and maintenance (Castro-Barros et al. 2015; Daelman et al. 2012; Kyung et al. 2015). In addition, the removal of total nitrogen (TN) and total phosphate (TP) in the activated sludge process exerts a great O_2 demand as high as 2 kg O_2 kg^{-1} chemical oxygen demand (COD) (Filipe 2011). Maintaining such a high concentration of dissolved O_2 by means of aeration in the conventional activated sludge process is unfeasible and has been proven to be expensive in providing a satisfactory solution. Perhaps the most efficient way to reduce the environmental impact of WWTPs, and to achieve sustainability, is to adopt a process which has low energy requirement, low carbon footprint, and low O_2 requirements. In this direction, efforts are being made to develop and adopt a novel strategy to utilize microalgae as an environmentally friendly and sustainable alternative to the energy-intensive WWTPs (Table 1.1). Algae and cyanobacteria use sunlight energy to carry out photosynthesis and contribute oxygen required for aerobic degradation, thereby sequestering nutrients such as TN, TP, organic carbon, and heavy metals (Chaturvedi et al. 2013; Goswami et al. 2021a). Moreover, these photoautotrophic organisms have the ability to capture atmospheric CO_2 and can assist in CO_2 abatement and mitigation of greenhouse gases (Goswami et al. 2020). This chapter critically examines the use of microalgae and cyanobacteria as an attractive emerging approach to integrate with wastewater treatment systems aimed toward the development of commercially viable technologies with multidisciplinary approaches.

TABLE 1.1
Advantages of Microalgae-Based Wastewater Treatment over Conventional Treatment Technology

Type of Pollutants	Conventional Treatment	Microalgae-Based WWT
Suspended solids	• Gravity settling	• Prior primary settling required due to poor light diffusion
Biodegradable organic pollutant	• Aerobic and anaerobic bacteria	• Aerobic degradation by algae-bacteria symbiosis
	• Aerobic degradation requires aeration by mechanical diffusers and pumps	• Molecular O_2 provided by algae
Nitrogen	• Chemical precipitation of organic substances • Microbial uptake	• Nutrient assimilation (nitrogen uptake NO_3) • Simultaneous nitrification/denitrification process in facultative oxidation ponds
	• Nitrification requires oxygenic environment; denitrification requires anoxygenic environment	
Phosphate	• Enhanced biological phosphorus removal (EBPR) (in the absence of NO_3 and O_2)	• Phosphate uptake by algal cells
Ammonia	• Ammonia stripping at alkaline pH by passing air stream	• Ammonia volatilization favored by an increase in pH and dissolved oxygen (DO)
Pathogens	• Physical (UV) • Chemical disinfection (chlorination, ozonation)	• High surface/volume ratio causes exposure to UV rays. • Pathogen disinfection by raising pH and DO
Heavy metals	• Chemical precipitation (addition of lime) • Chemical flocculation • Ion exchange • Adsorption • Ultrafiltration • Reverse osmosis	• Metal precipitation at high pH contributed by algae. • Active metal uptake into the cytoplasm • Metal adsorption on algae cells

1.2 WASTEWATER TREATMENT BY MICROALGAE

Algae are diverse, poly-phyletic, photosynthetic organisms and have dominated in freshwater, brackish water, marine, and wastewater streams, and even in extreme conditions of hot water springs and sub-zero snowy habitats (Wehr and Sheath 2015). Algal cells are biochemically composed of polysaccharides, proteins, lipids, omega-3 fatty acids, vitamins, pigments, enzymes, minerals, and secondary metabolites (Gupta et al. 2019; Michalak and Chojnacka 2015; Goswami et al. 2021b). Owing to the presence of a vast array of metabolites, algae are exploited in biotechnology, nutraceutical, pharmaceutical, cosmeceutical, health, and agriculture sectors (Mehariya et al. 2021a). Moreover, algae are known to have potential benefits in phycoremediation of wastewater owing to their CO_2 sequestration capability, nutrient removal, resource recovery, and oxygen production (Razzak et al. 2013; Wollmann et al. 2019; Goswami et al. 2021a). To produce 1 kg of dry algal biomass, almost 1.83 kg of atmospheric CO_2 is fixed (Chisti 2007). Microalgae are efficient in trapping solar energy and converting it to biomass with high CO_2 fixation rates that are significantly higher than land plants (Benedetti et al. 2018; Bhola et al. 2014). The growth rate of microalgae in liquid culture (such as wastewater) is better compared to land plants due to the metabolic flexibility of microalgae to shift from autotrophic to mixotrophic or heterotrophic conditions (Gupta et al. 2016a, 2016b). In addition, variations in environmental conditions and nutrient dynamics are better tolerated by algae due to robust cell physiology and stress-tolerating mechanisms. Algae are known for the biological removal of carbon, nitrogen, and phosphorus from wastewater effluents such as municipal, agricultural, brewery, refinery, and industrial effluents (Arun et al. 2020; Meng et al. 2020; Park et al. 2011; Tiron et al. 2015). Apart from removing N and P from wastewater, microalgae are also known for the removal of heavy metals present in significant amounts in industrial wastewater and acid mine drainage (AMD) by the synergistic effect of microalgae and bacterial consortium and biofilms (Abinandan et al. 2018; Kiran Marella et al. 2020). Often the eukaryotic microalgae and aerobic microalgae are grown simultaneously, since microalgae provide molecular oxygen employing photosynthesis and absorb CO_2 in return, released from heterotrophic bacteria (Flores-Salgado et al. 2021), while aerobic bacteria use molecular oxygen for heterotrophic metabolism and degrade organic matter. This process also eliminates the aeration cost, saving the energy requirements for intensive aeration, which itself accounts for 45–75% of total energy consumption in WWTPs (Rosso et al. 2008).

1.3 WASTEWATER TREATMENT BY CYANOBACTERIA

Cyanobacteria are prokaryotic microorganisms exhibiting various ecological and physiological functions. Ecologically, cyanobacteria are very well known for nutrient mineralization and nutrient recycling. Physiologically, cyanobacteria have high adaptability to changes in the environment such as variations in light, nutrients, temperature, and moisture conditions. Cyanobacteria can accumulate nitrogen (NH_4^+, NO_2^-, NO_3^-) and phosphate (PO_4^{3-}) and hence can sustain better growth in sewage or municipal wastewater. Moreover, simple growth requirements and plasticity to adopt variations in different effluents provide an opportunity to create low-cost, eco-friendly technologies. The strategy of integrating cyanobacteria in WWTPs creates a prospect to produce clean water with economically valuable byproducts (proteins, vitamins, enzymes, amino acids, biofertilizers, etc.) (Mehariya et al. 2021b). Cyanobacteria use CO_2 during photosynthesis and aid in achieving a carbon-neutral production process of WWTPs. Many cyanobacteria species such as *Arthrospira* sp. (Markou et al. 2012), *Oscillatoria* sp. (Kallarakkal et al. 2021), *Phormidium* sp. (Fagundes et al. 2019), *Anabena* sp. (Nayak et al. 2014), and *Nostoc* sp. (Devi and Parthiban 2020) have been used for treatment of wastewater not only for the removal of TN and TP but also for the reduction of COD and biochemical oxygen demand (BOD) (Table 1.2).

TABLE 1.2
Nutrient Removal Potential of Cyanobacteria and Microalgae in Different Wastewater Effluents

Microalgal Species	Wastewater Type	Wastewater Effluent	Biomass Productivity (g L^{-1} day^{-1})	N removal (mg L^{-1} day^{-1})	P removal (mg L^{-1} day^{-1})	References
Chlorella zofingiensis	Municipal wastewater	–	1.7	21	4.6	(Zhou et al. 2018)
Chlamydomonas reinhardtii	Wastewater	Anaerobically treated	0.38	–	0.56	(Pachés et al. 2020)
Chlorella vulgaris	Wastewater	Anaerobically treated	0.24	–	1.4	(Pachés et al. 2020)
Chlorella sp.	Municipal	Centrate	0.92	–	–	(Li et al. 2011)
Leptolyngbya sp.	Brewery wastewater (BW)	Pretreated	0.525	–	–	(Papadopoulos et al. 2020)
Leptolyngbya sp.	Municipal wastewater	Untreated	2.93	4.37	1.01	(Singh and Thakur 2015)
Leptolyngbya sp.	Agro-industrial wastewater	Aerobically treated	0.332	1.2	0.3	(Tsolcha et al. 2018)
Monoraphidium braunii	Wastewater	Anaerobically treated	0.65	–	2.63	(Pachés et al. 2020)
Muriellopsis sp.	Municipal	Centrate	1.13	47.5	3.8	(Morales-Amaral et al. 2015)
N. gaditana	Municipal	Centrate	0.4	35s	5.7	(Sepúlveda et al. 2015)
Scenedesmus obliquus	BW	–	0.2	0.4	9.5	(Ferreira et al. 2017)
Scenedesmus obliquus	Wastewater	Anaerobically treated	0.64	–	1.2	(Pachés et al. 2020)
Scenedesmus sp.	Municipal	Pretreated	0.073	–	–	(Sacristán de Alva et al. 2013)
Scenedesmus sp.	Municipal	–	0.13	8.85	0.97	(McGinn et al. 2012)

1.4 OPEN SYSTEMS

Open pond cultivation systems are widely adopted in the industry due to their high economic benefits and low operational capital (Benemann and Oswald 1996). Open systems are shallow ponds intended to treat wastewater through photosynthesis by microalgae and bacteria (aerobic and anaerobic) (Murthy 2011). Conventionally, the open pond systems are an adaptation of natural lagoons (waste stabilization ponds) where the agitation is provided by the wind. As industrialization and commercialization increased, open ponds were gradually adapted to artificial ponds, wherein mixing is ensured by a common stirrer or mechanical aerators to augment the oxygen supply, thus reducing the required size of the pond and hydraulic retention time (HRT) (Hendricks 2010; Owusu-Twum and Sharara 2020). Several modifications of open algal cultivation systems include stabilization ponds, oxidation ditches, artificial lagoons, raceway ponds (high-rate algal ponds, HRAPs), and revolving algal biofilm (RAB) systems (Ho and Goethals 2020). Open systems are known to provide a wide array of design options, owing to flexibility in the choice of construction materials, agitation techniques, and scalability of operation (Jerney and Spilling 2020). Although open pond systems have multiple advantages, there are several drawbacks to consider, as mentioned in Table 1.1, i.e. (1) difficulty in the regulation of crucial parameters such as pH, temperature, mixing rate, and light intensity; (2) CO_2 sparging, i.e., high CO_2 loss due to very short HRT, resulting in poor solubility of diffused CO_2; (3) seasonal variations often contributing to alteration in biomass production and nutrient removal rates, due to fluctuating weather conditions; (4) an increase in evaporation rate during the summer season due to high temperature, resulting in water losses; (5) a cardinal risk of cross-contamination associated with open environment, hampering the growth and biomass productivity, and even causing cultural collapse due to invasive algal/cyanobacterial species.

1.4.1 Stabilization Ponds/Oxidation Ditches/Lagoons

Stabilization ponds, oxidation ditches, or lagoons are recognized as some of the oldest forms of biological wastewater treatment and are known to have originated as early as 3000 years ago (Ho and Goethals 2020). In stabilization ponds, oxidation ditches, or lagoons, molecular oxygen acts as an oxidizing agent by virtue of accepting electrons (terminal electron acceptor). The biodegradable dissolved and suspended organic compounds are broken down by aerobic/anaerobic and heterotrophic bacteria (Figure 1.1). Oxygen is provided by two principal mechanisms: (1) by the growth of algae, which generates molecular oxygen by performing photosynthesis; and (2) by diffusion of atmospheric oxygen provided by power-driven surface aerators/diffused aerators or blowers. Depending upon the growth of bacteria, lagoons are classified as (1) anaerobic, (2) facultative, and (3) aerobic. In the anaerobic process, the biodegradable organic matter is metabolized by anaerobic bacteria whereby the volatile organic compounds are converted into CO_2 and CH_4. In several studies, the anaerobic co-digestion with algae is proven to be effective in increasing the bioconversion efficiency of CH_4. For example, the raw algae co-digested swine manure resulted in a cumulative increase in CH_4 yields from 163 to 245 mL CH_4/g

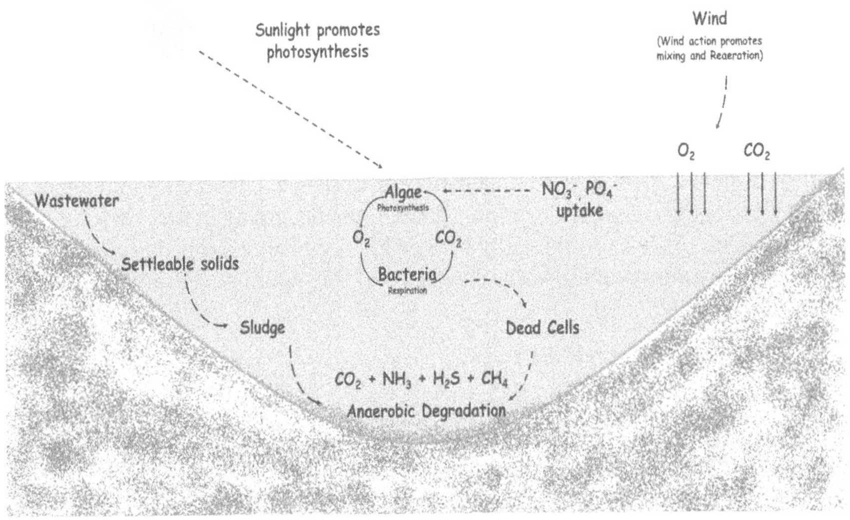

FIGURE 1.1 Phycoremediation mechanism of oxidation ponds for removal of nutrients from wastewater.

of volatile solid (González-Fernández et al. 2011). Mostly, anaerobic lagoons are used for the treatment of wastewater generated from concentrated animal feeding operations (CAFOs) (Wang et al. 2016). On the other hand, facultative lagoons provide the biodegradation of organic nutrients employing both anaerobic and aerobic processes. The bottom portion of the lagoon is dominated by anaerobic microorganisms, wherein the organic compounds are converted into CH_4 and CO_2 by facultative methanogenic bacteria, while the upper portion is aerobic owing to the presence of algae and air-liquid interface. Moreover, the presence of oxygen in the upper layer minimizes the odorous and volatile compounds by oxidation, thereby reducing the foul smell caused by the anaerobic process. Facultative lagoons exhibit a synergistic relationship among algae and bacteria (Humenik and Hanna 1971). Bacteria oxidize the organic compounds in the wastewater via anaerobic and aerobic metabolism, while algae provide the oxygen in the presence of sunlight. However, in the absence of sunlight, algae utilize the oxygen for heterotrophic metabolism and oxidize biodegradable organic matter. Hence, the average depth of facultative lagoons does not exceed above 2–3 meters, to allow maximum sunlight penetration and balance the organic degradation through adjusting the HRT and organic load (Wang et al. 2020). Another advantage of lagoons is inactivation and elimination of pathogens such as protozoan eggs, cyst, thermotolerant coliforms, enterobacteria (*Escherichia coli, Salmonella, Shigella*), anaerobic bacteria (*Clostridium*), and viruses such as coliphages (Locas et al. 2010; Pandey et al. 2016; Sharafi et al. 2012). The removal of the pathogen is caused by variations in the pH occurring due to autotrophic and heterotrophic conditions generated during day as well as at night. During daytime, the pH of the uppermost aerobic zone reaches as high as 10, since CO_2 produced by bacterial degradation is consumed by the algae. In contrast, in dark conditions, the pH drops below 7, as CO_2 is generated by the respiration of both algae and bacteria. Such fluctuations in the pH value and high HRT result in excellent pathogen removal, inactivation, and destruction (Dias et al. 2017). Apart from pathogen removal, facultative lagoons are known to remove nitrogen and phosphorus. The mechanism of nitrogen removal is assisted by nitrification, denitrification, and ammonia stripping phenomenon. Since facultative lagoon exhibits both oxygenic and anoxygenic zones, both nitrification and denitrification processes are favored simultaneously. Ammonia stripping is assisted by high pH (alkaline) occurring in the aerobic zone of the lagoon. Due to photosynthesis by algae, CO_2 is consumed, resulting in alkaline pH wherein the ammonium ions are converted to free ammonia (Butler et al. 2017).

1.4.2 Raceway Ponds (RWP)

Raceway ponds with paddle wheel agitation systems are widely used in the mass cultivation of algae since they tend to be the cheapest method of large-scale microalgal production. The primary design of RWP was first suggested by Oswald and co-workers at the University of California at Berkeley in the late 1950s (Oswald and Goleuke 1967). A typical RWP consists of a shallow pond with a depth of ~20–30 cm (Chisti 2016). The RWP facility is divided into several rectangular grids, with each grid having an oval-shaped channel (Figure 1.2). The movement,

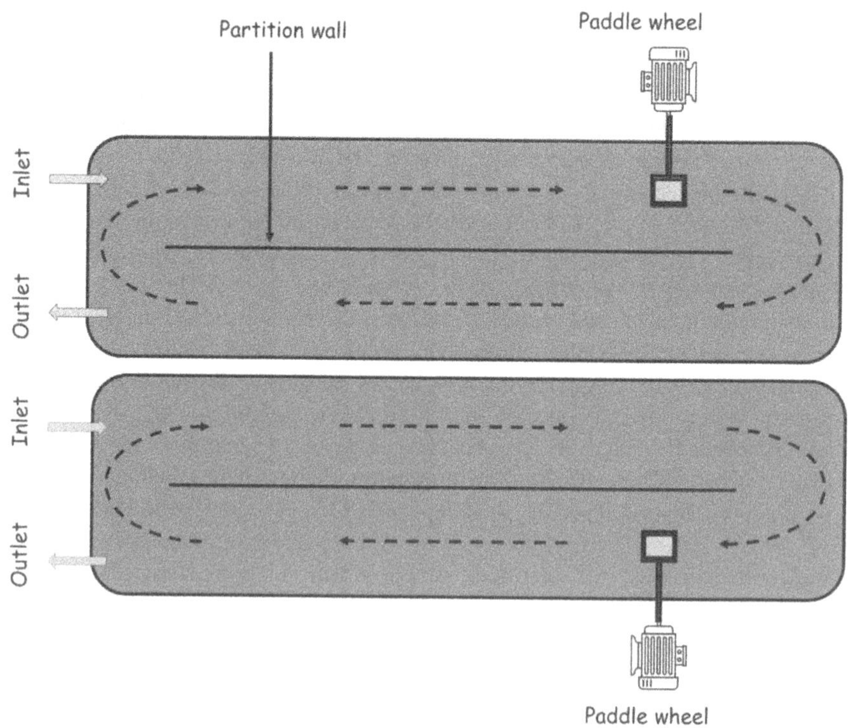

FIGURE 1.2 Raceway ponds with paddle wheel agitation systems.

circulation, and mixing of algae is facilitated by utilizing paddlewheels. The vertical mixing ensures algal cell proliferation due to recurrent exposure to light. Moreover, constant mixing ensures the prevention of sedimentation and bacterial/algae flocs, thereby maintaining diffusion of nutrients. In sizeable operations, often "dark zones" occur due to extensive long channels (Chen et al. 2016). Hence, balancing the flow rate, mass transfer is crucial in avoiding the accumulation of "dead/dark zones". To operate the RWP effectively, a balance of organic nutrient loading rate (OLR), HRT, pond depth, CO_2 supply, flow rate, and pH must always be optimized. RWP systems are cost-effective in comparison to WWTPs since they tend to utilize less space, are scalable, produce no unpleasant odor or fumes, consistently provide optimal nutrient removal, and are known to disinfect pathogens (Craggs et al. 2014).

1.4.3 Revolving Algal Biofilms (RAB)

RAB is a modification of rotating biological contractors (RBCs) used in wastewater treatment systems (Hassard et al. 2015). RAB uses vertically oriented rotating disks and belts partially submerged in the nutrient-rich liquid reservoir and partially exposed in the air (Figure 1.3) (Gross et al. 2015). Algae cells from the liquid broth are attached to the media on the conveyor belt often ridged and corrugated to increase the surface area (Zhao et al. 2018). Usually, the media on the conveyor belt are made up of inert material such as polyvinyl chloride (PVC). Aeration occurs by rotating action, whereby the media is exposed to air and light after contact with the wastewater, resulting in the exchange of CO_2 and O_2. The rotation of RAB leads to fluid mixing, convection of molecules due to pores present in the media and biofilm, nutrient and CO_2 diffusion across the biofilm, and successive product inter-exchange between the RAB reactor and the surroundings (Hassard et al. 2015). A biological slime or "biofilm" grows on the media which contains a consortium of cyanobacteria, microalgae, and aerobic bacteria. The nutrient removal capability of RAB is associated

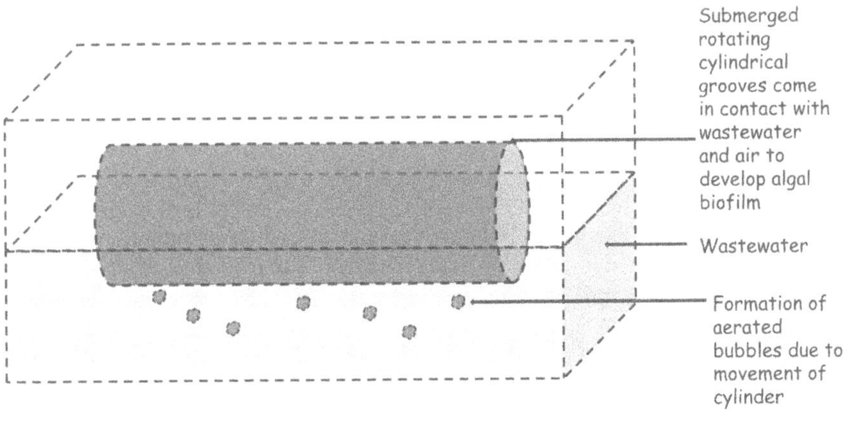

FIGURE 1.3 Revolving algal biofilms (RAB).

with the surface area of the media, quality, and concentration of the nutrients in the wastewater. The role of the algal-microbial consortium on biofilm also plays a crucial role in the removal rate (Roostaei et al. 2018). The extracellular polysaccharides (EPS) secreted by the algal-microbial consortia play a unique role in the creation of a diffusion barrier and act as an adsorbent. Moreover, EPS provide stability and enable a conducive microenvironment against the harsh external environment and establish microbial communities and promote interactions. EPS serve as a carbon source and energy reserve during water and nutrient stress conditions. As soon as the algal/cyanobacterial biofilm is developed, the symbiotic interaction between different autotrophic and heterotrophic microorganisms is established, thereby favoring degradation of organics and nutrient removal (Roeselers et al. 2007). The rotation of conveyor belt also creates gradients promoting microenvironment (aerobic, anaerobic, and anoxic conditions) within a single integrated system, resulting in diverse removal mechanisms (Dutta et al. 2007).

1.4.4 Photo Sequencing Batch Reactor (PSBR)

PSBRs are the modification of sequencing batch reactors (SBRs). The reactor uses the "fill and draw" operation wherein a single reactor is used to complete the entire steps of the activated sludge process (Di Iaconi et al. 2002). The reactor is filled for a certain duration along with constant exposure to light and subsequently aerated for a brief period. Aeration along with light illumination ensures the growth of autotrophic algae and cyanobacteria, resulting in increased oxygen. This eventually supports the growth of heterotrophic microbes that consume dissolved organic matter in the wastewater. Moreover, oxygen promotes the nitrification process by the conversion of NH_3 to NO_2^- and NO_3^-. NO_3^- is actively taken up by algae cells (Tang et al. 2018). Once the aeration step is complete, the sludge is settled down, while the treated water is drained (Ahirrao 2014). A small fraction of the sludge at the rector basin is retained for the next batch, which is rich in the population of algal, cyanobacterial, and heterotrophic microbial consortia (Arias et al. 2019). Retention of sludge helps in re-establishing the population of algal, cyanobacterial, and heterotrophic microbial consortia for the next batch of wastewater, thereby reducing the time for the next batch treatment.

1.5 CLOSED SYSTEMS

In a closed cultivation system, the algal cells are isolated from the atmosphere, often in closed containers. The closed cultivation systems are often rewarded with (1) high and stable growth rate due to better management of controlled conditions such as pH, temperature, HRT, mixing rate, and light intensity; (2) low water evaporation and CO_2 loss; (3) minimal weather dependency and impact of seasonal variations; (4) ease and flexibility in operation; (5) a lower risk of cross-contamination due to sterile environment; (5) low space requirement; (6) high quality and purity of the biomass; and (7) high reproductivity and predictability of production volume. Various closed cultivation systems are discussed in the following sections.

1.5.1 Photobioreactors (PBRs)

PBRs are closed vessels for phototrophic production, where light is provided by artificial means (Andersen 2005). A typical PBR has four phases: (1) solid phase (microalgal cells), (2) liquid phase (growth medium), (3) gaseous phase (CO_2 and O_2), and (4) light (electromagnetic wave). Based on the illumination, PBRs can be grouped into (a) flat plate, (b) tubular, and (c) column PBRs. Based on the flow of liquid, they are grouped into (a) stirred, (b) bubble column, and (c) air lift PBRs (Figure 1.4). PBRs, such as tubular or flat-panel PBR systems, are intended to provide effective light utilization, owing to the higher surface area to volume ratio, even in highly dense suspensions of wastewater and algal cells. Moreover, since PBRs offer a well-controlled environment and well-balanced CO_2 and O_2 transfer (in the case of bubble columns, flat-panel, and airlift PBRs), the growth rates and

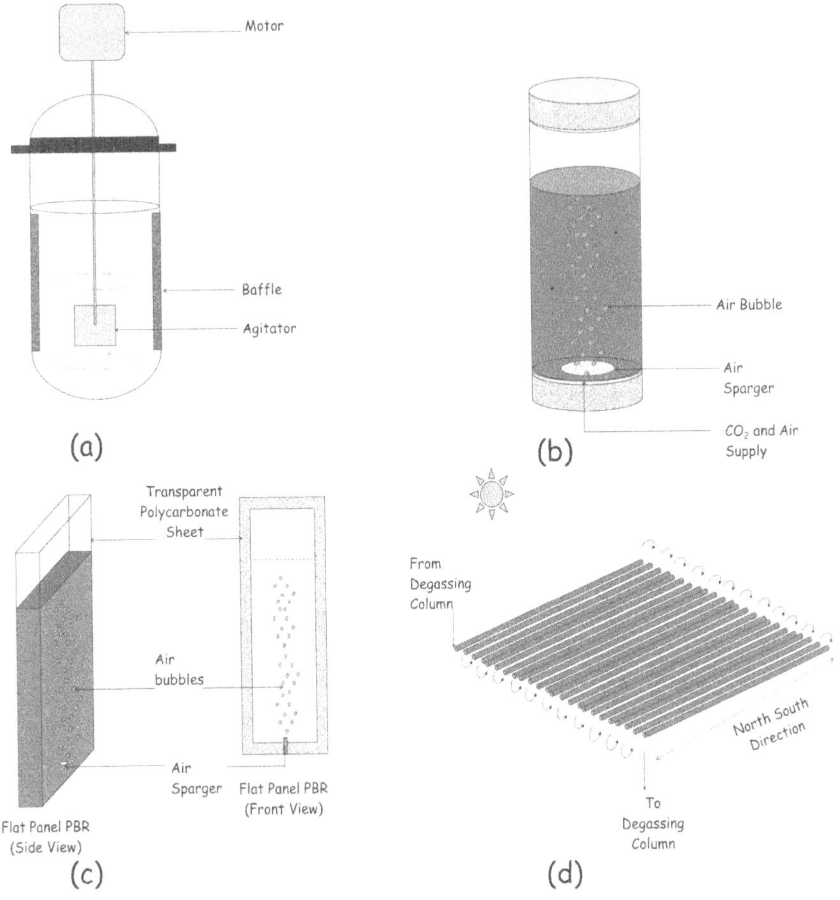

FIGURE 1.4 Types of photobioreactors: (a) Stirred tank PBRs, (b) Bubble column PBRs, (c) Flat panel PBRs, and (d) Horizontal tubular PBRs.

productivities are at par with open pond-based systems. However, the use of PBRs in wastewater treatment has been restricted due to material, construction, operation, and maintenance costs. Additionally, the production and treatment costs significantly increase on scale-up.

1.5.2 Immobilized Algae System

The concept of immobilizing algae has been used as early as the 1960s (Hallier and Park 1969). An immobilized system is defined as a system where the movement of the cell is restricted by natural or artificial means into the aqueous phase (Karel et al. 1985). Various techniques widely used to immobilize cells comprise aggregation, adsorption, confinement, and entrapment (Mallick 2002). Immobilization of algae in polymers (entrapment) such as alginate beads is also a widely used technique. Such immobilized algal cells are often used in fluidized bed reactor systems. In entrapment, algal cells are physically separated by the wastewater by means of immobilizing (trapping) the live cells within the polymer (Ca^{+2}-alginate) since the alginate pores are smaller than the microbes. The wastewater flows through these beads and contaminants are either absorbed and metabolized by algae cells or are adsorbed onto the alginate polymers, eventually providing cell growth and nutrient removal. The major advantage of using the immobilized system is a reduction in HRTs, option to recover the algae biomass, avoiding cell washouts, low space requirements, reuse of cells for next operations, enhanced cell resistance against unfavorable conditions (temperature, pH, and toxic compounds), protection against the damaging effects of photoinhibition, and ease of handling (de-Bashan and Bashan 2010). Immobilization permits higher cell loading as compared to suspended systems, resulting in higher removal rates (Whitton et al. 2018). Moreover, algal biomass recovery poses a significant challenge in large-scale bioreactors; in such operations, immobilized systems are becoming attractive alternatives (Goswami et al. 2021c). Although immobilization demonstrates a potential nutrient removal capability, there are major issues associated with the price of the polymer and dependency on the immobilization process. In this regard, various synthetic polymers such as acrylamide, polyurethane, polyvinyl, and resins are used. Apart from synthetic polymers, biological derivatives of algal polysaccharides such as alginate, carrageenan, agar, agarose, and fungi-derived chitosan are also explored (Eroglu et al. 2015). Two key applications of immobilization of algae cells include the removal of nutrients and biosorption of heavy metal from aqueous solutions (Ahmad et al. 2018, 2019; Mallick 2002).

1.5.3 Algal Membrane Photobioreactor (A-MPBR)

The A-MPBR technology is the amalgamation of a membrane filtration process such as microfiltration or ultrafiltration with an algal WWTP (phycoremediation) (Bilad et al. 2014). The primary step (phycoremediation) occurs whereby microalgae and microbes degrade pollutants in PBRs, while in the secondary step, purified water is filtered by a series of membranes (Luo et al. 2018). A-MPBRs is of two types: (1) internal/submerge type (also known as vacuum/gravity type), wherein the hollow

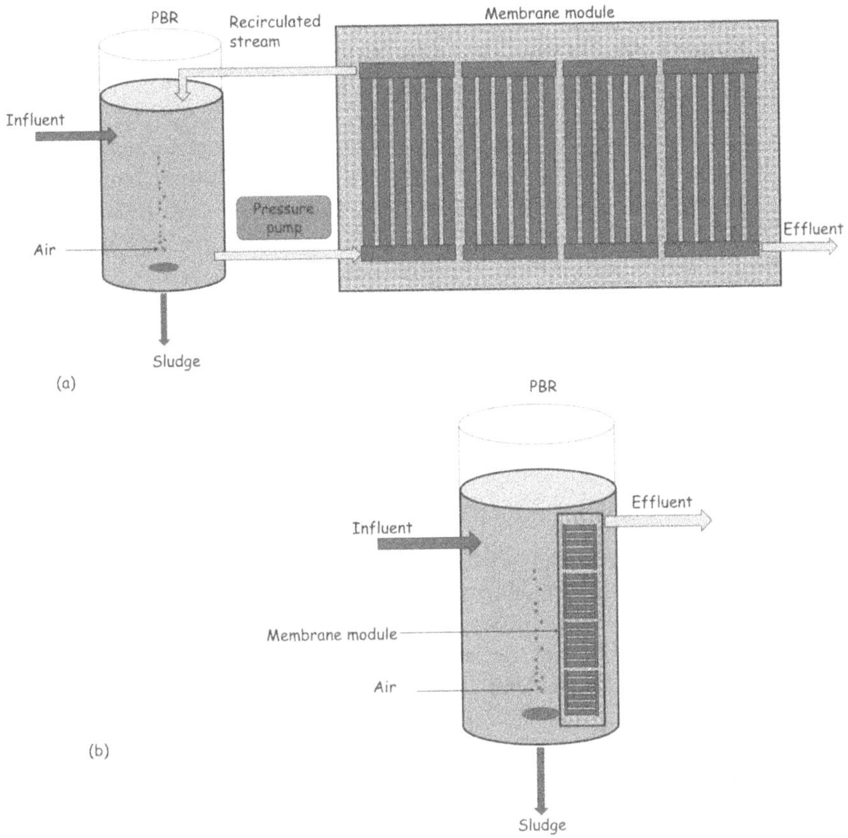

FIGURE 1.5 Configurations of algal membrane photobioreactor (A-MPBR): (a) Side-stream and (b) Immersed.

fiber or flat sheet membrane is submerged and is a constitutive part of the photobioreactor; and (2) external/side stream type (also known as pressure-driven), where membranes and photobioreactors both are separated by an inline pressure-pipe cartridge (Figure 1.5). The pressure pumps are used to separate treated water from the PBRs. The membranes can be flat sheets, tubular capillary tubes, or spiral wound membranes (Judd and Judd 2011). The membranes are accommodated in sections known as modules, or cassettes. The membrane's surface is continually scoured by passing air and encourage mixing. A-MPBRs are intended to enhance light utilization by providing a high surface/volume ratio, resulting in beneficial conditions for algal cell growth. The purpose of a membrane is to retain microalgae cells and reduce biomass washout, thereby segregating the control of solids retention time (SRT) and HRT (Honda et al. 2012). Such segregation or decoupling of HRT and SRT results in the production of high biomass as compared to that produced by PBRs. A-MPBRs have exhibited higher biomass productivity up to 3.5 times more than PBRs (Marbelia et al. 2014).

1.6 BIOTECHNOLOGICAL ADVANCEMENT TOWARD WASTEWATER TREATMENT: BETTER UNDERSTANDING WITH OMICS APPROACH

The advancement in biotechnology has inspired the emerging modern concepts such as "synthetic biology" and "systems biology." The synthetic biology branch aims for the creation/synthesis of biological parts. The biological system represents many functional proteins which may be synthesized from different genes expressed at different levels. For prediction of many proteins and gene expression, different omics strategies are used. For better and accurate predictions of disease conditions, many omics data are used. Overall, omics along with metabolic engineering are trying to develop a viable genetic circuit for redirecting the metabolic flux (Chaturvedi et al. 2021). For use of any organism in the different aspects of biology, two main approaches, viz. genomics and proteomics, are used (Choi et al. 2019).

1.6.1 Omics Approach in Wastewater Treatment

With the increase in demographic development, industrialization, and climate change, the need for pure/drinkable water is mounting (Kumar et al. 2018). Natural resources are decreasing due to erosion and high waste generation due to urbanization. Treatment facilities are not efficient for effectively treating the wastewater and removing hazardous chemicals from the available water (Agrawal et al. 2020; Kumar et al. 2019). With a process like eutrophication and contamination, the drinking water becomes unsafe, and proper disposal procedures are required for its better usability. There are many approaches for the treatment of this wastewater. However, a biological approach with the usage of the microbial commune can serve as an ideal cost-effective, environmentally sustainable, and eco-friendly approach for wastewater treatment (Medhi and Thakur 2018; Mishra et al. 2018; Singh and Thakur 2015; Agrawal et al. 2021a, 2021b). Among the biological entities, microalgae and macroalgae represent the most valuable tool for recycling wastewater with a generation of value-added goods (Patel et al. 2015). Algomics tools have enabled substantial strides in understanding the complexities of algal populations concerning metabolic responses in different wastewater conditions. The algomics approach has also led to the foundation of a new way for the modern wastewater remediation process (Pandey et al. 2019). The phytoremediation process involves the remediation of all forms of wastes (organic and inorganic ones). Among the microalgae, *Chlorella*, *Cyanobacteria*, *Oscillatoria sp.*, and *Scenedesmus sp.* represent an important group of species mostly studied in wastewater treatment (Jha et al. 2017). The use of these species for wastewater treatment led to significant removal of phosphorous and nitrogen content (Ye et al. 2016). At the genomic level, the study of all these species is restricted to a few reports, but the metagenomics approach using *Chlorella* species in lagoons is extensively studied, with reports of gene expression for optimal nitrogen and phosphorous removal (Ye et al. 2016). Further, genomic studies were conducted on two algal species: (1) *Cyanobacteria* and (2) *Nannochloropsis* (Lockhart and Winzeler 2000). Genome-wide analysis has indicated the presence of high T-terminals characteristically with independent guanine and cytosine (GC) content.

The independent GC content suggests that the utilization gap is very less in genetic material as characterized in the 41 genomes of the cyanobacteria (Prabha et al. 2017). The genome-wide analysis of *Nannochloropsis* indicates that two factors, transcription factor (TF) and transcription factor binding site (TFBS), usually regulate the triacylglycerol biosynthesis (Hu et al. 2014). Data integration and analysis has revealed significant precision in the identification of transcription factors regulating the triacylglycerol pathways in microalgal species like *Pavlova sp.*, *N. gaditana*, and *C. reinhardtii* (Thiriet-Rupert et al. 2016). Furthermore, the study of protein interactions at the genomic level revealed that the biochemical system is mostly regulated in microalgae at the genome level compared to the protein levels. Comparative genome analysis of different strains of cyanobacteria used in wastewater treatment showed that the natural pathways used for remediation of wastewater can also accommodate the pathways for biosynthesis of secondary metabolites. Based on the omics data analysis, the secondary metabolites were reported to have excellent therapeutic properties like anti-inflammatory, antiviral, and anti-cancer activities (Dittmann et al. 2015). Leao et al. (2017) also reported the genome-wide analysis of cyanobacteria and found a specific gene cluster having novel therapeutic metabolite activity with no match to reported compounds. Next-generation data of pulp and paper industries from Australia indicate that the algae growth is important for total suspended solids in the wastewater discharged from the paper industries. Approximately 19% of the sequences matched with different strains of cyanobacteria (Wilks and Scholes 2018). Multidimensional protein identification data of *C. reinhardtii* cultured in wastewater with a high amount of ammonia and nitrate reveals a total of two thousand proteins expressed. Most of the proteins were related to nitrogen assimilation and metabolism. Also, high levels of the enzyme utilized for biomolecule (carbohydrates, proteins, lipids) synthesis were expressed during the biomass augmentation of the *C. reinhardtii* species (Patel et al. 2015).

Transcriptomics data includes the form, formation, and utility of transcripts formed by any cell under an experimental and natural set of conditions. Transcriptomics will aid in giving better insights into the phenomena like intron rearrangement and formation of proteins via alternative splicing in the eukaryotic microalgae along with the coupled data of genomics. Transcriptomics data reveals that the microalgae strain *Chlorella protothecoides* consumes carbon sources more effectively for fatty lipid storage compare to the standard microalgal strain *Chlorella variabilis* (Gao et al. 2014). Further, data coupling of transcriptome and genome has served for better characterization of the transcription start site (TSS). The characterization of TSS has served as an important tool as it reveals that the photosynthetic pigment in response to the light observed by transcriptome data was like that of the cell behavior found in synthetic biology data (Liu et al. 2018). Data of TSS analysis was more precise in the predictability of targeted RNA expression in clustered regularly interspaced short palindromic repeats (CRISPR) experiments and analyzing the location of ribosome binding site (RBS) (Tan et al. 2018). Additionally, the comparative transcriptome discovered a gene upregulation pathway in photosynthesis covering photosystems I and II and contributing to effective wastewater treatment and biofuel production (Ban et al. 2019). Further combination of proteomics and transcriptomics has helped to gain insight for a microalgal response to various carbon and nitrogen sources with known probable gene targets for assimilation of the metabolic flux including the genes involved in the phosphorous and nitrogen removal pathways. Additionally, proteomics analysis of microalgal samples in wastewater processing has also opened the gates for the potential applications of biofuel production as a by-product of wastewater treatment. Combinatorial transcriptome and proteome analysis of *Picochlorum* species had indicated that regions essentially coding the genes responsible for the wastewater treatment consume more than one-half of the proteins expressed and the biological activity resembles the enzymes used in synthetic biology for inhibiting the pollutant-prompted stress conditions (Xiong et al. 2018). Metabolomics profiles of secondary metabolites confirmed that three filamentous cyanobacterial strains used in wastewater treatment were highly influenced by phosphate and nitrate contents (Crnkovic et al. 2018). Results indicate significant production of merocyclophane C production at elevated phosphate concentration with novel peptide synthesis. This metabolomics study further enhances the chances of resource recovery from different wastewater facilities.

1.7 CONCLUSION

The integration of algae and cyanobacteria with WWTP offers enormous potential and sustainable alternative toward resource recovery. This chapter explains the efficiency and capability of algae and cyanobacteria in WWTP in terms of nutrient removal and heavy metal sequestration, which leads to nullifying the environmental hazards and overall improved water quality. Consequently, the use of algae and cyanobacteria also provides the opportunity to carry out wastewater treatment operations with lower carbon and energy footprint as compared to conventional biological WWTPs. However, further research and deeper understanding is needed on approaches suitable for scale-up. Moreover, challenges such as strain selection, monoculture versus mixed culture (consortia), inoculum dose, nutrient load, HRT, effect of nitrogen and phosphate ratio, light and dark conditions, and effect of carbon source under real scenarios should be evaluated rigorously for better insight. In coming decades, the omics approaches will have a strong influence on industrial microbiology; hence, novel strain identification, improvement, and selection strategy should focus on algonomics, genomics, proteomics, and metabolomics. Such insights will help in understanding the cellular function keenly, and integrating such tools will be fruitful.

ABBREVIATIONS

AMD, acid mine drainage; CAFOs, concentrated animal feeding operations; COD, chemical oxygen demand; CRISPR, clustered regularly interspaced short palindromic repeats; DO, dissolved oxygen; EBPR, enhanced biological phosphorus removal; EPS, extracellular polysaccharides; GC, guanine and cytosine; HRAPs, high-rate algal ponds; HRT, hydraulic retention time; MPBR, membrane photobioreactor; OLR, organic nutrient loading rate; PBR, photobioreactors; PVC, polyvinyl chloride; RAB, revolving algal biofilms; RBCs, rotating biological contractors; RBS, ribosome binding site; RWPs, raceway ponds; SRT, solids retention time; TF, transcription factor; TFBS, transcription factor binding site; TN, total Nitrogen; TP, total phosphate; TSS, transcription start site; WWTPs, wastewater treatment processes

DECLARATION OF CONFLICT OF INTEREST

The authors declare that they have no known conflict of interest that could have appeared to influence the work reported in this chapter.

REFERENCES

Abinandan, S., Subashchandrabose, S. R., Venkateswarlu, K., and Megharaj, M. Microalgae–bacteria biofilms: a sustainable synergistic approach in remediation of acid mine drainage. *Appl. Microbiol. Biotechnol.* 2018. 102 (3): 1131–1144. doi: 10.1007/s00253-017-8693-7.

Agrawal, K., Bhatt, A., Bhardwaj, N., Kumar, B., Verma, P. Integrated approach for the treatment of industrial effluent by physico-chemical and microbiological process for sustainable environment. In: Combined Application of Physico-Chemical & Microbiological Processes for Industrial Effluent Treatment Plant. Springer, Singapore. 2020. pp. 119–143.

Agrawal, K., & Verma, P. Metagenomics: a possible solution for uncovering the "mystery box" of microbial communities involved in the treatment of wastewater. In: Wastewater Treatment Cutting Edge Molecular Tools, Techniques and Applied Aspects. Elsevier. 2021a. pp. 41–53.

Agrawal, K., & Verma, P. "Omics"—a step toward understanding of complex diversity of the microbial community. In: Wastewater Treatment Cutting Edge Molecular Tools, Techniques and Applied Aspects. Elsevier. 2021b. pp. 471–487.

Ahirrao, S. Zero Liquid Discharge Solutions. In V. V. Ranade and V. M. Bhandari (Eds.). Industrial Wastewater Treatment, Recycling and Reuse 2014. pp. 489–520.

Ahmad, A., Bhat, A. H., and Buang, A. Biosorption of transition metals by freely suspended and Ca-alginate immobilised with *Chlorella vulgaris*: kinetic and equilibrium modeling. *J. Clean. Prod.* 2018. 171: 1361–1375. doi: 10.1016/j.jclepro.2017.09.252.

Ahmad, A., Bhat, A. H., and Buang, A. Enhanced biosorption of transition metals by living *Chlorella vulgaris* immobilized in Ca-alginate beads. *Environ. Technol.* 2019. 40 (14): 1793–1809. doi: 10.1080/09593330.2018.1430171.

Andersen, R. A. Algal Culturing Techniques. Academic Press. 2005.

Angelakis, A. N., Koutsoyiannis, D., and Tchobanoglous, G. Urban wastewater and stormwater technologies in ancient Greece. *Water Res.* 2005. 39 (1): 210–220. doi: 10.1016/j.watres.2004.08.033.

Arias, D. M., Rueda, E., García-Galán, M. J., Uggetti, E., and García, J. Selection of cyanobacteria over green algae in a photo-sequencing batch bioreactor fed with wastewater. *Sci. Total Environ.* 2019. 653: 485–495. doi: 10.1016/j.scitotenv.2018.10.342.

Arun, S., Sinharoy, A., Pakshirajan, K., and Lens, P. N. L. Algae based microbial fuel cells for wastewater treatment and recovery of value-added products. *Renew. Sust. Energ. Rev.* 2020. 132: 110041. doi: 10.1016/j.rser.2020.110041.

Ban, S., Lin, W., Luo, Z., and Luo, J. Improving hydrogen production of *Chlamydomonas reinhardtii* by reducing chlorophyll content via atmospheric and room temperature plasma. Bioresour. Technol. 2019. 275: 425–429.

Benedetti, M., Vecchi, V., Barera, S., and Dall'Osto, L. Biomass from microalgae: The potential of domestication towards sustainable biofactories. *Microb. Cell Fact.* 2018. 17 (1): 173. doi: 10.1186/s12934-018-1019-3.

Benemann, J. R., and Oswald, W. J. Systems and economic analysis of microalgae ponds for conversion of CO_2 to biomass. Final report (DOE/PC/93204-T5). California Univ., Berkeley, CA (United States). Dept. of Civil Engineering. 1996. doi: 10.2172/493389.

Bhola, V., Swalaha, F., Ranjith Kumar, R., Singh, M., and Bux, F. Overview of the potential of microalgae for CO_2 sequestration. *Int. J. Environ. Sci. Technol.* 2014. 11 (7): 2103–2118. doi: 10.1007/s13762-013-0487-6.

Bilad, M. R., Discart, V., Vandamme, D., Foubert, I., Muylaert, K., and Vankelecom, I. F. J. Coupled cultivation and pre-harvesting of microalgae in a membrane photobioreactor (MPBR). *Bioresour. Technol.* 2014. 155: 410–417. doi: 10.1016/j.biortech.2013.05.026.

Butler, E., Hung, Y.-T., Suleiman Al Ahmad, M., Yeh, R. Y.-L., Liu, R. L.-H., and Fu, Y. P. Oxidation pond for municipal wastewater treatment. *Appl. Water Sci.* 2017. 7 (1): 31–51. doi: 10.1007/s13201-015-0285-z.

Castro-Barros, C. M., Daelman, M. R. J., Mampaey, K. E., van Loosdrecht, M. C. M., and Volcke, E. I. P. Effect of aeration regime on N_2O emission from partial nitritation-anammox in a full-scale granular sludge reactor. *Water Res.* 2015. 68: 793–803. doi: 10.1016/j.watres.2014.10.056.

Chaturvedi, V., Chandravanshi, M., Rahangdale, M., & Verma, P. An integrated approach of using polystyrene foam as an attachment system for growth of mixed culture of cyanobacteria with concomitant treatment of copper mine waste water. *J. Waste Manag.* 2013. 282798: 1–7. doi: https://doi.org/10.1155/2013/282798.

Chaturvedi, V., Goswami, R. K., Verma, P. Genetic engineering for enhancement of biofuel production in microalgae. In: Biorefineries: A Step Towards Renewable and Clean Energy. Springer. 2021. pp. 539–559.

Chen, Z., Zhang, X., Jiang, Z., Chen, X., He, H., and Zhang, X. Light/dark cycle of microalgae cells in raceway ponds: Effects of paddlewheel rotational speeds and baffles installation. *Bioresour. Technol.* 2016. 219: 387–391. doi: 10.1016/j.biortech.2016.07.108.

Chisti, Y. Biodiesel from microalgae. *Biotechnol. Adv.* 2007. 25 (3): 294–306. doi: 10.1016/j.biotechadv.2007.02.001.

Chisti, Y. Large-Scale Production of Algal Biomass: Raceway Ponds. In F. Bux and Y. Chisti (Eds.). Algae Biotechnology: Products and Processes 2016. pp. 21–40.

Choi, K. R., Jang, W. D., Yang, D., Cho, J. S., Park, D., and Lee, S. Y. Systems metabolic engineering strategies: integrating systems and synthetic biology with metabolic engineering. *Trends Biotechnol.* 2019. 37 (8): 817–837. doi: 10.1016/j.tibtech.2019.01.003.

Craggs, R., Park, J., Heubeck, S., and Sutherland, D. High rate algal pond systems for low-energy wastewater treatment, nutrient recovery and energy production. *N. Z. J. Bot.* 2014. 52 (1): 60–73. doi: 10.1080/0028825X.2013.861855.

Crnkovic, C. M., May, D. S., and Orjala, J. The impact of culture conditions on growth and metabolomic profiles of freshwater cyanobacteria. *J. Appl. Phycol.* 2018. 30: 375–384. doi: 10.1007/s10811-017-1275-3.

Daelman, M. R. J., van Voorthuizen, E. M., van Dongen, U. G. J. M., Volcke, E. I. P., and van Loosdrecht, M. C. M. Methane emission during municipal wastewater treatment. *Water Res.* 2012. 46 (11): 3657–3670. doi: 10.1016/j.watres.2012.04.024.

Davis, M. L. Water and Wastewater Engineering. Tata McGraw-Hill Education. 2017.

de-Bashan, L. E., and Bashan, Y. Immobilized microalgae for removing pollutants: review of practical aspects. *Bioresour. Technol.* 2010. 101 (6): 1611–1627. doi: 10.1016/j.biortech.2009.09.043.

Devi, T. E., and Parthiban, R. Hydrothermal liquefaction of *Nostoc ellipsosporum* biomass grown in municipal wastewater under optimized conditions for bio-oil production. *Bioresour. Technol.* 2020. 316: 123943. doi: 10.1016/j.biortech.2020.123943.

Di Iaconi, C., Lopez, A., Ramadori, R., Di Pinto, A. C., and Passino, R. Combined chemical and biological degradation of tannery wastewater by a periodic submerged filter (SBBR). *Water Res.* 2002. 36 (9): 2205–2214. doi: 10.1016/S0043-1354(01)00445-6.

Dias, D. F. C., Passos, R. G., and von Sperling, M. A review of bacterial indicator disinfection mechanisms in waste stabilisation ponds. *Rev. Environ. Sci. Biotechnol.* 2017. 16 (3): 517–539. doi: 10.1007/s11157-017-9433-2.

Dittmann, E., Gugger, M., Sivonen, K., and Fewer, D. P. Natural product biosynthetic diversity and comparative genomics of the cyanobacteria. Trends Microbiol. 2015. 23 (10): 642–652. doi: 10.1016/j.tim.2015.07.008.

Dutta, S., Hoffmann, E., and Hahn, H. H. Study of rotating biological contactor performance in wastewater treatment using multi-culture biofilm model. *Water Sci. Technol.* 2007. 55: (8–9): 345–353. doi: 10.2166/wst.2007.276.

Eroglu, E., Smith, S. M., and Raston, C. L. Application of Various Immobilization Techniques for Algal Bioprocesses. In N. R. Moheimani, M. P. McHenry, K. de Boer, and P. A. Bahri (Eds.). *B*iomass and Biofuels from Microalgae: Advances in Engineering and Biology 2015. pp. 19–44.

Fagundes, M. B., Falk, R. B., Facchi, M. M. X., Vendruscolo, R. G., Maroneze, M. M., Zepka, L. Q., Jacob-Lopes, E., and Wagner, R. Insights in cyanobacteria lipidomics: a sterols characterization from *Phormidium autumnale* biomass in heterotrophic cultivation. *Food Res. Int.* 2019. 119: 777–784. doi: 10.1016/j.foodres.2018.10.060.

Fardin, H. F., Hollé, A., Gautier, E., and Haury, J. Wastewater management techniques from ancient civilizations to modern ages: examples from South Asia. *Water Supply.* 2013. 13 (3): 719–726. doi: 10.2166/ws.2013.066.

Ferreira, A., Ribeiro, B., Marques, P. A. S. S., Ferreira, A. F., Dias, A. P., Pinheiro, H. M., Reis, A., and Gouveia, L. Scenedesmus obliquus mediated brewery wastewater remediation and CO_2 biofixation for green energy purposes. *J. Clean. Prod.* 2017. 165: 1316–1327. doi: 10.1016/j.jclepro.2017.07.232.

Filipe, C. D. M. Biological Wastewater Treatment. CRC Press. 2011. doi: 10.1201/b13775.

Flores-Salgado, G., Thalasso, F., Buitrón, G., Vital-Jácome, M., and Quijano, G. Kinetic characterization of microalgal-bacterial systems: contributions of microalgae and heterotrophic bacteria to the oxygen balance in wastewater treatment. *Biochem. Eng. J.* 2021. 165: 107819. doi: 10.1016/j.bej.2020.107819.

Gao, C., Wang, Y., Shen, Y., Yan, D., He, X., Dai, J., and Wu, Q. Oil accumulation mechanisms of the oleaginous microalga *Chlorella protothecoides* revealed through its genome, transcriptomes, and proteomes. *BMC Genomics.* 2014. 15: 582.

González-Fernández, C., Molinuevo-Salces, B., and García-González, M. C. Evaluation of anaerobic codigestion of microalgal biomass and swine manure via response surface methodology. *Appl. Energy.* 2011. 88 (10): 3448–3453. doi: 10.1016/j.apenergy.2010.12.035.

Goswami, R. K., Agrawal, K., Mehariya, S., Molino, A., Musmarra, D., Verma, P. Microalgae-Based Biorefinery for Utilization of Carbon Dioxide for Production of Valuable Bioproducts. In: Chemo-Biological Systems for CO_2 Utilization. Taylor & Francis. 2020. pp. 199–224.

Goswami, R. K., Mehariya, S., Verma, P, Lavecchia, R., Zuorro, A. Microalgae-based biorefineries for sustainable resource recovery from wastewater. *J. Water Process. Eng.* 2021a. 40: 101747 https://doi.org/10.1016/j.jwpe.2020.101747.

Goswami, R. K., Agrawal, K., Verma, P. An Overview of Microalgal Carotenoids: Advances in the Production and Its Impact on Sustainable Development Bioenergy Research: Evaluating Strategies for Commercialization and Sustainability. John Wiley & Sons. 2021b. pp. 105–128.

Goswami, R. K., Mehariya, S., Karthikeyan, O. P. K., Verma, P. Advanced microalgae-based renewable biohydrogen production systems: a review. *Bioresour. Technol.* 2021c. 320 (A): 124301 doi: https://doi.org/10.1016/j.biortech.2020.124301.

Gross, M., Mascarenhas, V., and Wen, Z. Evaluating algal growth performance and water use efficiency of pilot-scale revolving algal biofilm (RAB) culture systems. *Biotechnol. Bioeng.* 2015. 112 (10): 2040–2050. doi: 10.1002/bit.25618.

Gupta, P., Choi, H. J., and Lee, S. M. Enhanced nutrient removal from municipal wastewater assisted by mixotrophic microalgal cultivation using glycerol. *Environ. Sci. Pollut. Res.* 2016a. 23 (10): 10114–10123. doi: 10.1007/s11356-016-6224-1.

Gupta, P., Choi, H. J., Pawar, R. R., Jung, S. P., and Lee, S. M. Enhanced biomass production through optimization of carbon source and utilization of wastewater as a nutrient source. *J. Environ. Manage.* 2016b. 184: 585–595. doi: 10.1016/j.jenvman.2016.10.018.

Gupta, P., Rajput, M., Oza, T., Trivedi, U., and Sanghvi, G. Eminence of microbial products in cosmetic industry. *Nat. Prod. Bioprospect.* 2019. 9 (4): 267–278. doi: 10.1007/s13659-019-0215-0.

Gurung, K., Tang, W. Z., and Sillanpää, M. Unit energy consumption as benchmark to select energy positive retrofitting strategies for Finnish wastewater treatment plants (WWTPs): a case study of Mikkeli WWTP. *Environ. Process.* 2018. 5 (3): 667–681. doi: 10.1007/s40710-018-0310-y.

Hallier, U. W., and Park, R. B. Photosynthetic light reactions in chemically fixed *Anacystis nidulans, Chlorella pyrenoidosa*, and *Porphyridium cruentum. Plant Physiol.* 1969. 44 (4): 535–539. doi: 10.1104/pp.44.4.535.

Hassard, F., Biddle, J., Cartmell, E., Jefferson, B., Tyrrel, S., and Stephenson, T. Rotating biological contactors for wastewater treatment - a review. *Process Saf. Environ. Prot.* 2015. 94: 285–306. doi: 10.1016/j.psep.2014.07.003.

Hendricks, D. Fundamentals of Water Treatment Unit Processes: Physical, Chemical, and Biological. CRC Press. 2010.

Ho, L., and Goethals, P. L. M. Municipal wastewater treatment with pond technology: Historical review and future outlook. *Ecol. Eng.* 2020. 148: 105791. doi: 10.1016/j.ecoleng.2020.105791.

Honda, R., Boonnorat, J., Chiemchaisri, C., Chiemchaisri, W., and Yamamoto, K. Carbon dioxide capture and nutrients removal utilizing treated sewage by concentrated microalgae cultivation in a membrane photobioreactor. *Bioresour. Technol.* 2012. 125, 59–64. doi: 10.1016/j.biortech.2012.08.138.

Hu, J., Wang, D., Li, J., Jing, G., Ning, K., and Xu, J. Genome-wide identification of transcription factors and transcription-factor binding sites in oleaginous microalgae *Nannochloropsis. Sci. Rep.* 2014. 4 (1): 5454. doi: 10.1038/srep05454.

Humenik, F. J., and Hanna, G. P. Algal-bacterial symbiosis for removal and conservation of wastewater nutrients. *J. Water Pollut. Control. Fed.* 1971. 43 (4): 580–594.

Jansen, M. Water supply and sewage disposal at Mohenjo-Daro. *World Archaeol.* 1989. 21 (2): 177–192. doi: 10.1080/00438243.1989.9980100.

Jerney, J., and Spilling, K. Large Scale Cultivation of Microalgae: Open and Closed Systems. In K. Spilling (Ed.). Biofuels from Algae: Methods and Protocols 2020. pp. 1–8.

Jha, V., Puranik, S., and Purohit, H. J. Sequestration options for phosphorus in wastewater. In: Optimization and Applicability of Bioprocesses. Springer. 2017. pp. 115–140.

Judd, S., and Judd, C. Fundamentals. In: The MBR Book (2nd ed.). Butterworth-Heinemann. 2011. pp. 55–207. doi: 10.1016/B978-0-08-096682-3.10002-2.

Kallarakkal, K. P., Muthukumar, K., Alagarsamy, A., Pugazhendhi, A., and Naina Mohamed, S. Enhancement of biobutanol production using mixotrophic culture of *Oscillatoria* sp. in cheese whey water. *Fuel.* 2021. 284: 119008. doi: 10.1016/j.fuel.2020.119008.

Karel, S. F., Libicki, S. B., and Robertson, C. R. The immobilization of whole cells: engineering principles. *Chem. Eng. Sci.* 1985. 40 (8): 1321–1354. doi: 10.1016/0009-2509(85)80074-9.

Kiran Marella, T., Saxena, A., and Tiwari, A. Diatom mediated heavy metal remediation: a review. *Bioresour. Technol.* 2020. 305: 123068. doi: 10.1016/j.biortech.2020.123068.

Kumar, B., Agrawal, K., Bhardwaj, N., Chaturvedi, V., & Verma, P. Advances in Concurrent Bioelectricity Generation and Bioremediation Through Microbial Fuel Cells. In: Microbial Fuel Cell Technology for Bioelectricity. Springer, Cham. 2018. pp. 211–239.

Kumar, B., Agrawal, K., Bhardwaj, N., Chaturvedi, V., Verma, P. Techno-Economic Assessment of Microbe-Assisted Wastewater Treatment Strategies for Energy and Value-Added Product Recovery. In: Microbial Technology for the Welfare of Society. Springer. 2019. pp. 147–181.

Kyung, D., Kim, M., Chang, J., and Lee, W. Estimation of greenhouse gas emissions from a hybrid wastewater treatment plant. *J. Clean. Prod.* 2015. 95: 117–123. doi: 10.1016/j.jclepro.2015.02.032.

Leao, T., Castelão, G., Korobeynikov, A., Monroe, E. A., Podell, S., Glukhov, E., Allen, E. E., Gerwick, W. H. and Gerwick, L. Comparative genomics uncovers the prolific and distinctive metabolic potential of the cyanobacterial genus Moorea. *Proc. Natl. Acad. Sci. U.S.A.* 2017. 114 (12), 3198–3203. doi: 10.1073/pnas.1618556114.

Li, Y., Chen, Y. F., Chen, P., Min, M., Zhou, W., Martinez, B., Zhu, J., and Ruan, R. Characterization of a microalga Chlorella sp. well adapted to highly concentrated municipal wastewater for nutrient removal and biodiesel production. *Bioresour. Technol.* 2011. 102 (8): 5138–5144. doi: 10.1016/j.biortech.2011.01.091.

Liang, J., Cui, Y., Zhang, M., Chen, Z., Wang, S., and Li, X. Evalution of the fate of greenhouse gas emissions from the sludge mineralisation process in sludge treatment wetlands. *Ecol. Eng.* 2021. 159: 106124. doi: 10.1016/j.ecoleng.2020.106124.

Liu, Z., Zhang, J., Jin, J., Geng, Z., Qi, Q., and Liang, Q. Programming bacteria with light-sensors and applications in synthetic biology. *Front. Microbiol.* 2018. 9: 2692. doi: 10.3389/fmicb.2018.02692.

Locas, A., Martinez, V., and Payment, P. Removal of human enteric viruses and indicator microorganisms from domestic wastewater by aerated lagoons. *Can. J. Microbiol.* 2010. 56 (2): 188–194. doi: 10.1139/w09-124.

Lockhart, D. J., and Winzeler, E. A. Genomics, gene expression and DNA arrays. *Nature.* 2000. 405 (6788): 827–836. doi: 10.1038/35015701.

Luo, Y., Le-Clech, P., and Henderson, R. K. Assessment of membrane photobioreactor (MPBR) performance parameters and operating conditions. *Water Res.* 2018. 138: 169–180. doi: 10.1016/j.watres.2018.03.050.

Mallick, N. Biotechnological potential of immobilized algae for wastewater N, P and metal removal: a review. *Biometals.* 2002. 15 (4): 377–390. doi: 10.1023/A:1020238520948.

Marbelia, L., Bilad, M. R., Passaris, I., Discart, V., Vandamme, D., Beuckels, A., Muylaert, K., and Vankelecom, I. F. J. Membrane photobioreactors for integrated microalgae cultivation and nutrient remediation of membrane bioreactors effluent. *Bioresour. Technol.* 2014. 163: 228–235. doi: 10.1016/j.biortech.2014.04.012.

Markou, G., Chatzipavlidis, I., and Georgakakis, D. Cultivation of *Arthrospira (Spirulina) platensis* in olive-oil mill wastewater treated with sodium hypochlorite. *Bioresour Technol.* 2012. 112: 234–241. doi: 10.1016/j.biortech.2012.02.098.

McGinn, P. J., Dickinson, K. E., Park, K. C., Whitney, C. G., MacQuarrie, S. P., Black, F. J., Frigon, J. C., Guiot, S. R., and O'Leary, S. J. B. Assessment of the bioenergy and bioremediation potentials of the microalga *Scenedesmus* sp. AMDD cultivated in municipal wastewater effluent in batch and continuous mode. *Algal Res.* 2012. 1 (2): 155–165. doi: 10.1016/j.algal.2012.05.001

Medhi, K., and Thakur, I. S. Bioremoval of nutrients from wastewater by a denitrifier *Paracoccus denitrificans* ISTOD1. *Bioresour. Technol. Rep.* 2018. 1: 56–60.

Mehariya, S., Goswami, R. K., Karthikeysan, O. P., & Verma, P. Microalgae for high-value products: a way towards green nutraceutical and pharmaceutical compounds. Chemosphere. 2021a. 130553. doi: https://doi.org/10.1016/j.chemosphere.2021.130553

Mehariya, S., Kumar, P., Marino, T., Casella, P., Lovine, A., Verma, P., Musuma, Molino, A. Aquatic Weeds: A Potential Pollutant Removing Agent from Wastewater and Polluted Soil and Valuable Biofuel Feedstock, Bioremediation using weeds. Springer. 2021b. pp. 59–77.

Meng, F., Huang, W., Liu, D., Zhao, Y., Huang, W., Lei, Z., and Zhang, Z. Application of aerobic granules-continuous flow reactor for saline wastewater treatment: granular stability, lipid production and symbiotic relationship between bacteria and algae. *Bioresour. Technol.* 2020. 295: 122291. doi: 10.1016/j.biortech.2019.122291.

Michalak, I., and Chojnacka, K. Algae as production systems of bioactive compounds. *Eng. Life Sci.* 2015. 15(2): 160–176. doi: 10.1002/elsc.201400191.

Mishra, A., Medhi, K., Maheshwari, N., Srivastava, S., and Thakur, I. S., Biofuel production and phycoremediation by *Chlorella* sp. ISTLA1 isolated from landfill site. *Bioresour. Technol.* 2018. 253: 121–129. doi: 10.1016/j.biortech.2017.12.012.

Molinos-Senante, M., Hernández-Sancho, F., and Sala-Garrido, R. Economic feasibility study for wastewater treatment: a cost–benefit analysis. *Sci. Total Environ.* 2010. 408 (20): 4396–4402. doi: 10.1016/j.scitotenv.2010.07.014.

Morales-Amaral, M. del M., Gómez-Serrano, C., Acién, F. G., Fernández-Sevilla, J. M., and Molina-Grima, E. Production of microalgae using centrate from anaerobic digestion as the nutrient source. *Algal Res.* 2015. 9: 297–305. doi: 10.1016/j.algal.2015.03.018.

Murthy, G. S. Overview and Assessment of Algal Biofuels Production Technologies. In A. Pandey, C. Larroche, S. C. Ricke, C. G. Dussap, and E. Gnansounou (Eds.). Biofuels 2011. pp. 415–437.

Nayak, B. K., Roy, S., and Das, D. Biohydrogen production from algal biomass (Anabaena sp. PCC 7120) cultivated in airlift photobioreactor. *Int. J. Hydrog. Energy.* 2014. 39 (14): 7553–7560. doi: 10.1016/j.ijhydene.2013.07.120.

Oswald, W. J., and Goleuke, C. G. Large-scale production of algae (PB-296411). California Univ., Berkeley (USA). 1967. https://www.osti.gov/biblio/5398149.

Owusu-Twum, M. Y., and Sharara, M. A. Sludge management in anaerobic swine lagoons: a review. *J. Environ. Manag.* 2020. 271: 110949. doi: 10.1016/j.jenvman.2020.110949.

Pachés, M., Martínez-Guijarro, R., González-Camejo, J., Seco, A., and Barat, R. Selecting the most suitable microalgae species to treat the effluent from an anaerobic membrane bioreactor. Environ. Technol. 2020. 41 (3): 267–276. doi: 10.1080/09593330.2018.1496148

Pandey, P. K., Cao, W., Wang, Y., Vaddella, V., Castillo, A. R., Souza, A., and Rio, N. S. Simulating the effects of mesophilic anaerobic and aerobic digestions, lagoon system, and composting on pathogen inactivation. *Ecol. Eng.* 2016. 97: 633–641. doi: 10.1016/j.ecoleng.2016.10.047.

Pandey, A., Singh, M. P., Kumar, S., and Srivastava, S. Phycoremediation of persistent organic pollutants from wastewater: retrospect and prospects. In: Application of Microalgae in Wastewater Treatment. Springer. 2019. pp. 207–235.

Papadopoulos, K. P., Economou, C. N., Tekerlekopoulou, A. G., and Vayenas, D. V. Two-step treatment of brewery wastewater using electrocoagulation and cyanobacteria-based cultivation. *J. Environ. Manage.* 2020. 265: 110543. doi: 10.1016/j.jenvman.2020.110543.

Park, J. B. K., Craggs, R. J., and Shilton, A. N. Wastewater treatment high rate algal ponds for biofuel production. *Bioresour. Technol.* 2011. 102 (1): 35–42. doi: 10.1016/j.biortech.2010.06.158.

Patel, A. K., Huang, E. L., Low-Décarie, E., and Lefsrud, M. G. Comparative shotgun proteomic analysis of wastewater-cultured microalgae: nitrogen sensing and carbon fixation for growth and nutrient removal in *Chlamydomonas reinhardtii*. *J. Proteome Res.* 2015. 14 (8): 3051–3067. doi: 10.1021/pr501316h.

Prabha, R., Singh, D. P., Sinha, S., Ahmad, K., and Rai, A. Genome-wide comparative analysis of codon usage bias and codon context patterns among cyanobacterial genomes. *Mar. Genomics.* 2017. 32: 31–39. doi: 10.1016/j.margen.2016.10.001.

Razzak, S. A., Hossain, M. M., Lucky, R. A., Bassi, A. S., and de Lasa, H. Integrated CO_2 capture, wastewater treatment and biofuel production by microalgae culturing- a review. *Renew. Sust. Energ. Rev.* 2013. 27: 622–653. doi: 10.1016/j.rser.2013.05.063.

Roeselers, G., van Loosdrecht, M. C. M., and Muyzer, G. Heterotrophic pioneers facilitate phototrophic biofilm development. *Microb. Ecol.* 2007. 54 (3): 578–585. doi: 10.1007/s00248-007-9238-x.

Roostaei, J., Zhang, Y., Gopalakrishnan, K., and Ochocki, A. J. Mixotrophic microalgae biofilm: a novel algae cultivation strategy for improved productivity and cost-efficiency of biofuel feedstock production. *Sci. Rep.* 2018. 8 (1): 12528. doi: 10.1038/s41598-018-31016-1.

Rosso, D., Larson, L. E., and Stenstrom, M. K. Aeration of large-scale municipal wastewater treatment plants: State of the art. *Water Sci. Technol.* 2008. 57 (7): 973–978. doi: 10.2166/wst.2008.218.

Sacristán de Alva, M., Luna-Pabello, V. M., Cadena, E., and Ortíz, E. Green microalga *Scenedesmus acutus* grown on municipal wastewater to couple nutrient removal with lipid accumulation for biodiesel production. *Bioresour. Technol.* 2013. 146: 744–748. doi: 10.1016/j.biortech.2013.07.061

Sala-Garrido, R., Hernández-Sancho, F., and Molinos-Senante, M. Assessing the efficiency of wastewater treatment plants in an uncertain context: A DEA with tolerances approach. *Environ. Sci. Policy.* 2012. 18: 34–44. doi: 10.1016/j.envsci.2011.12.012.

Sepúlveda, C., Acién, F. G., Gómez, C., Jiménez-Ruíz, N., Riquelme, C., and Molina-Grima, E. Utilization of centrate for the production of the marine microalgae *Nannochloropsis gaditana*. *Algal Res.* 2015. 9: 107–116. doi: 10.1016/j.algal.2015.03.004.

Sharafi, K., Fazlzadehdavil, M., Pirsaheb, M., Derayat, J., and Hazrati, S. The comparison of parasite eggs and protozoan cysts of urban raw wastewater and efficiency of various wastewater treatment systems to remove them. *Ecol. Eng.* 2012. 44: 244–248. doi: 10.1016/j.ecoleng.2012.03.008.

Singh, J., and Thakur, I. S. Evaluation of cyanobacterial endolith *Leptolyngbya* sp. ISTCY101, for integrated wastewater treatment and biodiesel production: a toxicological perspective. *Algal Res.* 2015. 11: 294–303. doi: 10.1016/j.algal.2015.07.010.

Tan, X., Hou, S., Song, K., Georg, J., Klähn, S., Lu, X., and Hess, W. R. The primary transcriptome of the fast-growing cyanobacterium *Synechococcus elongatus* UTEX 2973. *Biotechnol. Biofuels.* 2018. 11: 218.

Tang, C. C., Tian, Y., Liang, H., Zuo, W., Wang, Z. W., Zhang, J., and He, Z. W. Enhanced nitrogen and phosphorus removal from domestic wastewater via algae-assisted sequencing batch biofilm reactor. *Bioresour. Technol.* 2018. 250: 185–190. doi: 10.1016/j.biortech.2017.11.028.

Thiriet-Rupert, S., Carrier, G., Chénais, B., Trottier, C., Bougaran, G., Cadoret, J. P., Schoefs, B., and Saint-Jean, B. Transcription factors in microalgae: genome-wide prediction and comparative analysis. *BMC Genomics.* 2016. 17: 282. doi: 10.1186/s12864-016-2610-9.

Tiron, O., Bumbac, C., Patroescu, I. V., Badescu, V. R., and Postolache, C. Granular activated algae for wastewater treatment. *Water Sci. Technol.* 2015. 71 (6): 832–839. doi: 10.2166/wst.2015.010.

Tsolcha, O. N., Tekerlekopoulou, A. G., Akratos, C. S., Antonopoulou, G., Aggelis, G., Genitsaris, S., Moustaka-Gouni, M., and Vayenas, D. V. A *Leptolyngbya*-based microbial consortium for agro-industrial wastewaters treatment and biodiesel production. *Environ. Sci. Pollut. Res.* 2018. 25 (18): 17957–17966. doi: 10.1007/s11356-018-1989-z.

Wang, Y., Fan, L., Crosbie, N., and Roddick, F. A. Photodegradation of emerging contaminants in a sunlit wastewater lagoon, seasonal measurements, environmental impacts and modelling. *Environ. Sci.: Water Res. Technol.* 2020. 6 (12): 3380–3390. doi: 10.1039/D0EW00527D.

Wang, M., Lee, E., Zhang, Q., and Ergas, S. J. Anaerobic co-digestion of swine manure and microalgae *Chlorella* sp.: experimental studies and energy analysis. *BioEnergy Res.* 2016. 9 (4): 1204–1215. doi: 10.1007/s12155-016-9769-4.

Wehr, J. D., and Sheath, R. G. Habitats of Freshwater Algae. In J. D. Wehr, R. G. Sheath, and J. P. Kociolek (Eds.). Freshwater Algae of North America (2nd ed.). 2015. pp. 13–74.
Whitton, R., Santinelli, M., Pidou, M., Ometto, F., Henderson, R., Roddick, F., Jarvis, P., Villa, R., and Jefferson, B. Tertiary nutrient removal from wastewater by immobilised microalgae: impact of wastewater nutrient characteristics and hydraulic retention time (HRT). *H2Open Journal*. 2018. 1 (1): 12–25. doi: 10.2166/h2oj.2018.008.
Wilks, R., and Scholes, E. DNA sequencing of pulp and paper wastewater treatment systems to inform process analysis. *Appita J.: J. Tech. Assoc. Australian New Zealand Pulp Paper Ind.* 2018. 71: 157.
Wollmann, F., Dietze, S., Ackermann, J.-U., Bley, T., Walther, T., Steingroewer, J., and Krujatz, F. Microalgae wastewater treatment: biological and technological approaches. *Eng. Life Sci.* 2019. 19 (12): 860–871. doi: 10.1002/elsc.201900071.
Xiong, J. Q., Kurade, M. B., Jeon, B. H. Can microalgae remove pharmaceutical contaminants from water? *Trends Biotechnol.* 2018. 36 (1): 30–44. doi: 10.1016/j.tibtech.2017.09.003.
Ye, J., Song, Z., Wang, L., and Zhu, J. Metagenomic analysis of microbiota structure evolution in phytoremediation of a swine lagoon wastewater. *Bioresour. Technol.* 2016. 219: 439–444. doi: 10.1016/j.biortech.2016.08.013.
Zhao, X., Kumar, K., Gross, M. A., Kunetz, T. E., and Wen, Z. Evaluation of revolving algae biofilm reactors for nutrients and metals removal from sludge thickening supernatant in a municipal wastewater treatment facility. *Water Res.* 2018. 143: 467–478. doi: 10.1016/j.watres.2018.07.001.
Zhou, W., Wang, Z., Xu, J., and Ma, L. Cultivation of microalgae *Chlorella zofingiensis* on municipal wastewater and biogas slurry towards bioenergy. *J. Biosci. Bioeng.* 2018. 126 (5): 644–648. doi: 10.1016/j.jbiosc.2018.05.006.

2 Wastewater Utilization as Growth Medium for Seaweed, Microalgae and Cyanobacteria, Defined as Potential Source of Human and Animal Services

Silvia Lomartire, Diana Pacheco, Glácio Souza Araújo, João C. Marques, Leonel Pereira, and Ana M. M. Gonçalves

CONTENTS

DOI: 10.1201/9781003155713-2

2.1 INTRODUCTION

Seaweeds, microalgae and cyanobacteria are potential sources of nutrients and bioactive compounds used in different fields, as shown in Figure 2.1. Seaweed cultivation is implied for food, pharmaceutical and cosmetics industries, along with agriculture fertilizers. In 2018, almost 600 different aquatic species used in biotechnological industry (Reid et al. 2019) were cultivated in 200 countries (FAO 2018). Microalgae and cyanobacteria are microscopic organisms with eukaryotic and prokaryotic cells, respectively, and both contain a huge quantity of bioactive compounds used for commercial use (Shanab et al. 2012; García et al. 2017).

Recent works show the most abundant microalgal classes used for widespread applications (Mehariya et al. 2021a; Goswami et al. 2021a) such as food and feed (Yaakob et al. 2014; Liu and Chen 2016; Bleakley and Hayes 2017), nutraceutical products (Yan et al. 2016; Bilal et al. 2017; Wells et al. 2017, Mehariya et al. 2021b), and biotechnological products (Chew et al. 2017; Odjadjare et al. 2017) are cyanobacteria, Chlorophyta, Bacillariophyta and Chrysophyceae (García et al. 2017).

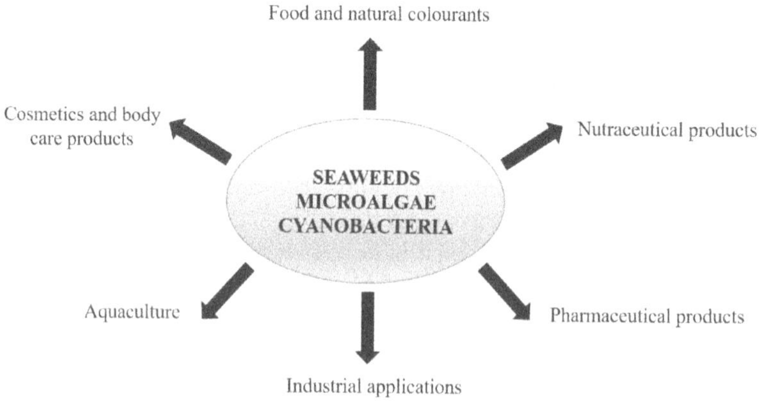

FIGURE 2.1 Products obtained by seaweeds, microalgae and cyanobacteria bioactive compounds.

Microalgae are cultivated for their polyunsaturated fatty acids (PUFAs), which have a good flavour and also do not contain cholesterol and heavy metals or polychlorobiphenyls (Robles Medina et al. 1995). Moreover, PUFAs are considered a source of antioxidant compounds (Li et al. 2007; Natrah et al. 2007; Hajimahmoodi et al. 2010; Rodriguez-Garcia and Guil-Guerrero 2008; Chacón-Lee and González-Mariño 2010; Lee et al. 2010). Scientists have been researching on microalgal antioxidant properties to find replacements of synthetic antioxidants (Goiris et al. 2012), particularly due to the presence of carotenoids that have a significant antioxidant action (Jahnke 1999; Takaichi 2011; Goswami et al. 2021b).

Seaweed cultivation is usually carried out in a natural environment, where the biomass production is increased ten times (Lüning 1990), but seaweeds are also used in fish aquaculture; numerous studies show that the presence of seaweed in fish aquaculture tanks or open sea fish cages (Buschmann et al. 2001; Troell et al. 1999; Chopin et al. 1999; Neori et al. 2000; Hernández et al. 2002) help the removal of heavy nutrients present in fed aquaculture. Neori et al. (1996) demonstrated that more than 90% of nutrients discharged from farm fish were removed by seaweed.

The oldest production system for microalgae is open pond system, where microalgae can grow in their natural environment (Agrawal et al. 2020). Nevertheless, this system has some limitations such as control of the cultivation temperature during the diurnal cycle or contamination due to other species (Kadir et al. 2018). To overcome the disadvantages of the open pond system, photobioreactor (PBR) systems have been used. The benefits of PBRs are multiple, since cultivation could be maintained in different conditions; they prevent production of other algal strain and contamination from other organisms (Salama et al. 2017; Safi et al. 2014); closed PBR systems can increase the biomass production and prevent the water loss by evaporation (Posten 2009). However, PBRs are quite expensive with regard to the profit that comes from selling microalgae (Posten 2009). The presence of a hybrid cultivation system helps reduce microalgae costs; it consists of transfer of a portion of microalgal cells from the PBR to open ponds. In this way, the lipid content of microalgal cells increases (Kadir et al. 2018). Furthermore, costs may be reduced by using wastewater as a growth medium for cultivation (Kumar et al. 2019); a good growth medium for microalgae could be piggery wastewater due to the presence of nitrogen and phosphate in high quantity (Kadir et al. 2018) that will be absorbed by the microalgae species (De-Bashan and Bashan 2010; Hoffmann 1998; Mallik 2002) like *Chlorella* (Chlorophyta), which was treated with wastewater to have an increase in biomass production. Results showed that *Chlorella* absorbs nutrients present in wastewater for enhanced biomass and lipid yield (Rawat et al. 2011). This chapter focuses on the utility of seaweed, microalgae and cyanobacteria in the food, pharmaceutical, and industrial sectors and the use of wastewater as a potential growth medium for their cultivation.

2.2 CORRELATION BETWEEN BIOLOGICAL TOOLS AND PRODUCTION OF SERVICES FOR HUMANS AND ANIMALS

Being single-celled, filamentous or colonial organisms, microalgae may have a prokaryotic or eukaryotic cell structure (De-Bashan and Bashan 2010). Prokaryotes embrace the cyanobacteria, also called blue-green algae, while eukaryotes belong mainly to the phyla Bacillariophyta (diatoms), Miozoa (dinoflagellates), Haptophyta, Coccolithophyceae (coccolithophorids), Cryptophyta, Cryptophyceae (cryptomonads), Prasinophytina and Chlorophyta (green algae) (Bicudo and Menezes 2006).

Microalgae are organisms that make up most of the phytoplankton and have a fundamental position in the ecosystem, since they can provide nutriment for several organisms. Commercially, they are essential in the production of zooplankton, larvae and juveniles of crustaceans and fish. In addition, microalgae have been used for years as a human ingredient in food and recipes (Milledge 2011), and some species are still included in human diets (Lourenço 2006; Das et al. 2014).

Aphanizomenon flosaquae, *Arthrospira platensis* (formerly *Spirulina platensis*) (cyanobacteria) *Chlorella* sp., *Dunaliella salina* and *Dunaliella tertiolecta* (Chlorophyta) possess a high protein content and nutrition values; for this reason, these species are largely available in the market in the food section (Soletto et al. 2005; Rangel-Yagui et al. 2004).

Biotechnology applied to algae has resulted in major advances, especially over the past three decades. Some genera of microalgae such as *Botryococcus*, *Dunaliella*, *Haematococcus* (Chlorophyta), *Chlorella* and *Arthrospira* (cyanobacteria) are mainly cultivated to produce proteins and pigments like astaxanthin and β-carotene, glycerol, biofuels, drugs and other fine chemicals compounds (Spolaore et al. 2006).

Algae, in general, produce polysaccharides that have aroused great interest because of some of their physicochemical properties, making them quite useful for industrial purposes. The use of these polysaccharides is common in industries as thickeners, stabilizers and emulsifiers. Cultures of microalgae as useful sources for different products (fuel, animal and human food, etc.), as well as to obtain pharmaceutical products have been investigated for over 40 years. In addition, microalgae cultures as a source of other valuable bioactive compounds, such as lipids, enzymes, proteins and pigments, have also become quite evident (You and Barnett 2004).

Some species of microalgae can produce high amounts of biomass and accumulate large amounts of compounds with biological activity, which can be used in distinct applications: in the industrial, food, health and energy sectors (Lee and Han 2015). The algal composition varies according to environmental conditions, with species rich in proteins, polysaccharides, pigments and fatty acids (George et al. 2014).

These parameters directly affect biomass production and the biosynthesis of lipids, polysaccharides, proteins or pigments (Radmann and Costa 2008). Thus, the nutritional mode, cultivation system and strategy will depend on the cultivated species and the final bio-product that we want to extract (Valduga et al. 2009).

Under heterotrophic conditions, growth occurs in the absence of light and microalgae use organic carbon through aerobic respiration. This method has shown efficiency in producing bioactive compounds for further commercialization in cosmetic, pharmaceutical and food supplement products (Perez-Garcia et al. 2015). In mixotrophic growth, microalgae assimilate organic carbon, such as glucose or glycerol, through aerobic respiration (Kang et al. 2004).

2.2.1 Use in Aquaculture

Microalgae are commonly used in aquaculture for feeding molluscs post-larvae, fish and crustaceans due to the ease of cultivation, their small size, marked growth rate and high content of PUFAs (Derner et al. 2006; Guedes and Malcata 2012).

Several researches came out with the importance of using immunostimulants to prevent diseases, such as nutritional factors lactoferrin and vitamins B and C, in addition to hormonal factors such as prolactin. Other compounds extracted from various organisms, such as chitin polysaccharides obtained from the crustacean exoskeleton, β-glucan extracted from the cell walls of fungi and yeasts and sulphated polysaccharides (SP) of algae, have also shown immune-stimulating effects. These compounds mainly activate phagocytosis and antibacterial activity of defence cells against pathogens and their metabolites (Sakai et al. 1992; Sakai 1999; Campa-Córdova et al. 2002; Farias et al. 2004; Huang, Zhou, and Zhang 2006).

In addition to the immune-stimulating effect, SP from seaweed are also known to have other bioactivities, such as antiviral, anticoagulant, anti-thrombotic and anti-nociceptive activities (Hayashi et al. 1996; Farias et al. 2001; Pereira 2018). Some studies report that the administration of an optimal concentration of a certain compound with immunostimulatory activity is extremely important to result in an effective response. Some reports have shown that low concentrations have a better effect than higher ones (Boonyaratpalin et al. 1995; Park and Jeong 1996).

The administration of immunostimulants must be well monitored, as some of them can cause adverse effects on animals under experimentation. Hauton and Smith (2004) reported that it is very important to use an ideal dose, whether administered through immersion baths or orally, since an exaggerated dose may cause some risk of toxicity, especially when administered in the long term (Hauton and Smith 2004). Thus, the administration of an optimal dose of these compounds is extremely important to obtain an effective response, and several reports have shown that low doses have a better effect than higher doses (Araújo et al. 2008).

2.3 SEAWEED AS A POTENTIAL SOURCE FOR FOOD INDUSTRY, NUTRACEUTICAL AND PHARMACEUTICAL PRODUCTS

Seaweeds are widely used in industrial, nutraceutical and pharmaceutical applications. Table 2.1 includes the most used seaweeds in those fields.

2.3.1 Industrial Applications of Seaweed

Like microalgae, the biomass of macroalgae shows a high amount of energy, bio-syngas, bio-oil, biohydrogen and bio-char, confirming their potential as a raw material for biofuels (Bhardwaj et al. 2020; Goswami et al. 2021c; Mehariya et al. 2021c). Macroalgae and microalgae chemical compounds can influence the features of the product, such as pH, viscosity, density, higher heating value (HHV) and their structure. By far, the characteristics of these organisms allow them to be used as a raw material for biofuels production, which can be commercialized in the future (Lee et al. 2020).

TABLE 2.1
Seaweeds Used Most in Biotechnological Applications

	Seaweed	Reference
Industrial applications	*Carpophyllum flexuosum* *Carpophyllum plumosum* *Ecklonia radiata* *Undaria pinnatifida*	Zhang et al. (2020)
	Kappaphycus alvarezii	Rudke et al. (2020) Khatri et al. (2019)
	Solieria robusta *Gracilaria corticata var. cylindrica* *Gracilaria corticata*	Khatri et al. (2019)
	Gelidium corneum	Méndez et al. (2019)
Nutraceutical applications	*Neoporphyra haitanensis*	(Wei et al. 2015)
	Sargassum polycystum *Padina boryana* *Turbinaria ornata*	Fauziee et al. (2020)
	Sarconema filiforme filiforme *Turbinaria conoides* *Ulva compressa*	Venkatesan et al. (2019)
Pharmaceutical applications	*Ulva* sp. *Gracilaria* sp.	Kazir et al. (2019)
	Macrocystis pyrifera *Lessonia vadosa*	Casoni et al. (2020)
	Padina australis *Sargassum aquifolium*	Chia et al. (2018)
	Ulva lactuca	Leston et al. (2015)

Wang et al. (2020) cite that low costs of producing biofuels from macroalgae can be achieved through the reduction in costs of each processing step (initial design, harvesting, pre-treatment, transport, selection of raw material and conversion process). The authors comment that, in general, a balance must be reached among the production of bioenergy from macroalgae, the cost of biofuel and the environmental impact caused by this activity (Wang et al. 2020).

All the processes involved in biorefinery aim to produce several compounds or products from microalgal biomass at the same time. Among four species of brown algae (Ochrophyta, Phaeophyceae), *Carpophyllum flexuosum*, *C. plumosum*, *Ecklonia radiata* and *Undaria pinnatifida* investigated by Zhang et al. (2020), extraction of carbohydrates, mannitol, pigments and phlorotannins was performed. The authors obtained yields, calculated based on the dry weight of each product in each species of kelp, of 3.4–9.8% (pigments), 22.2–30.7% (mannitol), 0.1–5.1% (phlorotannins), 5.2–15.5% (alginates), 12.2–18.5% (other carbohydrates) and 13.5–19.5% (residual algae). The results indicated that brown algae is a potential candidate for use in biorefineries in order to produce biomaterials, adding value and bioenergy (Zhang et al. 2020).

Rudke et al. (2020) mention that the biomass of the red macroalga (Rhodophyta) *Kappaphycus alvarezii* is useful in the production of chlorophyll, β-carotene, essential amino acids and phytohormones. The authors suggest the possibility, in the future, to extract ethanol, pigments and protein, fertilizers, and further compounds from this red macroalgae (Rudke et al. 2020).

A solid residue from macroalgae biomass generated by industrial agar production was used by Ferrera-Lorenzo et al. (2014) as a precursor to pyrolysis for obtaining coal, oil and gas. Chemical analysis of algae showed that its high carbon, hydrogen and nitrogen content, together with its low ash content, becomes a potential precursor to activated carbon. The authors also observed that the fraction of oil, through the gas chromatography (GC) technique, showed compounds such as phenols, pyrroles and furans. The gas fraction had a high syngas content, increasing its high heating value (Ferrera-Lorenzo et al. 2014).

Khatri et al. (2019) studied the alternative use of red algae biomass in the production of biofuel. The authors verified that there is an intricate network of polymers in the cell walls of the macroalgae *K. alvarezii* involved in the production of oligosaccharide units, as it is for the red macroalgae (Rhodophyta) *Solieria robusta*, *Gracilaria corticata* var. *cylindrica* and *G. corticata*. The cell wall of marine algae showed similarity with the polymers found in the cell wall of more developed plants, which indicated the potential of using the biomass of these marine algae for biofuel production (Khatri et al. 2019).

Studying the residues of the agar-agar industry, Méndez et al. (2019) used this product as a raw material for obtaining hydrocarbons through hydrothermal carbonization, varying temperature and time, for the red macroalga *Gelidium corneum* (formerly *Gelidium sesquipedale*). The authors observed that the carbon content and the heating value increased with hydrothermal carbonization (Méndez et al. 2019).

Gullón et al. (2021) cite the incorporation of marine algae in meat products to avoid salt addition and obtaining healthier foods. The authors comment that some challenges must be overcome, such as proper selection of kelp according to the characteristics desired in the final products, as well as the dosages to be used. In addition, it is necessary to carefully evaluate the effects of the addition of algae on sensory, nutritional, physical-chemical and microbiological properties (Gullón et al. 2021).

Larson et al. (2021) researched 74 women from coastal villages in South Sulawesi, Indonesia, and found out that seaweed is considered as a big source of income; therefore, the growth of these algae also has an important social and economic impact. Part of the income obtained from algae was fundamental in creating concrete changes in five of the ten most important contributors to well-being: transport, education, housing, basic needs, other needs, etc. No negative changes were associated with the algae cultivation activity; indeed, the index of satisfaction increased in all the villages studied (Larson et al. 2021).

2.3.2 Nutraceutical Applications of Seaweeds

With regard to macroalgae, fucoxanthin, a carotenoid found in brown algae (Matsuno 2001), is consumed to prevent diseases and illnesses such as obesity, heart complication, diabetes, hypertension and cancer, besides disorders associated with inflammation (Miyashita et al. 2020).

The red seaweed *Neoporphyra haitanensis* (formerly *Pyropia haitanensis*) has a hydrocolloid that shows activity against bacteria, related to the presence of 1,8-dihydroxy-anthraquinone that communicate with the wall and cell membrane leading to cell deconstruction and cytoplasm leakage and, thus, strongly inhibiting cell growth (Wei et al. 2015).

Hydrocolloids have several benefits for human health, specifically a therapeutic potential in preventing and treating diseases such as hypertension, diabetes, gut problems and tumors (Manzoor et al. 2020).

Actually, researches are focused on combining different and new techniques for extracting hydrocolloids to overcome all the limitations that singular techniques present. These industrial processes will require further investigation and optimization due to the high number of parameters involved in the extraction of hydrocolloids. This will allow the incorporation of green chemistry in the hydrocolloid industry (Gomez et al. 2020).

Fauziee et al. (2020) reported the properties and bioactivities of natural polysaccharides extracted from Phaeophyceae (also called brown algae) from Malaysia, aiming at the development of high-value functional ingredients. The authors extracted alginate, fucoidan and laminarin from the brown macroalgae *Sargassum polycystum, Padina boryana and Turbinaria ornata. P. boryana* registered significantly ($p \leq 0.05$) higher carbohydrate content (74.78±1.63%) with higher fucoidan yield (Fpad=1.59±0.16%), while *T. ornata* contained significantly ($p \leq 0.05$) higher alginate yield (Atur=105.19±3.45%). Meanwhile, laminarin extracts had a significantly higher total phenolic content ($p \leq 0.05$) (Lsar=43.29±0.43 mg GAE g^{-1}) and superoxide anion elimination activity (Lsig=21.7±3.6%) (Fauziee et al. 2020).

Water-soluble polysaccharides of three Indian algae were studied by Venkatesan et al. (2019); they found that the red seaweed *Sarconema filiforme filiforme* had a higher content of glucose, protein, moisture and essential minerals, in addition to antioxidant power. For this macroalga, the polysaccharides also showed high larvicidal activity against the mosquito *Anopheles stephensi* (0.255567 mg L^{-1}) followed by the macroalga *Turbinaria conoides* for *Aedes aegypti* (0.174348 mg L^{-1}) and the green seaweed (Chlorophyta) *Ulva compressa* (formerly *Enteromorpha compressa*) for *Culex quinquefasciatus* (0.1111175 mg L^{-1}). In conclusion, Venkatesan et al. (2019) showed that the polysaccharides of the three macroalgae can be used not only as alternative and healthy food and nutraceutical source but also as a mosquitocide agent.

Barbosa et al. (2019) described the mechanisms related to the antiallergic and anti-inflammatory potential of phlorotannins and phlorotannin-rich natural compounds extracted from marine macroalgae, still little explored, with a broad range of activity and low toxicity (Barbosa et al. 2019).

Bioactive peptides and polysaccharides extracted from marine macroalgae have several biological activities, where the most favorable ones show antihypertensive, antioxidant, anticoagulant, anti-inflammatory, anti-tumor and immunostimulatory activities. These biological activities help in the development of functional food (Lafarga et al. 2020).

Seaweed has been used as a functional food in commercial products as a stabilizer, thickening agent, emulsifier, texture modifier and for phytochemicals enriched with vitamins and dietary fiber (Ganesan et al. 2019).

Gullón et al. (2020) cited the importance of using seaweed and/or seaweed extracts in meat products as a strategy for reformulating these products, increasing shelf life, or increasing nutritional, textural, organoleptic and sensory aspects for health promotion. Usually, this reformulation aims to replace some components present in meat products that are harmful to consumers, such as biocompounds, to modify the level of fat and prevent oxidative deterioration (Gullón et al. 2020).

2.3.3 Pharmaceutical Products From Seaweeds

Kazir et al. (2019) developed a particular system to produce a high-protein concentrate from the macroalgae *Ulva* sp. and *Gracilaria* sp. suitable for food implication and to study the digestibility, amino acid composition and antioxidant properties of algal protein concentrates (APCs). The authors found APC levels of 70% and 86% for *Ulva* sp. and *Gracilaria* sp., respectively. The authors also found that APCs exhibited antioxidant activity through polyphenolic compounds. Finally, these two macroalgae appear to be suitable and sustainable for human nutrition thanks to their content of essential amino acids, good digestibility and antioxidant properties (Kazir et al. 2019).

Guilherme et al. (2020) determined the amount of selenium in Antarctic Phaeophyceae macroalgae (*Adenocystis utricularis*, *Cystosphaera jacquinotii*, *Ascoseira mirabilis*, *Desmarestia anceps* and *Himantothallus grandifolius*) and Rhodophyta (*Georgiella confluens*, *Curdiea racovitzae* and *Iridaea cordata*) through voltammetric method. The authors also evaluated the effect of the matrix and the influence of copper concentration. After microwave-aided sample digestion, the limit of quantification was 5.21×10^{-3} mg kg^{-1} for the macroalgae *G. confluens* to 9.85×10^{-3} mg kg^{-1} for *I. cordata*. Quantification of Se was performed in the concentration range of 0.23 mg kg^{-1} for the macroalga *C. jacquinotii* to 1.22 mg kg^{-1} for *A. mirabilis* (Guilherme et al. 2020).

Casoni et al. (2020) used the sea brown macroalgae *Macrocystis pyrifera* and *Lessonia vadosa* for the production of sorbitol. This product is transformed into isosorbide, a molecule which can be used as a medication (dinitrate isosorbide) for cardiovascular issues (Casoni et al. 2020).

Chia et al. (2018) used a system to isolate phlorotannin, an active bio-compound indicated for its beneficial effect on human health, which could be used in pharmaceutical and food applications. Using 2-propanol/ammonium sulfate, the greatest recovery of phlorotannin was 76.1% and 91.67% with a purification factor of 2.49 and 1.59 for the brown macroalgae *Padina australis* and *Sargassum aquifolium* (formerly *Sargassum binderi*), respectively, in a system with good recycling of compounds, therefore, ecological (Chia et al. 2018).

Leston et al. (2015) presented two methods of determination (MS/MS and UV) for sulfathiazole, an antibiotic that causes risk to natural ecosystems, based on liquid chromatography, for the green seaweed *Ulva lactuca*. The authors obtained values of 2.79 ng g^{-1} and 2.83 μg g^{-1} for the MS/MS and UV methods, respectively, showing that this species of macroalga has great potential as a biological indicator of contamination (Leston et al. 2015).

2.3.4 Therapeutical Applications of Seaweeds

Thanigaivel et al. (2016) cite that the use of macroalgae bioactive compounds is an alternative way to control the spread of diseases in the different cultivations of aquatic organisms. Marine macroalgae are readily available and are the most suitable, in addition to being harmless to beneficial bacteria such as probiotics. These immunostimulants induce antipathogenic activity and may be used to avoid the use of synthetic antibiotics that cause resistance to organisms. The authors stated that the bioactive compounds encapsulated in the feed in the form of granules can be used to treat bacterial infections in fish and crustacean cultivations (Thanigaivel et al. 2016).

Marine macroalgae, such as *Fucus spiralis* and *Carpodesmia amentacea* (formerly *Cystoseira stricta*), can be a novelty as a source of antioxidants and phytochemicals, and could be used in the cosmetic, pharmaceutical and food industries for stress, disorder and free radical induction (Grina et al. 2020).

Leódido et al. (2017) mention that a part of the sulfated polysaccharide extracted from the macroalgae *Gracilaria intermedia* reduced gastrointestinal transit through anticholinergic mechanisms, minimizing diarrhea induced by the bacteria *Escherichia coli* and prevented weight loss in mice. In addition, the fraction of the extracted sulfated polysaccharide did not induce any signs of toxicity, thus being a possible candidate for the treatment of diarrhea (Leódido et al. 2017).

Okimura et al. (2020) studied the effect of a sulfated polysaccharide extracted from the edible brown seaweed *Ascophyllum nodosum* against intranasal infection by the bacterium *Streptococcus pneumoniae*. Control mice not treated with the polysaccharide began to die after seven days and 80% died fourteen days after infection. Oral assumption of the polysaccharide before and after bacterial infection resulted in a notable increase in the survival rate of the mice, with 90% at the lowest dose (167 mg kg^{-1} of body weight day^{-1}) and 100% at high dosage (500 mg kg^{-1} of body weight day^{-1}), surviving 14 days post-infection. The authors suggest that ascophyllan administered orally has a therapeutic effect on infection by *S. pneumoniae*, activating the host's defense system (Okimura et al. 2020).

Poulose et al. (2020) formulated a photoprotective cream using melanin produced by the bacterium *Halomonas venusta* isolated from the marine sponge *Callyspongia* sp. and a concentrate of the red seaweed *Gelidium spinosum*, which enhanced antioxidant and wound healing properties. The antimicrobial effect can reduce the emergence of drug-resistant bacteria and collateral effects of synthetic creams. When applied on the skin, the cream provided a cooling effect and immediate disappearance without the formation of a white or oily film (Poulose et al. 2020).

In brown algae, there is a large number and variety of bioactive compounds, such as phloroglucinol, fucoxanthin and fucoidan, which are among the most abundant and with promising anticancer activity. Antioxidant activity, inhibition of cell proliferation, induction of cell death, and suppression both of metastasis and angiogenesis are some of the anticancer mechanisms those three compounds possess, along with modulating breast cancer (Pádua et al. 2015).

2.4 MICROALGAE AS A POTENTIAL SOURCE FOR FOOD INDUSTRY, NUTRACEUTICAL AND PHARMACEUTICAL PRODUCTS

In this part of the chapter, the uses of microalgae in food, nutraceutical and pharmaceutical industries are analyzed, enhancing the most used strains of microalgae for achieving a healthy lifestyle and wellness.

2.4.1 Microalgae's Applications in Food Industry

Microalgae have the ability to reproduce in high concentration and in different types of environment, producing valuable products such as proteins and lipids (Borowitzka 2013). For this reason, they have a large commercial value and many companies have been commercializing microalgae as food or functional constituent of food (Pulz and Gross 2004).

The Food and Drug Administration (FDA) of the United States defines algal products as safe food. *Arthrospira platensis* (cyanobacteria), *Haematococcus lacustris* (formerly *Haematococcus pluvialis*), *Chlorella vulgaris*, (Chlorophyta), *Dunaliella salina* and *Crypthecodinium cohnii* (Miozoa) are included in the food sources registered into the Generally Recognized As Safe (GRAS) category (García et al. 2017).

In the second half of the 1990s has begun the use of microalgae for bioproducts of species that easily grow in outdoor ponds (Craggs et al. 2013): *C. vulgaris*, *A. platensis* (*Spirulina*) and the production of β-carotene from *D. salina* (Varshney et al. 2015).

Cyanotech (Hawaii) started its cultivation of the algae *H. lacustris* in the late 1990s, as a source of the carotenoid astaxanthin principally used in nutraceutical human products and also as a colouring agent for cultivated salmonids fish (Haque et al. 2016; Panis and Carreon 2016).

Microalgae have been consumed by populations since ancient times: Aztecs used the cyanobacteria *Arthrospira platensis* and *A. maxima* from Lake Toxcoco (Mexico) as complement food; Spanish historians described local fishermen collecting huge masses from lakes of blue-green colour for baking cakes (Matos 2019).

Chlorella and *Spirulina* are still available in the market since they are among the most nutritional food recognized by man and used for feeding animals. The production of *Spirulina* reaches more than 12,000 tons per year (70% of the production is in China, India and Taiwan), used in healthy drinks, cereal bars, biological soups, cookies, etc. (Andrade 2018; Santos et al. 2016), while *Chlorella* has a production of 500 tons per year (Draaisma et al. 2013).

There are other species used as animal food; for instance, there is widespread use in aquaculture of the genera *Tetraselmis* (Chlorophyta), *Isochrysis*, *Pavlova* (Haptophyta), *Phaeodactylum*, *Chaetoceros*, *Skeletonema*, *Thalassiosira* (Bacillariophyta) and *Nannochloropsis* (Eustigmatophyceae) (García et al. 2017).

The reason for considering microalgae as a source of food finds its base in the high protein content of some species (e.g., 55–70% for *S. platensis* and 42–55% for *C. vulgaris* per dry matter), plus the proven digestibility of these proteins. Studies show that the protein solubility (PS) was statistically the highest and the same for *C. vulgaris* at 84% and *Micractinium conductrix* (formerly *Micractinium reisseri*) at 78%, while PS was the lowest for *Nannochloropsis bacillaris* at 64%, and intermediate for *Tetracystis* sp. at 73% (Tibbetts et al. 2016).

Microalgal pigments like phycocyanin (PC) and phycoerythrin (PE) have an important commercial role not only in the food sector but also in biomedical and pharmaceutical areas (Borowitzka 2013).

Among the microalgal pigments produced, a few species of *Dunaliella* are widely researched for β-carotene production, used as a natural food colourant (Saha and Murray 2018).

Before considering microalgae as an ingredient of food, it is fundamental to execute a series of toxicological test, since some species of microalgae present toxic compounds produced by themselves (toxins) or obtained by accumulation (as heavy metals) (Matos 2017).

It is proven that – among others – *Spirulina*, *Isochrysis galbana*, *H. lacustris*, *C. vulgaris*, *D. salina*, *Phaeodactylum tricornutum* and *P. purpureum* do not produce toxins and are available in the market for aquaculture uses and food supplements (Matos 2019).

2.4.2 Nutraceutical Applications of Microalgae

The nutraceutical industry establishes a connection between the nutrients found in food products and their advantages to health, like prevention or treatment of diseases (Kalra 2003).

Microalgae are rich in compounds useful for functional aliments, such as proteins, peptides and amino acids (Galasso et al. 2019).

A global interest for nutraceutical and functional food showed up after these substances were recommended for health enhancement; consequently, more than 470 food products were commercially available (Bishop and Zubeck 2012).

Marine peptides have a healthy effect on organisms, since they prevent diseases like hypertension, cancer, antimutagenic, anti-tyrosinase, anticoagulant and also exert antioxidant properties (Vo et al. 2013; Rizzello et al. 2016). Moreover, proteins of marine origin have been shown to be beneficial to the vascular system and the heart (Sheih et al. 2009; Ejike et al. 2017; Suetsuna and Chen 2001).

Vitamins are important for health as enzyme cofactors, and some vitamins also show antioxidant activity. Since there are some vitamins that our body cannot synthetize, we need to consume them by external elements. Microalgae represent a source of vitamins: pro-vitamin A, vitamin C, vitamin E, and a few vitamins of the B group (e.g., B12) (Galasso et al. 2019); some strains also produce vitamin D (Jäpelt and Jakobsen 2013).

Another group of bioactive molecules synthesized by microalgae are mycosporine-like amino acids (MAAs) (Llewellyn and Airs 2010), which protect cells from UV-induced damage and the double bonds provide the cells with photostability and antioxidant response (Lawrence et al. 2017). Furthermore, from microalgae, it is possible to extract eicosapentaenoic acid (EPA) and docosahexanoic acid (DHA) (Pulz and Gross 2004), a unique vegetarian source for them.

Considering the huge diversity of microalgae, the main species known for nutraceutical potential are *Botryococcus braunii* (Sathasivam et al. 2019), *Chlorella vulgaris*, *Haematococcus lacustris* (Chlorophyta) (Molino et al. 2018), also known for antioxidant and antibacterial potential (Santhakumaran et al. 2020), and *Dunaliella salina* (Maadane et al. 2015).

Among the microalgae present in the market, there are *C. vulgaris*, *D.salina*, *H. lacustris* (Chlorophyta) and *Nannochloropsis oculata* (Eustigmatophyceae) (Saha and Murray 2018) or *Porphyridium purpureum* (Rhodophyta), which are commercialized for protein (28–39%), lipids (9–14%) and polysaccharides (40–57%) (Velea et al. 2011).

Chlorella is popularly produced and present in the market as a healthy aliment especially in the U.S.A and Asia. This microalga showed production of vitamin B12 and has the ability to remove problematic metals and pesticides (Shim et al. 2008; Arimoto-Kobayashi et al. 1998). It has been well-documented that *Chlorella* is involved in diverse antitumor (Tanaka et al. 1990), immune-pharmacol antioxidant, anti-inflammatory (Guzmán et al. 2003) and antimicrobial functions (Hasegawa et al. 1989). Moreover, it has the ability to regulate blood pressure, cholesterol level, heal rapidly and reinforce the immune system (Bewicke and Potter 1984).

Dunaliella salina (dominant in nutraceutical production) is famous for the production of β-carotene, protein and glycerol, easily extracted through its thin cell wall (Bental et al. 1988; Murthy et al. 2005). Studies revealed that β-carotene prevents various types of tumors (lung, pancreas, stomach, colon, cervix, rectum, prostate, breast and ovarian) (van Poppel and Goldbohm 1995). Carotenoids from *Dunaliella* are potent free radical scavengers that reduce levels of lipid peroxidation and enzyme inactivation, thereby restoring enzyme activity (Murthy et al. 2005). Likewise, β-carotene from *Dunaliella* is useful for enhancing intracellular communication and immune response (Hughes et al. 1997).

Haematococcus (predominantly *H. lacustris*) is mostly produced in the U.S.A and Israel (Spolaore et al. 2006), and its bioactive compounds are used for biotechnological purposes (Guerin et al. 2003).

This microalga is predominantly used for astaxanthin production, a carotenoid pigment that showed antioxidant activity stronger than β-carotene and more effective than vitamin E (Miki 1991; Naguib 2000).

It has antioxidant (Fassett and Coombes 2012), anti-cancer (Bertram 1999), anti-inflammatory and anti-bacterial activities (Bennedsen et al. 1999). Pigment is important for the protection of immune response, its antioxidant properties, enhancing the vision and protection against solar radiations (Jyonouchi et al. 1995).

2.4.3 Pharmaceutical Applications of Microalgae

Some organisms possess a variety of properties due to their elements; they could be useful against oxidation, inflammation, diabetic issues, fungal, viral, bacterial and parasitic infections, depending on the compounds they produce. This peculiarity of some organisms, like microalgae, permits the use of these molecules for potential pharmaceutical properties (Saha and Murray 2018).

For instance, polysaccharides from *Chlorella vulgaris*, *C. stigmatophora*, *Scenedesmus quadricauda* (Chlorophyta) and *Phaeodactylum tricornutum* (Bacillariophyta) have anti-inflammatory, immunomodulatory and antioxidant properties (Guzmán et al. 2003; Mohamed 2008). Gürlek et al. (2020) investigated the extracts of *Galdieria sulphuraria* (Rhodophyta, Cyanidiophyceae), *Ettlia carotinosa*, *Neochloris texensis*, *Mychonastes homosphaera* (formerly *Chlorella minutissima*), *Chlorella vulgaris*, *Stichococcus bacillaris* (Chlorophyta), *Crypthecodinium cohnii* (Miozoa) and *Schizochytrium limacinum* (fungi) to attest their antioxidant activities. For the first time, not only was *S. bacillaris'* antioxidant capacity determined but also the radical scavenging activity (RSA) of this species was detected with high value. In the future, *S. bacillaris* could be used in biotechnological applications (Gürlek et al. 2020).

The pigments phycocyanin and PE, as citated above, are not only widely used in the alimentary industry but also in the medical field thanks to the anti-inflammatory, anticancer, antiviral, neuroprotective and hepatoprotective properties. Recently, they have already been used in some cancer therapies (Matos 2019), and their fluorescence properties suggest the use of these molecules for diagnostic applications (Matos 2017). Moreover, the biological activities in microalgae are due to proteins or peptides as well (Brasil et al. 2017; Fan et al. 2014).

Microalgae are also involved in the production of antibiotics; the first antibacterial compound found in a microalga was from *Chlorella* by Pratt et al. (1944); the antibacterial compound named chlorellin, basically a mix of fatty acids, shows activity against gram-bacteria (positive and negative) (Pratt et al. 1944). Bloom of marine microalgae *Phaeocystis pouchetii* has been shown to produce acrylic acid, an antibiotic substance (Eberlein et al. 1985).

Among microalgae, there are some strains that have a relevant antimycotic activity, such as *Chlorella vulgaris* and the cyanobacteria *Anabaena flosaquae*, *Nostoc oryzae* (formerly *Anabaena oryzae*), *Desmonostoc muscorum* (formerly *Nostoc muscorum*), *Nostoc humifusum*, *Oscillatoria* sp., *Leptolyngbya fragilis* (formerly *Phormedium fragile*), *Arthrospira platensis* and *Wollea saccata* (Singh et al. 2017). These organisms reduce the fungal mycelium of *Cercospora beticola* (fungi), which causes disease in leaves of plants that produce sugar. Anyway, the most functional antifungal is the cyanobacteria *Arthrospira platensis* (Singh et al. 2017).

The use of microalgal proteins is also related to the cosmetic industry; proteins and derivatives of *A. platensis* and *Porphyridium purpureum* provide nutrients to skin and hair, and may be used in the production of cosmetics such as as chemicals emulsifiers and foaming products for personal care (body lotions, shampoos, soap, hair and shaving foam) (Matos 2019).

2.4.4 Companies That Produce Microalgae-Based Products

The use of microalgae is convenient in food economy because not only do they give some advantages that standard synthetic and traditional molecules do not have, but also the cost of obtaining microalgae-based molecules is less than the synthetic ones.

Due to the high rate of proliferation, the ability to grow in extreme environment and the easy way to be genetically manipulated, microalgae are ideal for industrial purposes (Chaturvedi et al. 2021; Ghosh et al. 2016; Vazquez-Villegas et al. 2018).

A good commercial example of phycocyanin extracted by *A. platensis* is presented by Lina Blue, a Japanese company. This company produces a blue non-toxic powder that tastes sweet (Matos 2019).

Microlife Nutrition, an Italian company, has developed *Spirulina* as an ingredient for rice, pasta, sfogliatine and grissini. Moreover, *Spirulina* powder was used by Microlife Nutrition to produce wafer with orange flavour, chocolate bars and biscuits (Fradique et al. 2010), meanwhile French companies produce pasta from *Spirulina* (Fradique et al. 2010).

The microalga *Haematococcus*, already cited for the extraction of astaxanthin, is utilized for products by Cynotech Corporation, Parry Nutraceuticals, BioReal, Inc., Fuji Health Science, Valensa International, Alga Technologies and Aquasearch Inc. (Bishop and Zubeck 2012). Furthermore, *Chlorella* is distributed by Lucky Vitamin™, Prime Chlorella™ Distribution Inc., Sun Chlorella, HerbMark, Puritan's Pride and more (Bishop and Zubeck 2012). The dominant companies in the production of *Dunaliella* are located in Israel, China, U.S.A, and Australia and include Betatene, Western Biotechnology, AquaCarotene LTD, Cyanotech Corp., Nature Beta Technologies and more (Del Campo et al. 2007), which mainly produce *Dunaliella* for β-carotene production.

2.5 CYANOBACTERIA AS A POTENTIAL SOURCE FOR FOOD INDUSTRY, NUTRACEUTICAL AND PHARMACEUTICAL PRODUCTS

Cyanobacteria are part of an ancient group of gram-negative prokaryotes that are very common in every environment. They have a strong adaptation and can live in extreme conditions influenced by high UV radiation, extreme temperatures and with low presence of nutrients (Tandeau de Marsac and Houmard 1993).

Cyanobacteria are involved in biotechnological applications; their biomolecules are approved for drugs and clinical trials (Gerwick and Moore 2012). Due to that, cyanobacteria are considered for the extraction of elements that could be used in food, nutraceutical and pharmaceutical industries (Abed et al. 2009; Vijayakumar and Menakha 2015; Bermejo et al. 1997; Saini et al. 2018).

Bioactive compounds produced by cyanobacteria have various properties which offer health benefits; indeed they are very commonly found in the market (Gupta et al. 2013).

This part of the chapter reports the uses of cyanobacteria in food, nutraceutical and pharmaceutical industries, and Table 2.2 shows the microalgae and cyanobacteria mostly used in biotechnological applications.

2.5.1 CYANOBACTERIA'S APPLICATION IN FOOD INDUSTRY

Aquatic environments are considered the greatest source of bioactive compounds used in biotechnological applications (Freitas et al. 2012). Some cyanobacteria

TABLE 2.2
Microalgae and Cyanobacteria Most Used in Biotechnological Applications

	Microalgae	Reference	Cyanobacteria	References
Food industry	*Haematococcus lacustris* *Chlorella vulgaris* *Crypthecodinium cohnii*	García et al. (2017); Varshney et al. (2015); Haque et al. (2016); Panis and Carreon (2016); Varshney et al. (2015)	*Dolichospermum circinale* *Microcystis pulverea* *Nostoc punctiforme, Nodosilinea bijugata*	Sharma et al. (2011)
	Dunaliella salina	Draaisma et al. (2013); Varshney et al. (2015)	*Arthrospira platensis*	Mishra et al. (2010).
			Nostoc flagelliforme	Gao (1998)
	Isochrysis galbana	(Matos 2019)	*Nostoc sphaeroides*	Qiu et al. (2002)
	Phaeodactylum tricornutum *Pennisetum purpureum*			
Nutraceutical applications	*Botryococcus braunii*	Sathasivam et al. (2019)	*Geitlerinema sp.,*	Renugadevi et al. (2018)
			Arthrospira platensis	Mishra et al. (2010)
	Chlorella vulgaris *Haematococcus lacustris* *Dunaliella salina*	Molino et al. (2018); Spolaore et al. (2006) Maadane et al. (2015); Bental et al. (1988); Murthy et al. (2005)	*Trichormus variabilis* *Aphanotece microscopica*	Di Pippo et al. (2013); Gismondi et al. (2016) Manzoni Maroneze et al. (2019).
			Microcoleus autumnalis	
	Nannochloropsis oculate	Saha and Murray (2018)		
	Porphyridium purpureum	Velea et al. (2011)		

rich in nutrients and bioactive compounds are used in the food industry, such as *Dolichospermum circinale* (formerly *Anabaena hassallii*), *Microcystis pulverea*, *Spirulina*, *Nostoc punctiforme* and *Nodosilinea bijugata* (formerly *Phormidium bijugatum*) (Sharma et al. 2011).

Records demonstrate that cyanobacterial biomass was used as food since ancient times; historical data proves that the cyanobacterium *Arthrospira* (commercially known as *Spirulina*) was consumed by the Aztecs (Mexico) and in Africa long ago (Grosshagauer et al. 2020). It is considered as "super food" because of the huge number of active components produced: 50–70% protein and fatty polyunsaturated acids (1.5–2%), lipids (5–6%) as well as various vitamins (Grosshagauer et al. 2020).

The genus *Nostoc* is well-known since ancient times as well; *Nostoc flagelliforme* has been present in Chinese cuisine since the Jin Dynasty (265–316 A.D.) (Gao 1998). This cyanobacterium is called "Facai" in China, which means "hair vegetable", from its similarity with black hair; it is also a symbol of fortune because the pronunciation of the word "Facai" refers to a word which means "to be fortunate and get rich" (Gao 1998). The genus *Nostoc* may be used as complete food or as supplement for diets as well (Manolidi et al. 2018). *Nostoc sphaeroides* is a peculiar species found in spheroid colony shape (Huang et al. 1998), which is also used as an ingredient in traditional Chinese recepies (Qiu et al. 2002).

Cyanobacteria are largely used in the food market (Mantzouki et al. 2018; Carmichael and Boyer 2016), especially in Asia and South America. Nowadays, more than 70 countries commercialize cyanobacterial-based products (Merel et al. 2013).

Algae, microalgae and cyanobacteria are used as natural food colourants; their colour is due to the presence of phycobiliproteins (PBPs), PE and PC, which are coloured water-soluble proteins. The characteristic blue colour of cyanobacteria is due to the presence of PC pigment (Hu et al. 2018), widely used as natural blue colour in food and make up (Hemlata and Fatma 2018; Moraes et al. 2011).

For example, PC from *Spirulina* has been added in sweets, yogurts, ice creams, candies, soft and healthy drinks, green tea, jellies, etc. (Mishra et al. 2010).

PBPs of isolated Nostocaceae strains are used as colorants as well (Galetovic et al. 2020).

Nevertheless, PE has been mostly used for fluorescence techniques in medical field, more than in the alimentary industry (Galetovic et al. 2020).

2.5.2 Nutraceutical Applications of Cyanobacteria

Cyanobacterial PBPs, PE and PC, as already mentioned, are very abundant in cyanobacteria and not only are they used in the food industry but also for nutraceutical purpose for their antioxidant properties. The antioxidant property is fundamental to assure the wellness of our health; it has the ability to delay or inhibit the oxidation process and also defend the organisms from free radicals (Renugadevi et al. 2018).

Among the genus rich in PC is *Geitlerinema* sp., whose pigment shows antioxidant properties (Renugadevi et al. 2018) determined by the phosphomolybdenum method (Prieto et al. 1999; Aktumsek et al. 2013). Results indicate that PBPs and methanol found in strains of cyanobacteria from the Atacama Desert also have antioxidant properties similar to fruits (e.g., pineapple, berries, passion-fruits, etc.) (Kuskoski et al. 2005), which make those cyanobacteria good candidates for biotechnological applications in the nutritional, pharmaceutical and cosmetic industries (Galetovic et al. 2020).

The use of cyanobacterium *Arthrospira platensis*, known as *Spirulina*, has been suggested in diets of patients who suffer from malnutrition, immune diseases and neural issues. Although the high presence of bioactive compounds has been proved, studies about the antiviral effects of *Spirulina* are still required in order to better understand what help they can offer in pharmaceutical applications (Mishra et al. 2014).

The production of PUFAs and exopolymeric substances (EPS) from the native strain of the cyanobacteria *Trichormus variabilis* has been shown useful in nutraceutical products and biomaterials (Di Pippo et al. 2013; Gismondi et al. 2016). Apart from the production of PUFAs and EPS, *T. variabilis* is a good source of linoleic acid and α-linolenic acid, essential nutritional components that show specific functions in the human metabolism (Koller et al. 2014), such as in the regulation of oxygen and electron transport and membrane fluidity (Funk 2001; Cardozo et al. 2007). Moreover, linoleic and α-linolenic acid may be effective in cardiac protection (Judé et al. 2006) and cancer prevention (Rose and Connolly 1999; Xu and Qian 2014; Molfino et al. 2017), type 2 diabetes, inflammation and obesity (Cardozo et al. 2007; García et al. 2017; Lee et al. 2016), suggesting the use of *T. variabilis* as a low-cost source of PUFA for nutraceutical applications (Bellini et al. 2018).

Furthermore, cyanobacteria are the richest source of vitamins. For example, *Spirulina* is found to be rich in vitamin B12 (two to six times richer than raw beef liver) and vitamin E, with 20 g of *Spirulina* fulfilling the daily body requirements of Vitamin B12 and 70% of B1 (thiamine), 50% of Vitamin B2 (riboflavin) and 12% of Vitamin B3 (niacin) (Switzer 1982). It also contains other nutrients, including B complex vitamins, beta-carotene, vitamin E, manganese, zinc, copper, iron, selenium and γ linolenic acid (an essential fatty acid) (Gupta et al. 2013).

Besides, cyanobacterial carotenoid esters accumulated in non-stressed conditions may provide an extra value to the importance of carotenoids for the production of functional food (Manzoni Maroneze et al. 2019). Three carotenoid esters useful for this purpose are identified in *Aphanotece microscopica* and *Microcoleus autumnalis* (formerly *Phormidium autumnale*) (Manzoni Maroneze et al. 2019).

2.5.3 Pharmaceutical Applications of Cyanobacteria

Cyanobacteria are a great source of bioactive compounds that exhibit properties used in the pharmacological field (Moore 1996; Beltran and Neilan 2000) such as against microbial activities, immunosuppressant, tumor, HIV, bacterial activities, fungal activities, inflammatory diseases, tuberculosis and viral activities (Gademann and Portmann 2008).

Furthermore, cyanobacteria produce many substances that express antifungal, antibiotic and anticarcinogens activities (Dittmann et al. 2001).

The chemical structures and bioactive compounds of cyanobacteria could be helpful in finding new antitumoral agents, which do not have the side effects of the current treatments (e.g., chemotherapy) (Vijayakumar and Menakha 2015).

Among the bioactive compounds with anticancer properties, borophycin is obtained from *Nostoc carneum* (formerly *Nostoc spongiaeforme*), and it showed high cytotoxicity for cancer (Davidson 1995; Gupta et al. 2012; Hemscheidt et al. 1994; Torres et al. 2014). Apratoxin A, which is a potential cytotoxic metabolite, has been discovered to be a powerful cancer cell cytotoxin used in cancer treatments (Grinberg et al. 2002). Cylindrospermopsin is a cytotoxin (Runnegar et al. 2002) that presents neurotoxic (Kiss et al. 2002) effects and it is a carcinogen (Humpage et al. 2000). The species identified as cylindrospermopsin producers are: *Raphidiopsis raciborskii* (formerly *Cylindrospermopsis raciborskii*), *Chrysosporum ovalisporum* (formerly *Aphanizomenon ovalisporum*), *Aphanizomenon flosaquae*, *Umezakia natans*, *Raphdiopsis curvata*, *Chrysosporum bergii* (formerly *Anabaena bergii*), *Anabaena lapponica* and *Microseira wollei* (formerly *Lyngbya wollei*) (Seifert et al. 2007); dragonamide C and dragonamide D, isolated from a cyanobacteria *Lyngbya* sp. (Gunasekera et al. 2008) show anticancerogenic activity as well (Singh et al. 2017).

Studies show that phytol, a fundamental component of chlorophyll and also a cofactor for photosynthesis in cyanobacteria (Nowicka and Kruk 2010; Kiyota et al. 2014), can be helpful in improving the immune system against early stages of tumor (Mukund et al. 2014); phytol also presents anti-inflammatory properties (Shimizu and Tomoo 1994; Silva-Stenico et al. 2011).

Cyanobacteria also express antiviral properties; Asthana et al. (2006) conducted studies on the cyanobacteria *Fischerella* sp., which show an active action against *Mycobacterium tuberculosis*, *Enterobacter aerogenes*, *Staphylococcus aureus*, *Pseudomonas aeruginosa*, *Salmonella typhi* and *Escherichia coli* (Asthana et al. 2006).

Cyanobacterial PBPs and pigments are extensively used in bioindustry for coloration of food and cosmetic products; furthermore, some pigments – like carotenoids – act even as antioxidants, protecting our organisms from UV radiations. These compounds may implement the immunological response and also prevent the risk of cardiovascular diseases, cancer, multiple sclerosis, etc. (Larsson et al. 2007). Carotenoids have been discovered in huge quantities in the *Anabaena* strains (Nayak et al. 2007).

Cyanobacteria are also used to produce antibiotics from compounds such as phenols and fatty acids (Kamenarska et al. 2009); thermophilic cyanobacteria (*Oscillatoria subbrevis*, *Oscillatoria tenuis*, *Pseudanabaena limnetica*, formerly *Oscillatoria limnetica*), *Achroonema angustatum* (formerly *Oscillatoria angusta*), *Phormidium articulatum* (formerly *Oscillatoria articulata*), *Synechocystis aquatilis* and *Cyanobacterium cedrorum* (formerly *Synechococcus cerdrorum*) show a response against *Bacillus subtilis* and *Bacillus pumilus* bacterial activities. Cultivating species of *Phormidium*, antimicrobial compounds capable of arresting the reproduction of both gram-positive and gram-negative bacteria, have also been discovered (Fish and Codd 1994). Furthermore, *Phormidium's* antimicrobial properties are expressed against fungal species as well, like *Candida albicans* and *Cladosporium resinae*, which are immune to others cyanobacteria strains (Patel et al. 2019).

Cyanobacteria are also used in the field of cosmetics, for example, for the production of sunscreens that contain biopterin glucose, an UV-A-absorbing chromophore (Matsunaga et al. 1993), or scytonemin, a carotenoid that has showed antioxidant activity in many tests and also has UV absorption properties (Takamatsu et al. 2003).

Among antioxidant biological compounds produced by cyanobacteria, lycopene is considered one of the powerful antioxidants, followed by tocopherol, carotene and lutein. These biological compounds can be also used as protection by UV radiations (Singh et al. 2012). *Wollea vaginicola* (formerly *Anabaena vaginicola*) has been discovered as a good source of lycopene and is a good candidate for anti-aging agent production (Hashtroudi et al. 2013).

MAAs also contribute in the protection by UV radiations; they guarantee a limited but significant protection, for example, MAAs present in *Nostoc commune* absorb photons that reach cells (Scherer et al. 1988).

Well-known enterprises made investments in microalgal-based cosmetics products (e.g., Louis Vuitton Moët Hennessy, Paris, France, and Danial Jouvance, Carnac, France) (Mourelle et al. 2017).

As already mentioned, PC, phycocyanobilins and phycoerythrobilins are largely used in the food industry, but also could be interesting to involve them in the production of make up products (Mourelle et al. 2017).

2.6 METHODS OF CULTIVATING MACROALGAE, MICROALGAE AND CYANOBACTERIA

As previously shown, algae and cyanobacteria can be a wide source of bioactive compounds with several daily and biotechnological applications. Marine resources such as macroalgae, microalgae and cyanobacteria could be a useful tool to provide food, feed and pharmacological products to the humanity (Bjerregaard et al. 2016). However, to upscale algal biomass production and optimize the production of bioactive compounds, it is important to explore less expensive and sustainable cultivation methods and strategies (Pacheco et al. 2020).

2.6.1 Macroalgae Cultivation

Macroalgae aquaculture is a current practice that aims to reduce environmental pressure on natural ecosystems feedstock. In fact, macroalgae cultivation contributes to reduce the risk of coastal eutrophication. The Table 2.2. give a list of microalgae and cyanobacteria used in biotechnological applications, promoting ecological quality and may also create opportunities to develop innovative products to several industries (Torres et al. 2019). Hence, this practice is beneficial in several areas, namely environmental, economic and societal (Pérez-Lloréns et al. 2020; Chojnacka 2012; Salehi et al. 2019).

For instance, there are several strategies and methods that can be employed in macroalgae aquaculture. Besides that, different parameters must be considered in order to achieve good results (Emblemsvåg et al. 2020). Globally, China, Japan, Korea, Indonesia and Philippines are the major macroalgae producers (Goecke et al. 2020; Ferdouse et al. 2018) over the years. This practice has been evolved over the world, such as in Europe (Peteiro et al. 2016; Goecke et al. 2020), Africa (Msuya

2011) and in the whole of America (Augyte et al. 2017; Camus et al. 2018; Pellizzari and Reis 2011).

Due to environmental concerns in macroalgae production, geographical location is a determining factor (Emblemsvåg et al. 2020). For this reason, macroalgae production can be conducted through onshore, offshore or nearshore techniques (Hafting et al. 2012).

Onshore or land-based cultivation offers semi-controlled conditions and is carried in closed systems, such as tanks, ponds or raceways (Hafting et al. 2012). In this technique, it is possible to control several biotic parameters, namely nutrient concentration, pH, photoperiod or aeration rate, in order to achieve higher biomass yields or to provoke a stimulus in the macroalgae aimed at a certain compound production (Cotas et al. 2019; Hafting et al. 2015).

Contrary to the prior technique, offshore macroalgae cultivation is conducted on the coastline. Therefore, there are several strategies that can be employed, such as line cultivation (off-bottom, submerged hanging line or long-line), net cultivation, floating raft or rock-based aquaculture (Buschmann et al. 2017; Sudhakar et al. 2018; Hafting et al. 2012; Reid et al. 2020; Grote 2019; Kim et al. 2014; Peteiro et al. 2014). In these types of macroalgae aquaculture techniques, the cultures are exposed to several biotic and abiotic factors, which are not easy to control and will affect the biomass yield, such as predation, epiphytism, nutrient concentration, temperature variation, solar exposure and intensity, storms and maritime currents (Fernand et al. 2017; Fletcher 1995; Vairappan et al. 2008; Buck et al. 2018). In addition, a nearshore technique can be adopted, consisting of macroalgae cultivation in estuaries or in locations near the coast (Liu and Su 2017).

Then, the same cultivation techniques can be employed in offshore culture. For instance, these water bodies offer protection against sea waves and the proximity to the coastline is also an economic advantage due to lower transport costs (Grote 2019).

2.6.2 Microalgae and Cyanobacteria Cultivation

Marine or freshwater microalgae and cyanobacteria cultivation techniques and strategies have been evolved in order to optimize not only their growth rate and biomass increase, but also their bioactive compound production (Plaza et al. 2009; Di Lena et al. 2019; Spolaore et al. 2006; Goswami et al. 2020). *Haematococcus lacustris* (formerly known as *Haematococcus pluvialis*) and *Arthrospira* sp. (commonly known as *Spirulina* sp.) are the most produced, respectively (FAO 2014). However, microalgae and cyanobacteria aquaculture is still residual when compared with macroalgae (FAO 2014).

Therefore, these microorganisms can be cultured in open (i.e., ponds, lakes, tanks or raceways) (Chen et al. 2017; Satyanarayana et al. 2011) or closed systems (Figure 2.2) (i.e., plastic bags, tubular, flat-plate or tubular PBRs) (Nwoba et al. 2016; Fazal et al. 2018; Barsanti and Gualtieri 2014).

In resemblance to macroalgae cultivation techniques, in microalgae and cyanobacteria aquaculture, closed culture systems also offer a higher control of production parameters. Alternatively, in open systems, there are several biological and physical-chemicals factors that can be difficult to cope with. Although the operation in PBRs

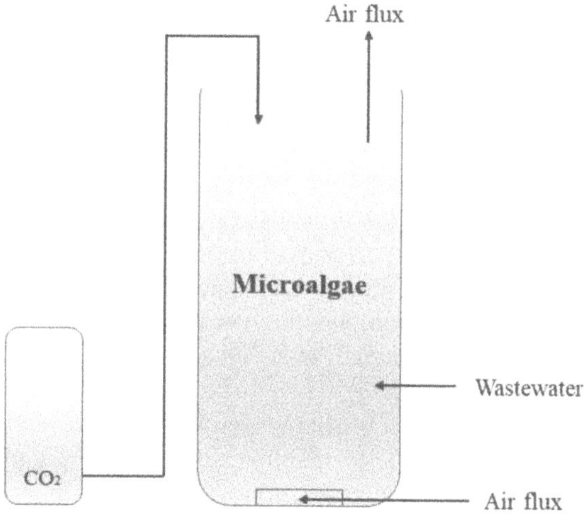

FIGURE 2.2 Schematic diagram of a PBR for microalgal cultivation (inspired by Gao et al. 2016).

is expensive due to the high electrical costs (Molazadeh et al. 2019), the scientific community has been doing efforts to improve PBR design in order to reduce the associated costs.

Particularly in open systems, the co-existence of several bacteria, fungi and other microalgae and/or cyanobacteria species often takes place, which can positively or negatively affect the growth of the species of interest (Arora et al. 2019). Nevertheless, the co-culture method has been studied, in order to understand if the interaction between different microorganisms can help in achieving the final goal (Quijano et al. 2017). Moreover, it is necessary to consider the similarity between the growth requirements of the cultured species (Magdouli et al. 2016). Concurrently, promising research has been focused on how to prevent microalgae cultivation contamination and to reduce harvesting costs (Mallik 2002). In this context, several researches have been developed the immobilization method, where the microalgae or cyanobacteria are entrapped in a polymeric matrix that allows gas and nutrient exchange (Eroglu et al. 2012; Ruiz-Marin et al. 2010; De-Bashan and Bashan 2010).

2.7 RURAL AND INDUSTRIAL WASTEWATER APPLICATION AS POTENTIAL GROWTH SUBSTRATE

Nowadays, aquatic resources have been the focus of several researchers for the quest of natural and novel bioactive compounds (Biris-Dorhoi et al. 2020). Algae and cyanobacteria can contribute to this quest and can also be cultivated through environmentally friendly techniques. Concurrently, the cultivation of these photosynthetic organisms is beneficial to the environment, due to their carbon dioxide biosequestration capacity and oxygen production (Laurens et al. 2020). Nevertheless, it is necessary to consider that the growth substrate needs to satisfy the algal or cyanobacteria growth requirements and the production of compounds of interest (Azma et al. 2011).

Variations in culture conditions, such as temperature (Meinerz et al. 2009), light intensity (Guedes et al. 2010) and composition of nutrients in the medium (Yang et al. 2011) are decisive factors and further translate into biomass production and influence synthesis of lipids, proteins, carbohydrates or pigments (Radmann and Costa 2008). In this context, it is important to refer that there are specific culture media for certain groups and/or species of microalgae and cyanobacteria (Volkmann et al. 2008). The culture medium selection will depend on the choice of compound of interest. The compound of intrest also defines the use of a specific extraction method, whether for proteins, lipids, carbohydrates or pigments (Mulbry et al. 2009; Valduga et al. 2009).

Thus, for each alga and cyanobacteria species, the productivity and the biochemical composition of the cells strongly depend on the type of cultivation and the nutritional profile of the culture medium (Guedes et al. 2011). For the purpose of lowering the costs associated with the culture medium acquisition, researchers have investigated the use of wastewater as a growth medium (Chaturvedi et al. 2013; Jard et al. 2013; Wei et al. 2013; Wu et al. 2015). With wastewater application in algae and cyanobacteria cultivation, is possible to promote a circular economy (Sudhakar et al. 2018; Kumar et al. 2019; Kumar and Verma 2021)

2.7.1 Macroalgae

Wastewater from fish farming aquaculture can be used as macroalgae growth media, since these waters often have a high inorganic load (i.e., nitrite, nitrate, phosphorous) (Ramli et al. 2020). In this context, macroalgae acts as a biofilter and can transform wastewater into a useful resource for these commercially valuable photosynthetic organisms (Neori and Nobre 2012). Afterwards, these macroalgae can be used as a natural fertilizer for further agriculture application. Alternatively, this biomass may also be used as a supplement in animal alimentation (i.e., fish, cow) (Fleurence et al. 2012; Biris-Dorhoi et al. 2020).

Macroalgae rich nutritional composition and *in vitro* assays already demonstrated that the supplementation at low concentrations of animal feed with macroalgae can fulfil nutritional requirements of several animals and be a health promoter ingredient (Makkar et al. 2016).

Laminarin, a SP extracted from brown macroalgae *Laminaria* spp., has promoted piglet gastrointestinal health (Gahan et al. 2009).

However, researchers found that developing a fish feed supplement was possible by culturing the red macroalgae *Agarophyton vermiculophyllum* (previously known as *Gracilaria vermiculophylla*), *Porphyra dioica* and the green macroalgae *Ulva* spp. through integrated multitrophic aquaculture (IMTA). Thus, it was observed that the inclusion of 10% of *P. dioica* and *Ulva* spp. in fish feed had no harmful impacts on fish (*Oreochromis niloticus*) juveniles, while the same concentration of *A. vermiculophyllum* caused a lower growth and lower ingestion yields (Silva et al. 2015).

Nevertheless, as macroalgae tend to bioaccumulate metals, minerals and other compounds, the chemical composition analysis is recommended before its application in food and feed (Makkar et al. 2016).

2.7.2 Microalgae and Cyanobacteria

Microalgae and cyanobacteria can significantly contribute to agro-industrial wastewater treatment, reducing the organic and inorganic load, while producing biomass that can be further applied to several biotechnological applications (Christenson and Sims 2011).

Microalgae can produce highly valuable primary and secondary metabolites used in therapies that require anti-cancer, antioxidant, anti-inflammatory and anti-fungal activities (Shanab et al. 2012; Kothari et al. 2017).

Yet, just a few species have been researched for their bioactivities. Moreover, there is a lack of information in how culture conditions could change the activity of microalgae bioactive compounds (Lauritano et al. 2016).

For another perspective, researchers cultivated *Chlorella* sp., *Scenedesmus* sp. and diatom isolated from an anaerobic digestion piggery effluent. Thus, the consortia of these three species growing in anaerobic digestion piggery effluent in a helical tubular PBR was more efficient. Moreover, the biochemical profile of these microalgae was demonstrated to be suitable for animal feed or biodiesel manufacturing (Nwoba et al. 2016).

For another perspective, some cyanobacteria, such as *Arthrospira* sp. (Campbell et al. 1982), *Synechococcus* sp. (Miyake et al. 1996) and *Synechocystis* sp. (Wu et al. 2001), produce energy storage compounds, namely poly-b-hydroxybutyrate (PHB) granules. This natural compound has thermoplastic properties and can be applied in the development of natural and biodegradable plastics.

Furthermore, microalgae can be a useful tool for developing natural fertilizers due to their micronutrient and macronutrient concentrations, which are essential for plant growth (Shaaban 2001; Ronga et al. 2019; Faheed and Fattah 2008).

In fact, it is possible to manipulate microalgae and cyanobacteria biochemical profile through the variation of cultivation parameters (Markou and Georgakakis 2011).

However, for utilizing wastewater as a growth media for microalgae and cyanobacteria cultivation, the chemical variation of wastewater is a major bottleneck, which will imply differences in biochemical profile of the cultured species (Christenson and Sims 2011).

2.8 CONCLUSION

Nowadays, seaweeds, microalgae and cyanobacteria are gaining more attention in food, nutraceutical and pharmaceutical industries due to their properties. As previously shown, these marine organisms have bioactive compounds useful for humans and animals' health; they work on prevention of diseases and can be used to sustain a healthy lifestyle due to their activities against inflammation, viruses and bacteria.

In this chapter, we focused on the most common species of seaweed, microalgae and cyanobacteria used for production of food, nutraceutical and pharmaceutical products.

Furthermore, this chapter analyzed the different methods of cultivation of seaweed, microalgae and cyanobacteria and enhance the application of wastewater as a growth medium in order to reduce costs associated with culture medium acquisition and to recycle the wastewater that comes out from rural industries. For instance, wastewater fish farming aquaculture can be used as macroalgae growth media, while agro-industrial wastewater can contribute as a medium for microalgae and cyanobacteria.

In conclusion, using wastewater growth medium for seaweeds, microalgae and cyanobacteria, it is possible not only to cultivate them but also to have reduction in production cost of natural products for food, nutraceutical and pharmaceutical industry.

ACKNOWLEDGEMENTS

This work is financed by national funds through FCT - Foundation for Science and Technology, I.P., within the scope of the projects UIDB/04292/2020 granted to MARE - Marine and Environmental Sciences Centre and UIDP/50017/2020+UIDB/50017/2020 (by FCT/MTCES) granted to CESAM - Centre for Environmental and Marine Studies. Silvia Lomartire thanks to the project PORBIOTA - E-Infrastructure Portuguese Information and Research in Biodiversity (POCI-01-0145-FEDER-022127) which financed this research, supported by Competitiveness and Internationalization Operational Programme and Regional Operational Programme of Lisbon, through FEDER, and by the Portuguese Foundation for Science and Technology (FCT), through national funds (OE). Diana Pacheco thanks to PTDC/BIA-CBI/31144/2017 - POCI-01 project -0145-FEDER-031144 - MARINE INVADERS, co-financed by the ERDF through POCI (Operational Program Competitiveness and Internationalization) and by the Foundation for Science and Technology (FCT, IP). Ana M. M. Gonçalves acknowledges University of Coimbra for the contract IT057-18-7253.

ABBREVIATIONS

APCs, algal protein concentrates; DHA, docosahexanoic acid; EPA, eicosapentaenoic acid; EPS, exopolymeric substances; FDA, Food and Drug Administration; GC, gas chromatography; GRAS, Generally Recognized As Safe; HHV, higher heating value; MAAs, mycosporine- like amino acids; PBR, photobioreactor; PBPs, phycobiliproteins; PC, phycocyanin; PE, phycoerythrin; PHB, poly-b-hydroxybutyrate; PUFA, polyunsaturated fatty acid; PS, protein solubility; RSA, radical scavenging activity; SP, sulphated polysaccharides; UV, ultraviolet

CONFLICT OF INTEREST STATEMENT

The authors declare no conflict of interest.

REFERENCES

Abed, R. M. M., S. Dobretsov, and K. Sudesh. 2009. "Applications of Cyanobacteria in Biotechnology." *Journal of Applied Microbiology* 106 (1): 1–12. doi:10.1111/j.1365-2672.2008.03918.x.

Agrawal, Komal, Ankita Bhatt, Nisha Bhardwaj, Bikash Kumar, and Pradeep Verma. 2020. "Algal Biomass: Potential Renewable Feedstock for Biofuels Production–Part I." In *Biofuel Production Technologies: Critical Analysis for Sustainability*, pp. 203–237. Springer, Singapore doi: https://doi.org/10.1007/978-981-13-8637-4_8

Aktumsek, Abdurrahman, Gokhan Zengin, Gokalp Ozmen Guler, Yavuz Selim Cakmak, and Ahmet Duran. 2013. "Antioxidant Potentials and Anticholinesterase Activities of Methanolic and Aqueous Extracts of Three Endemic Centaurea L. Species." *Food and Chemical Toxicology* 55: 290–296. doi:10.1016/j.fct.2013.01.018.

Andrade, Lidiane M. 2018. "Chlorella and Spirulina Microalgae as Sources of Functional Foods, Nutraceuticals, and Food Supplements; an Overview." *MOJ Food Processing & Technology* 6 (1): 45–58. doi:10.15406/mojfpt.2018.06.00144.

Araújo, Glacio Souza, Wladimir Ronald Lobo Farias, José Ariévilo Gurgel Rodrigues, Valeska Martins Torres, and Grazielle da Costa Pontes. 2008. "Administração Oral Dos Polissacarídeos Sulfatados Da Rodofícea Gracilaria Caudata Na Sobrevivência de Pós-Larvas de Tilápia." *Revista Ciencia Agronomica* 39 (4): 548–554.

Arimoto-Kobayashi, Sakae, Naomi Inada, Hiromi Nakano, Haruki Rai, and Hikoya Hayatsu. 1998. "Iron-Chlorophyllin-Mediated Conversion of 3-Hydroxyamino-1-Methyl-5H-Pyrido[4,3-b]Indole (Trp-P-2(NHOH)) into Its Nitroso Derivative." *Mutation Research* 400 (1–2): 259–269. doi:10.1016/S0027-5107(98)00033-5.

Arora, Neha, Alok Patel, Juhi Mehtani, Parul A. Pruthi, Vikas Pruthi, and Krishna Mohan Poluri. 2019. "Co-Culturing of Oleaginous Microalgae and Yeast: Paradigm Shift towards Enhanced Lipid Productivity." *Environmental Science and Pollution Research* 26 (17): 16952–16973. doi:10.1007/s11356-019-05138-6.

Asthana, Ravi K., Arunima Srivastava, Arvind M. Kayastha, Gopal Nath, and Sureshwar P. Singh. 2006. "Antibacterial Potential of γ-Linolenic Acid from Fischerella Sp. Colonizing Neem Tree Bark." *World Journal of Microbiology and Biotechnology* 22 (5): 443–448. doi:10.1007/s11274-005-9054-8.

Augyte, Simona, Charles Yarish, Sarah Redmond, and Jang K. Kim. 2017. "Cultivation of a Morphologically Distinct Strain of the Sugar Kelp, Saccharina Latissima Forma Angustissima, from Coastal Maine, USA, with Implications for Ecosystem Services." *Journal of Applied Phycology* 29 (4): 1967–1976. doi:10.1007/s10811-017-1102-x.

Azma, Mojtaba, Mohd Shamzi Mohamed, Rosfarizan Mohamad, Raha Abdul Rahim, and Arbakariya B. Ariff. 2011. "Improvement of Medium Composition for Heterotrophic Cultivation of Green Microalgae, Tetraselmis Suecica, Using Response Surface Methodology." *Biochemical Engineering Journal* 53 (2): 187–195. doi:10.1016/j.bej.2010.10.010.

Barbosa, Mariana, Graciliana Lopes, Paula B. Andrade, and Patrícia Valentão. 2019. "Bioprospecting of Brown Seaweeds for Biotechnological Applications: Phlorotannin Actions in Inflammation and Allergy Network." *Trends in Food Science and Technology* 86: 153–171. doi:10.1016/j.tifs.2019.02.037.

Barsanti, Laura, and Paolo Gualtieri. 2014. *Algae: Anatomy, Biochemistry, and Biotechnology*. Boca Raton, FL: CRC Press.

Bellini, Erika, Matteo Ciocci, Saverio Savio, Simonetta Antonaroli, Dror Seliktar, Sonia Melino, and Roberta Congestri. 2018. "Trichormus Variabilis (Cyanobacteria) Biomass: From the Nutraceutical Products to Novel EPS-Cell/Protein Carrier Systems." *Marine Drugs* 16 (9). 298. doi:10.3390/md16090298.

Beltran, E. C., and B. A. Neilan. 2000. "Geographical Segregation of the Neurotoxin-Producing Cyanobacterium Anabaena Circinalis." *Applied and Environmental Microbiology* 66 (10): 4468–4474. doi:10.1128/AEM.66.10.4468-4474.2000.

Bennedsen, Mads, Xin Wang, Roger Willén, Torkel Wadström, and Leif Percival Andersen. 1999. "Treatment of H. Pylori Infected Mice with Antioxidant Astaxanthin Reduces Gastric Inflammation, Bacterial Load and Modulates Cytokine Release by Splenocytes." *Immunology Letters* 70 (3): 185–189. doi:10.1016/S0165-2478(99)00145-5.

Bental, Michal, Hadassa Degani, and Mordhay Avron. 1988. "23 Na-NMR Studies of the Intracellular Sodium Ion Concentration in the Halotolerant Alga Dunaliella Salina." *Plant Physiology* 87 (4): 813–817. doi:10.1104/pp.87.4.813.

Bermejo, Ruperto, Eva M. Talavera, José M. Alvarez-Pez, and Juan C. Orte. 1997. "Chromatographic Purification of Biliproteins from Spirulina Platensis. High-Performance Liquid Chromatographic Separation of Their α and β Subunits." *Journal of Chromatography A* 778 (1–2): 441–450. doi:10.1016/S0021-9673(97)00577-3.

Bertram, John S. 1999. "Carotenoids and Gene Regulation." *Nutrition Reviews* 57 (6): 182–191. doi:10.1016/S0899-9007(00)00248-3.

Bewicke, D., and B. A. Potter. 1984. *Chlorella: The Emerald Food.* Berkley, CA: Ronin Publishing.

Bhardwaj, Nisha, Komal Agrawal, and Pradeep Verma. 2020. "Algal Biofuels: An Economic and Effective Alternative of Fossil Fuels." In *Microbial Strategies for Techno-economic Biofuel Production*, pp. 59–83. Springer doi: https://doi.org/10.1007/978-981-15-7190-9_7

Bicudo, Carlos E. De M., and Mariângela Menezes. 2006. Gêneros de Algas de Águas Continentais Do Brasil (Chave Para Identificação e Descrições).

Bilal, Muhammad, Tahir Rasheed, Ishtiaq Ahmed, and Hafiz M.N. Iqbal. 2017. "High-Value Compounds from Microalgae with Industrial Exploitability - A Review." *Frontiers in Bioscience - Scholar* 9 (3): 319–342. doi:10.2741/s490.

Biris-Dorhoi, Elena Suzana, Delia Michiu, Carmen R. Pop, Ancuta M. Rotar, Maria Tofana, Oana L. Pop, Sonia A. Socaci, and Anca C. Farcas. 2020. "Macroalgae—A Sustainable Source of Chemical Compounds with Biological Activities." *Nutrients* 12 (10): 1–23. doi:10.3390/nu12103085.

Bishop, West M., and Heidi M. Zubeck. 2012. "Evaluation of Microalgae for Use as Nutraceuticals and Nutritional Supplements." *Journal of Nutrition & Food Sciences* 2 (5). doi:10.4172/2155-9600.1000147.

Bjerregaard, Rasmus, Diego Valderrama, Neil Sims, Ricardo Radulovich, James Diana, Mark Capron, John Forster, et al. 2016. *Seaweed Aquaculture for Food Security, Income Generation and Environmental Health in Tropical Developing Countries. Seaweed Aquaculture for Food Security, Income Generation and Environmental Health in Tropical Developing Countries.* World Bank, Washington, DC. doi:10.1596/24919.

Bleakley, Stephen, and Maria Hayes. 2017. "Algal Proteins: Extraction, Application, and Challenges Concerning Production." *Foods* 6 (5): 33. doi:10.3390/foods6050033.

Boonyaratpalin, S., M. Boonyaratpalin, K. Supamattaya, and Y. Yoride. 1995. *Effects of Peptidoglucan (PG) on Growth, Survival, Immune Responses, and Tolerance to Stress in Black Tiger Shrimp Penaeus Monodon. Diseases in Asian Aquaculture*, Edited by J. R. Shariff, M. Subasighe, R. P. Arthur.

Borowitzka, Michael A. 2013. "High-Value Products from Microalgae-Their Development and Commercialisation." *Journal of Applied Phycology* 25 (3): 743–756. doi:10.1007/s10811-013-9983-9.

Brasil, Bruno dos Santos Alves Figueiredo, Félix Gonçalves de Siqueira, Thaís Fabiana Chan Salum, Cristina Maria Zanette, and Michele Rigon Spier. 2017. "Microalgae and Cyanobacteria as Enzyme Biofactories." *Algal Research* 25 (April): 76–89. doi:10.1016/j.algal.2017.04.035.

Buck, Bela H., Max F. Troell, Gesche Krause, Dror L. Angel, Britta Grote, and Thierry Chopin. 2018. "State of the Art and Challenges for Offshore Integrated Multi-Trophic Aquaculture (IMTA)." *Frontiers in Marine Science* 5 (May): 1–21. doi:10.3389/fmars.2018.00165.

Buschmann, A. H., M. Troell, and N. Kautsky. 2001. "Integrated Algal Farming: A Review." *Cahiers de Biologie Marine* 42 (1–2): 83–90.

Buschmann, Alejandro H., Carolina Camus, Javier Infante, Amir Neori, Álvaro Israel, María C. Hernández-González, Sandra V. Pereda, et al. 2017. "Seaweed Production: Overview of the Global State of Exploitation, Farming and Emerging Research Activity." *European Journal of Phycology* 52 (4): 391–406. doi:10.1080/09670262.2017.1365175.

Campa-Córdova, A. I., N. Y. Hernández-Saavedra, R. De Philippis, and F. Ascencio. 2002. "Generation of Superoxide Anion and SOD Activity in Haemocytes and Muscle of American White Shrimp (Litopenaeus Vannamei) as a Response to β-Glucan and Sulphated Polysaccharide." *Fish and Shellfish Immunology* 12 (4): 353–366. doi:10.1006/fsim.2001.0377.

Campbell, J., S. E. Stevens, and D. L. Balkwill. 1982. "Accumulation of Poly-β-Hydroxybutyrate in Spirulina Platensis." *Journal of Bacteriology* 149 (1): 361–363. doi:10.1128/jb.149.1.361-363.1982.

Del Campo, José A., Mercedes García-González, and Miguel G. Guerrero. 2007. "Outdoor Cultivation of Microalgae for Carotenoid Production: Current State and Perspectives." *Applied Microbiology and Biotechnology* 74 (6): 1163–1174. doi:10.1007/s00253-007-0844-9.

Camus, Carolina, Javier Infante, and Alejandro H. Buschmann. 2018. "Overview of 3 Year Precommercial Seafarming of Macrocystis Pyrifera along the Chilean Coast." *Reviews in Aquaculture* 10 (3): 543–559. doi:10.1111/raq.12185.

Cardozo, Karina H. M., Thais Guaratini, Marcelo P. Barros, Vanessa R. Falcão, Angela P. Tonon, Norberto P. Lopes, Sara Campos, et al. 2007. "Metabolites from Algae with Economical Impact." *Comparative Biochemistry and Physiology Part C: Toxicology & Pharmacology* 146 (1–2): 60–78. doi:10.1016/j.cbpc.2006.05.007.

Carmichael, Wayne W., and Gregory L. Boyer. 2016. "Health Impacts from Cyanobacteria Harmful Algae Blooms: Implications for the North American Great Lakes." *Harmful Algae* 54: 194–212. doi:10.1016/j.hal.2016.02.002.

Casoni, Andrés I., Fernando D. Ramos, Vanina Estrada, and M. Soledad Diaz. 2020. "Sustainable and Economic Analysis of Marine Macroalgae Based Chemicals Production - Process Design and Optimization." *Journal of Cleaner Production* 276: 122792. doi:10.1016/j.jclepro.2020.122792.

Chacón-Lee, T. L., and G. E. González-Mariño. 2010. "Microalgae for 'Healthy' Foods-Possibilities and Challenges." *Comprehensive Reviews in Food Science and Food Safety* 9 (6): 655–675. doi:10.1111/j.1541-4337.2010.00132.x.

Chaturvedi, Venkatesh, Monika Chandravanshi, Manoj Rahangdale, and Pradeep Verma. 2013. "An integrated approach of using polystyrene foam as an attachment system for growth of mixed culture of cyanobacteria with concomitant treatment of copper mine waste water." *Journal of Waste Management* 2013 (1): 1–7. doi:10.1155/2013/282798.

Chaturvedi, Venkatesh, Rahul Kumar Goswami, and PradeepVerma. 2021. "Genetic Engineering for Enhancement of Biofuel Production in Microalgae." In *Biorefineries: A Step Towards Renewable and Clean Energy*, pp. 539–559. Springer doi: https://doi.org/10.1007/978-981-15-9593-6_2

Chen, Yong, Li-ping Sun, Zhi-hui Liu, Greg Martin, and Zheng Sun. 2017. "Integration of Waste Valorization for Sustainable Production of Chemicals and Materials via Algal Cultivation." *Topics in Current Chemistry* 375 (6): 89. doi:10.1007/s41061-017-0175-y.

Chew, Kit Wayne, Jing Ying Yap, Pau Loke Show, Ng Hui Suan, Joon Ching Juan, Tau Chuan Ling, Duu Jong Lee, and Jo Shu Chang. 2017. "Microalgae Biorefinery: High Value Products Perspectives." *Bioresource Technology* 229: 53–62. doi:10.1016/j.biortech.2017.01.006.

Chia, Shir Reen, Pau Loke Show, Siew Moi Phang, Tau Chuan Ling, and Hwai Chyuan Ong. 2018. "Sustainable Approach in Phlorotannin Recovery from Macroalgae." *Journal of Bioscience and Bioengineering* 126 (2): 220–225. doi:10.1016/j.jbiosc.2018.02.015.

Chojnacka, Katarzyna. 2012. "Biologically Active Compounds in Seaweed Extracts - the Prospects for the Application." *The Open Conference Proceedings Journal* 3 (1): 20–28. doi:10.2174/1876326x01203020020.

Chopin, Thierry, Charles Yarish, Robert Wilkes, Ellen Belyea, Shan Lu, and Arthur Mathieson. 1999. "Developing Porphyra/Salmon Integrated Aquaculture for Bioremediation and Diversification of the Aquaculture Industry." *Journal of Applied Phycology* 11 (5): 463–472. doi:10.1023/A:1008114112852.

Christenson, Logan, and Ronald Sims. 2011. "Production and Harvesting of Microalgae for Wastewater Treatment, Biofuels, and Bioproducts." *Biotechnology Advances* 29 (6): 686–702. doi:10.1016/j.biotechadv.2011.05.015.

Cotas, João, Artur Figueirinha, Leonel Pereira, and Teresa Batista. 2019. "The Effect of Salinity on Fucus Ceranoides (Ochrophyta, Phaeophyceae) in the Mondego River (Portugal)." *Journal of Oceanology and Limnology* 37 (3): 881–891. doi:10.1007/s00343-019-8111-3.

Craggs, Rupert J., Tryg J. Lundquist, and John R. Benemann. 2013. "Wastewater Treatment and Algal Biofuel Production." *Applied Phycology*. doi:10.1007/978-94-007-5479-9.

Das, P., S. C. Manda, S. K. Bhagabati, M. S. Akhtar, and S. K. Singh. 2014. "Important Live Food Organisms and Their Role in Aquaculture." In *Frontiers in Aquaculture*, edited by Sukham, pp. 69–86. Delhi: Narendra Publishing House.

Davidson, Bradley S. 1995. "New Dimensions in Natural Products Research: Cultured Marine Microorganisms." *Current Opinion in Biotechnology* 6 (3): 284–291. doi:10.1016/0958-1669(95)80049-2.

De-Bashan, Luz E., and Yoav Bashan. 2010. "Immobilized Microalgae for Removing Pollutants: Review of Practical Aspects." *Bioresource Technology* 101 (6): 1611–1627. doi:10.1016/j.biortech.2009.09.043.

Derner, R. B., S. Ohse, M. Villela, S. M. Carvalho, and R. Fett. 2006. "Microalgas, Produtos e Aplicações." *Ciência Rural* 36 (6): 1959–1967.

Dittmann, E., B. A. Neilan, and T. Börner. 2001. "Molecular Biology of Peptide and Polyketide Biosynthesis in Cyanobacteria." *Applied Microbiology and Biotechnology* 57 (4): 467–473. doi:10.1007/s002530100810.

Draaisma, René B., René H. Wijffels, P. M. Slegers, Laura B. Brentner, Adip Roy, and Maria J. Barbosa. 2013. "Food Commodities from Microalgae." *Current Opinion in Biotechnology* 24 (2): 169–177. doi:10.1016/j.copbio.2012.09.012.

Eberlein, K., M. T. Leal, K. D. Hammer, and W. Hickel. 1985. "Dissolved Organic Substances during a Phaeocystis Pouchetii Bloom in the German Bight (North Sea)." *Marine Biology* 89 (3): 311–316. doi:10.1007/BF00393665.

Ejike, Chukwunonso E.C.C., Stephanie A. Collins, Nileeka Balasuriya, Andrew K. Swanson, Beth Mason, and Chibuike C. Udenigwe. 2017. "Prospects of Microalgae Proteins in Producing Peptide-Based Functional Foods for Promoting Cardiovascular Health." *Trends in Food Science and Technology* 59: 30–36. doi:10.1016/j.tifs.2016.10.026.

Emblemsvåg, Jan, Nina Pereira Kvadsheim, Jon Halfdanarson, Matthias Koesling, Bjørn Tore Nystrand, Jan Sunde, and Céline Rebours. 2020. "Strategic Considerations for Establishing a Large-Scale Seaweed Industry Based on Fish Feed Application: A Norwegian Case Study." *Journal of Applied Phycology* 32 (6). doi:10.1007/s10811-020-02234-w.

Eroglu, Ela, Vipul Agarwal, Michael Bradshaw, Xianjue Chen, Steven M. Smith, Colin L. Raston, and K. Swaminathan Iyer. 2012. "Nitrate Removal from Liquid Effluents Using Microalgae Immobilized on Chitosan Nanofiber Mats." *Green Chemistry* 14 (10): 2682–2685. doi:10.1039/c2gc35970g.

Faheed, Fayza, and Zeinab Abd-El Fattah. 2008. "Effect of Chlorella Vulgaris as Bio-Fertilizer on Growth Parameters and Metabolic Aspects of Lettuce Plant." *Agriculture and Social Sciences* 4: 165–69.

Fan, Xiaodan, Lu Bai, Liang Zhu, Li Yang, and Xuewu Zhang. 2014. "Marine Algae-Derived Bioactive Peptides for Human Nutrition and Health." *Journal of Agricultural and Food Chemistry* 62 (38): 9211–9222. doi:10.1021/jf502420h.

FAO. 2014. "The State of World Fisheries and Aquaculture - Opportunities and Challenges." Rome.

FAO. 2018. "The State of World Fisheries and Aquaculture: Meeting the Sustainable Development Goals." Rome.

Farias, W. R. L., R. A. Nazareth, and P. A. S. Mourão. 2001. "Dual Effects of Sulfated D-Galactans from the Red Algae Botryocladia Occidentalis Preventing Thrombosis and Inducing Platelet Aggregation." *Thrombosis and Haemostasis* 86 (6): 1540–1546. doi:10.1055/s-0037-1616760.

Farias, W. R. L., H. J. Rebouças, V. M. Torres, J. A. G. Rodrigues, G. C. Pontes, F. H. O. Silva, and A. H. Sampaio. 2004. "Enhancement of Growth in Tilapia Fingerlings (Oreochromis Niloticus) by Sulfated D- Galactans Extracted from Marine Algae." *Revista Ciência Agronômica* 35: 189–195.

Fassett, Robert G., and Jeff S. Coombes. 2012. "Astaxanthin in Cardiovascular Health and Disease." *Molecules* 17 (2): 2030–2048. doi:10.3390/molecules17022030.

Fauziee, Nur Afifah Mohd, Lee Sin Chang, Wan Aida Wan Mustapha, Adibi Rahiman Md Nor, and Seng Joe Lim. 2020. "Functional Polysaccharides of Fucoidan, Laminaran and Alginate from Malaysian Brown Seaweeds (*Sargassum Polycystum, Turbinaria Ornata* and *Padina Boryana*)." *International Journal of Biological Macromolecules* 167: 1135–1145. doi:10.1016/j.ijbiomac.2020.11.067.

Fazal, Tahir, Azeem Mushtaq, Fahad Rehman, Asad Ullah Khan, Naim Rashid, Wasif Farooq, Muhammad Saif Ur Rehman, and Jian Xu. 2018. "Bioremediation of Textile Wastewater and Successive Biodiesel Production Using Microalgae." *Renewable and Sustainable Energy Reviews* 82 (February): 3107–3126. doi:10.1016/j.rser.2017.10.029.

Ferdouse, Fatima, Susan Løvstad Holdt, Rohan Smith, Pedro Murúa, and Zhengzyong Yang. 2018. "The Global Status of Seaweed Production, Trade and Utilization." *FAO Globefish Research Programme* 124 (120).

Fernand, Francois, Alvaro Israel, Jorunn Skjermo, Thomas Wichard, Klaas R. Timmermans, and Alexander Golberg. 2017. "Offshore Macroalgae Biomass for Bioenergy Production: Environmental Aspects, Technological Achievements and Challenges." *Renewable and Sustainable Energy Reviews* 75 (October): 35–45. doi:10.1016/j.rser.2016.10.046.

Ferrera-Lorenzo, N., E. Fuente, I. Suárez-Ruiz, R. R. Gil, and B. Ruiz. 2014. "Pyrolysis Characteristics of a Macroalgae Solid Waste Generated by the Industrial Production of Agar-Agar." *Journal of Analytical and Applied Pyrolysis* 105: 209–216. doi:10.1016/j.jaap.2013.11.006.

Fish, S. A., and G. A. Codd. 1994. "Bioactive Compound Production by Thermophilic and Thermotolerant Cyanobacteria (Blue-Green Algae)." *World Journal of Microbiology & Biotechnology* 10 (3): 338–341. doi:10.1007/BF00414875.

Fletcher, Robert L. 1995. "Epiphytism and Fouling in Gracilaria Cultivation: An Overview." *Journal of Applied Phycology* 7 (3): 325–333. doi:10.1007/BF00004006.

Fleurence, Joël, Michèle Morançais, Justine Dumay, Priscilla Decottignies, Vincent Turpin, Mathilde Munier, Nuria Garcia-Bueno, and Pascal Jaouen. 2012. "What Are the Prospects for Using Seaweed in Human Nutrition and for Marine Animals Raised through Aquaculture?" *Trends in Food Science and Technology* 27 (1): 57–61. doi:10.1016/j.tifs.2012.03.004.

Fradique, Mónica, Ana Paula Batista, M. Cristiana Nunes, Lúisa Gouveia, Narcisa M. Bandarra, and Anabela Raymundo. 2010. "Incorporation of *Chlorella Vulgaris* and *Spirulina Maxima* Biomass in Pasta Products. Part 1: Preparation and Evaluation." *Journal of the Science of Food and Agriculture* 90 (10): 1656–1664. doi:10.1002/jsfa.3999.

Freitas, Ana C., Dina Rodrigues, Teresa A. P. Rocha-Santos, Ana M. P. Gomes, and Armando C. Duarte. 2012. "Marine Biotechnology Advances towards Applications in New Functional Foods." *Biotechnology Advances* 30 (6): 1506–1515. doi:10.1016/j.biotechadv.2012.03.006.

Funk, C. D. 2001. "Prostaglandins and Leukotrienes: Advances in Eicosanoid Biology." *Science* 294 (5548): 1871–1875. doi:10.1126/science.294.5548.1871.

Gademann, Karl, and Cyril Portmann. 2008. "Secondary Metabolites from Cyanobacteria: Complex Structures and Powerful Bioactivities." *Current Organic Chemistry* 12 (4): 326–341. doi:10.2174/138527208783743750.

Gahan, D. A., M. B. Lynch, J. J. Callan, J. T. O'Sullivan, and J. V. O'Doherty. 2009. "Performance of Weanling Piglets Offered Low-, Medium- or High-Lactose Diets Supplemented with a Seaweed Extract from *Laminaria Spp.*" *Animal* 3 (1): 24–31. doi:10.1017/S1751731108003017.

Galasso, Christian, Antonio Gentile, Ida Orefice, Adrianna Ianora, Antonino Bruno, Douglas M. Noonan, Clementina Sansone, Adriana Albini, and Christophe Brunet. 2019. "Microalgal Derivatives as Potential Nutraceutical and Food Supplements for Human Health: A Focus on Cancer Prevention and Interception." *Nutrients* 11 (6): 1–22. doi:10.3390/nu11061226.

Galetovic, Alexandra, Francisca Seura, Valeska Gallardo, Rocfo Graves, Juan Cortés, Carolina Valdivia, Javier Nuñez, et al. 2020. "Use of Phycobiliproteins from Atacama Cyanobacteria as Food Colorants in a Dairy Beverage Prototype." *Foods* 9 (2): 1–13. doi:10.3390/foods9020244.

Ganesan, Abirami R., Uma Tiwari, and Gaurav Rajauria. 2019. "Seaweed Nutraceuticals and Their Therapeutic Role in Disease Prevention." *Food Science and Human Wellness* 8 (3): 252–263. doi:10.1016/j.fshw.2019.08.001.

Gao, Feng, Chen Li, Zhao Hui Yang, Guang Ming Zeng, Li Juan Feng, Jun zhi Liu, Mei Liu, and Hui wen Cai. 2016. "Continuous Microalgae Cultivation in Aquaculture Wastewater by a Membrane Photobioreactor for Biomass Production and Nutrients Removal." *Ecological Engineering* 92: 55–61. doi:10.1016/j.ecoleng.2016.03.046.

Gao, Kunshan. 1998. "Chinese Studies on the Edible Blue-Green Alga, *Nostoc Flagelliforme*: A Review." *Journal of Applied Phycology* 10 (1): 37–49. doi:10.1023/A:1008014424247.

García, José L., Marta de Vicente, and Beatriz Galán. 2017. "Microalgae, Old Sustainable Food and Fashion Nutraceuticals." *Microbial Biotechnology* 10 (5): 1017–1024. doi:10.1111/1751-7915.12800.

George, Basil, Imran Pancha, Chahana Desai, Kaumeel Chokshi, Chetan Paliwal, Tonmoy Ghosh, and Sandhya Mishra. 2014. "Effects of Different Media Composition, Light Intensity and Photoperiod on Morphology and Physiology of Freshwater Microalgae *Ankistrodesmus Falcatus* - A Potential Strain for Bio-Fuel Production." *Bioresource Technology* 171: 367–374. doi:10.1016/j.biortech.2014.08.086.

Gerwick, William H., and Bradley S. Moore. 2012. "Lessons from the Past and Charting the Future of Marine Natural Products Drug Discovery and Chemical Biology." *Chemistry and Biology* 19 (1): 85–98. doi:10.1016/j.chembiol.2011.12.014.

Ghosh, Ashmita, Saumyakanti Khanra, Madhumanti Mondal, Gopinath Halder, O. N. Tiwari, Supreet Saini, Tridib Kumar Bhowmick, and Kalyan Gayen. 2016. "Progress toward Isolation of Strains and Genetically Engineered Strains of Microalgae for Production of Biofuel and Other Value Added Chemicals: A Review." *Energy Conversion and Management* 113: 104–118. doi:10.1016/j.enconman.2016.01.050.

Gismondi, Alessandra, Francesca Di Pippo, Laura Bruno, Simonetta Antonaroli, and Roberta Congestri. 2016. "Phosphorus Removal Coupled to Bioenergy Production by Three Cyanobacterial Isolates in a Biofilm Dynamic Growth System." *International Journal of Phytoremediation* 18 (9): 869–876. doi:10.1080/15226514.2016.1156640.

Goecke, Franz, Gunnar Klemetsdal, and Åshild Ergon. 2020. "Cultivar Development of Kelps for Commercial Cultivation — Past Lessons and Future Prospects" 8 (February). doi:10.3389/fmars.2020.00110.

Goiris, Koen, Koenraad Muylaert, Ilse Fraeye, Imogen Foubert, Jos De Brabanter, and Luc De Cooman. 2012. "Antioxidant Potential of Microalgae in Relation to Their Phenolic and Carotenoid Content." *Journal of Applied Phycology* 24 (6): 1477–1486. doi:10.1007/s10811-012-9804-6.

Gomez, L. P., C. Alvarez, M. Zhao, U. Tiwari, J. Curtin, M. Garcia-Vaquero, and B. K. Tiwari. 2020. "Innovative Processing Strategies and Technologies to Obtain Hydrocolloids from Macroalgae for Food Applications." *Carbohydrate Polymers* 248: 116784. doi:10.1016/j.carbpol.2020.116784.

Goswami, Rahul Kumar, Komal Agrawal, Sanjeet Mehariya, Antonio Molino, Dino Musmarra, and PradeepVerma. 2020. "Microalgae-Based Biorefinery for Utilization of Carbon Dioxide for Production of Valuable Bioproducts." In *Chemo-Biological Systems for CO_2 Utilization*, pp. 199–224. Taylor & Francis doi: https://doi.org/10.1201/9780429317187

Goswami, Rahul Kumar, Sanjeet Mehariya, Pradeep Verma, Roberto Lavecchia, Antonio Zuorro. 2021a. "Microalgae-Based Biorefineries for Sustainable Resource Recovery from Wastewater." *Journal of Water Process Engineering* 40: 101747 doi:10.1016/j.jwpe.2020.101747.

Goswami, Rahul Kumar, Komal Agrawal, Pradeep Verma. 2021b. "An Overview of Microalgal Carotenoids: Advances in the Production and Its Impact on Sustainable Development." In: *Bioenergy Research: Evaluating Strategies for Commercialization and Sustainability*, pp. 105–128. John Wiley & Sons.

Goswami, Rahul Kumar, Sanjeet Mehariya, Obulisamy Parthiba Karthikeysan, and Pradeep Verma. 2021c. "Advanced microalgae-based renewable biohydrogen production systems: A review." *Bioresource Technology*, 320 (A): 124301 doi:10.1016/j.biortech.2020.124301

Grina, Fatiha, Zain Ullah, Erhan Kaplaner, Abderrahman Moujahid, Rabiaa Eddoha, Boubker Nasser, Pınar Terzioğlu, et al. 2020. "In Vitro Enzyme Inhibitory Properties, Antioxidant Activities, and Phytochemical Fingerprints of Five Moroccan Seaweeds." *South African Journal of Botany* 128: 152–160. doi:10.1016/j.sajb.2019.10.021.

Grinberg, Michal, Rachel Sarig, Yehudit Zaltsman, Dan Frumkin, Nicholas Grammatikakis, Eitan Reuveny, and Atan Gross. 2002. "TBID Homooligomerizes in the Mitochondrial Membrane to Induce Apoptosis." *Journal of Biological Chemistry* 277 (14): 12237–12245. doi:10.1074/jbc.M104893200.

Grosshagauer, Silke, Klaus Kraemer, and Veronika Somoza. 2020. "The True Value of Spirulina." *Journal of Agricultural and Food Chemistry* 68 (14): 4109–4115. doi:10.1021/acs.jafc.9b08251.

Grote, Britta. 2019. "Recent Developments in Aquaculture of Palmaria Palmata (Linnaeus) (Weber & Mohr 1805): Cultivation and Uses." *Reviews in Aquaculture* 11 (1): 25–41. doi:10.1111/raq.12224.

Guedes, A. C., and F. X. Malcata. 2012. *Nutritional Value and Uses of Microalgae in Aquaculture.* Edited by Z. A. Muchlisin. IntechOpen. doi: https://doi.org/10.5772/30576

Guedes, A. Catarina, Helena M. Amaro, and F. Xavier Malcata. 2011. "Microalgae as Sources of High Added-Value Compounds-a Brief Review of Recent Work." *Biotechnology Progress* 27 (3): 597–613. doi:10.1002/btpr.575.

Guedes, A. Catarina, Luís A. Meireles, Helena M. Amaro, and F. Xavier Malcata. 2010. "Changes in Lipid Class and Fatty Acid Composition of Cultures of Pavlova Lutheri, in Response to Light Intensity." *Journal of the American Oil Chemists' Society* 87 (7): 791–801. doi:10.1007/s11746-010-1559-0.

Guerin, Martin, Mark E. Huntley, and Miguel Olaizola. 2003. "Haematococcus Astaxanthin: Applications for Human Health and Nutrition." *Trends in Biotechnology* 21 (5): 210–216. doi:10.1016/S0167-7799(03)00078-7.

Guilherme, Juliana Rocha, Bruna Pacheco, Bruno Meira Soares, Claudio Martin Pereira de Pereira, Pio Colepicolo, and Daiane Dias. 2020. "Phaeophyceae and Rhodophyceae Macroalgae from the Antarctic: A Source of Selenium." *Journal of Food Composition and Analysis* 88: 103430. doi:10.1016/j.jfca.2020.103430.

Gullón, Beatriz, Mohammed Gagaoua, Francisco J. Barba, Patricia Gullón, Wangang Zhang, and José M. Lorenzo. 2020. "Seaweeds as Promising Resource of Bioactive Compounds: Overview of Novel Extraction Strategies and Design of Tailored Meat Products." *Trends in Food Science and Technology* 100: 1–18. doi:10.1016/j.tifs.2020.03.039.

Gullón, Patricia, Gonzalo Astray, Beatriz Gullón, Daniel Franco, Paulo Cezar Bastianello Campagnol, and José M. Lorenzo. 2021. "Inclusion of Seaweeds as Healthy Approach to Formulate New Low-Salt Meat Products." *Current Opinion in Food Science* 40: 20–25. doi:10.1016/j.cofs.2020.05.005.

Gunasekera, Sarath P., Cliff Ross, Valerie J. Paul, Susan Matthew, and Hendrik Luesch. 2008. "Dragonamides C and D, Linear Lipopeptides from the Marine Cyanobacterium Brown Lyngbya Polychroa." *Journal of Natural Products* 71 (5): 887–90. doi:10.1021/np0706769.

Gupta, Charu, Dhan Prakash, Amar P. Garg, and Sneh Gupta. 2012. "Whey Proteins: A Novel Source of Bioceuticals." *Middle East Journal of Scientific Research* 12 (3): 365–375. doi:10.5829/idosi.mejsr.2012.12.3.64227.

Gupta, Vishal, Sachitra Kumar Ratha, Anjuli Sood, Vidhi Chaudhary, and Radha Prasanna. 2013. "New Insights into the Biodiversity and Applications of Cyanobacteria (Blue-Green Algae)-Prospects and Challenges." *Algal Research* 2 (2): 79–97. doi:10.1016/j.algal.2013.01.006.

Gürlek, Ceren, Çağla Yarkent, Ayşe Köse, Başak Tuğcu, Ilgın Kımız Gebeloğlu, Suphi Öncel, and Murat Elibol. 2020. "Screening of Antioxidant and Cytotoxic Activities of Several Microalgal Extracts with Pharmaceutical Potential." *Health and Technology* 10 (1): 111–117. doi:10.1007/s12553-019-00388-3.

Guzmán, S., A. Gato, M. Lamela, M. Freire-Garabal, and J. M. Calleja. 2003. "Anti-Inflammatory and Immunomodulatory Activities of Polysaccharide from Chlorella Stigmatophora and Phaeodactylum Tricornutum." *Phytotherapy Research* 17 (6): 665–670. doi:10.1002/ptr.1227.

Hafting, Jeff T., James S. Craigie, Dagmar B. Stengel, Rafael R. Loureiro, Alejandro H. Buschmann, Charles Yarish, Maeve D. Edwards, and Alan T. Critchley. 2015. "Prospects and Challenges for Industrial Production of Seaweed Bioactives." *Journal of Phycology* 51 (5): 821–837. doi:10.1111/jpy.12326.

Hafting, Jeff T., Alan T. Critchley, M. Lynn Cornish, Scott A. Hubley, and Allan F. Archibald. 2012. "On-Land Cultivation of Functional Seaweed Products for Human Usage." *Journal of Applied Phycology* 24 (June): 385–392. doi:10.1007/s10811-011-9720-1.

Hajimahmoodi, Mannan, Mohammad Ali Faramarzi, Najmeh Mohammadi, Neda Soltani, Mohammad Reza Oveisi, and Nastaran Nafissi-Varcheh. 2010. "Evaluation of Antioxidant Properties and Total Phenolic Contents of Some Strains of Microalgae." *Journal of Applied Phycology* 22 (1): 43–50. doi:10.1007/s10811-009-9424-y.

Haque, Fatima, Animesh Dutta, Mahendra Thimmanagari, and Yi Wai Chiang. 2016. "Intensified Green Production of Astaxanthin from Haematococcus Pluvialis." *Food and Bioproducts Processing* 99. *Institution of Chemical Engineers*: 1–11. doi:10.1016/j.fbp.2016.03.002.

Hasegawa, Takashi, Kuniaki Tanaka, Kimiko Ueno, Sugi Ueno, Masao Okuda, Yasunobu Yoshikai, and Kikuo Nomoto. 1989. "Argumentatio of the Resistance against Escherichia Coli by Oral Administration of a Hot Water Extract of Chlorella Vulgaris in Rats." *International Society for Lmmunopharrnacology* 11 (8): 971–976.

Hashtroudi, Mehri Seyed, Zeinab Shariatmadari, Hossein Riahi, and Alireza Ghassempour. 2013. "Analysis of Anabaena Vaginicola and Nostoc Calcicola from Northern Iran, as Rich Sources of Major Carotenoids." *Food Chemistry* 136 (3–4): 1148–1153. doi:10.1016/j.foodchem.2012.09.055.

Hauton, Chris, and Valerie J. Smith. 2004. "In Vitro Cytotoxicity of Crustacean Immunostimulants for Lobster (Homarus Gammarus) Granulocytes Demonstrated Using the Neutral Red Uptake Assay." *Fish and Shellfish Immunology* 17 (1): 65–73. doi:10.1016/j.fsi.2003.12.003.

Hayashi, Kyoko, Toshimitsu Hayashi, and Ichiro Kojima. 1996. "A Natural Sulfated Polysaccharide, Calcium Spirulan, Isolated from Spirulina Platensis: In Vitro and Ex Vivo Evaluation of Anti-Herpes Simplex Virus and Anti-Human Immunodeficiency Virus Activities." *AIDS Research and Human Retroviruses* 12 (15): 1463–1471. doi:10.1089/aid.1996.12.1463.

Hemlata, Sumbul Afreen, and Tasneem Fatma. 2018. "Extraction, Purification and Characterization of Phycoerythrin from Michrochaete and Its Biological Activities." *Biocatalysis and Agricultural Biotechnology* 13: 84–89. doi:10.1016/j.bcab.2017.11.012.

Hemscheidt, Thomas, Melany P. Puglisi, Linda K. Larsen, Gregory M.L. Patterson, Richard E. Moore, Jorge L. Rios, and Jon Clardy. 1994. "Structure and Biosynthesis of Borophycin, a New Boeseken Complex of Boric Acid from a Marine Strain of the Blue-Green Alga Nostoc Linckia." *Journal of Organic Chemistry* 59 (12): 3467–3471. doi:10.1021/jo00091a042.

Hernández, I., J. F. Martínez-Aragón, J. L. Pérez-Lloréns, R. Vázquez, and J. J. Vergara. 2002. "Biofiltering Efficiency in Removal of Dissolved Nutrients by Three Species of Estuarine Macroalgae Cultivated with Sea Bass (Dicentrarchus Labrax) Waste Waters 1. Phosphate." *Journal of Applied Phycology* 14 (5): 375–384. doi:10.1023/A:1022134701273.

Hoffmann, James P. 1998. "Wastewater Treatment with Suspended and Nonsuspended Algae." *Journal of Phycology* 34 (5): 757–763. doi:10.1046/j.1529-8817.1998.340757.x.

Hu, Jianjun, Dillirani Nagarajan, Quanguo Zhang, Jo Shu Chang, and Duu Jong Lee. 2018. "Heterotrophic Cultivation of Microalgae for Pigment Production: A Review." *Biotechnology Advances* 36 (1): 54–67. doi:10.1016/j.biotechadv.2017.09.009.

Huang, Xuxiong, Hongqi Zhou, and Hui Zhang. 2006. "The Effect of Sargassum Fusiforme Polysaccharide Extracts on Vibriosis Resistance and Immune Activity of the Shrimp, Fenneropenaeus Chinensis." *Fish and Shellfish Immunology* 20 (5): 750–757. doi:10.1016/j.fsi.2005.09.008.

Huang, Zebo, Yongding Liu, Berit Smestad Paulsen, and Dag Klaveness. 1998. "Studies on Polysaccharides from Three Edible Species of Nostoc (Cyanobacteria) with Different Colony Morphologies: Comparison of Monosaccharide Compositions and Viscosities of Polysaccharides from Field Colonies and Suspension Cultures." *Journal of Phycology* 34 (6): 962–968. doi:10.1046/j.1529-8817.1998.340962.x.

Hughes, David A., Anthony J. A. Wright, Paul M. Finglas, Abigael C. J. Peerless, Angela L. Bailey, Sian B. Astley, Andrew C. Pinder, and Susan Southon. 1997. "The Effect of β-Carotene Supplementation on the Immune Function of Blood Monocytes from Healthy Male Nonsmokers." *Journal of Laboratory and Clinical Medicine* 129 (3): 309–317. doi:10.1016/S0022-2143(97)90179-7.

Humpage, Andrew R., Michael Fenech, Philip Thomas, and Ian R. Falconer. 2000. "Micronucleus Induction and Chromosome Loss in Transformed Human White Cells Indicate Clastogenic and Aneugenic Action of the Cyanobacterial Toxin, Cylindrospermopsin." *Mutation Research* 472 (1–2): 155–161. doi:10.1016/S1383-5718(00)00144-3.

Jahnke, Leland S. 1999. "Massive Carotenoid Accumulation in Dunaliella Bardawil Induced by Ultraviolet-A Radiation." *Journal of Photochemistry and Photobiology B: Biology* 48 (1): 68–74. doi:10.1016/S1011-1344(99)00012-3.

Jäpelt, Rie B., and Jette Jakobsen. 2013. "Vitamin D in Plants: A Review of Occurrence, Analysis, and Biosynthesis." *Frontiers in Plant Science* 4 (May): 1–21. doi:10.3389/fpls.2013.00136.

Jard, G., H. Marfaing, H. Carrère, J. P. Delgenes, J. P. Steyer, and C. Dumas. 2013. "French Brittany Macroalgae Screening: Composition and Methane Potential for Potential Alternative Sources of Energy and Products." *Bioresource Technology* 144: 492–498. doi:10.1016/j.biortech.2013.06.114.

Judé, Sébastien, Sébastien Roger, Eric Martel, Pierre Besson, Serge Richard, Philippe Bougnoux, Pascal Champeroux, and Jean Yves Le Guennec. 2006. "Dietary Long-Chain Omega-3 Fatty Acids of Marine Origin: A Comparison of Their Protective Effects on Coronary Heart Disease and Breast Cancers." *Progress in Biophysics and Molecular Biology* 90 (1–3): 299–325. doi:10.1016/j.pbiomolbio.2005.05.006.

Jyonouchi, H., S. Sun, Y. Tomita, and M. D. Gross. 1995. "Astaxanthin, a Carotenoid without Vitamin A Activity, Augments Antibody Responses in Cultures Including T-Helper Cell Clones and Suboptimal Doses of Antigen." *Journal of Nutrition* 125 (10): 2483–2492. doi:10.1093/jn/125.10.2483.

Kadir, Wan Nadiah Amalina, Man Kee Lam, Yoshimitsu Uemura, Jun Wei Lim, and Keat Teong Lee. 2018. "Harvesting and Pre-Treatment of Microalgae Cultivated in Wastewater for Biodiesel Production: A Review." *Energy Conversion and Management* 171 (June): 1416–1429. doi:10.1016/j.enconman.2018.06.074.

Kalra, Ekta K. 2003. "Nutraceutical - Definition and Introduction." *AAPS PharmSci* 5 (3): 1–2. doi:10.1208/ps050325.

Kamenarska, Zornitsa, Julia Serkedjieva, Hristo Najdenski, Kamen Stefanov, Iva Tsvetkova, Stefka Dimitrova-Konaklieva, and Simeon Popov. 2009. "Antibacterial, Antiviral, and Cytotoxic Activities of Some Red and Brown Seaweeds from the Black Sea." *Botanica Marina* 52 (1): 80–86. doi:10.1515/BOT.2009.030.

Kang, Ruijuan, Jing Wang, Dingji Shi, Wei Cong, Zhaoling Cai, and Fan Ouyang. 2004. "Interactions between Organic and Inorganic Carbon Sources during Mixotrophic Cultivation of Synechococcus Sp." *Biotechnology Letters* 26 (18): 1429–1432. doi:10.1023/B:BILE.0000045646.23832.a5.

Kazir, Meital, Yarden Abuhassira, Arthur Robin, Omri Nahor, Jincheng Luo, Alvaro Israel, Alexander Golberg, and Yoav D. Livney. 2019. "Extraction of Proteins from Two Marine Macroalgae, Ulva Sp. and Gracilaria Sp., for Food Application, and Evaluating Digestibility, Amino Acid Composition and Antioxidant Properties of the Protein Concentrates." *Food Hydrocolloids* 87: 194–203. doi:10.1016/j.foodhyd.2018.07.047.

Khatri, Kusum, Mangal S. Rathore, Surabhi Agrawal, and Bhavanath Jha. 2019. "Sugar Contents and Oligosaccharide Mass Profiling of Selected Red Seaweeds to Assess the Possible Utilization of Biomasses for Third-Generation Biofuel Production." *Biomass and Bioenergy* 130 (October): 105392. doi:10.1016/j.biombioe.2019.105392.

Kim, Jang K., George P. Kraemer, and Charles Yarish. 2014. "Field Scale Evaluation of Seaweed Aquaculture as a Nutrient Bioextraction Strategy in Long Island Sound and the Bronx River Estuary." *Aquaculture* 433: 148–156. doi:10.1016/j.aquaculture.2014.05.034.

Kiss, T., Á. Vehovszky, L. Hiripi, A. Kovács, and L. Vörös. 2002. "Membrane Effects of Toxins Isolated from a Cyanobacterium, Cylindrospermopsis Raciborskii, on Identified Molluscan Neurones." *Comparative Biochemistry and Physiology Part C: Toxicology & Pharmacology* 131 (2): 167–176. doi:10.1016/S1532-0456(01)00290-3.

Kiyota, Hiroshi, Yukiko Okuda, Michiho Ito, Masami Yokota Hirai, and Masahiko Ikeuchi. 2014. "Engineering of Cyanobacteria for the Photosynthetic Production of Limonene from CO2." *Journal of Biotechnology* 185: 1–7. doi:10.1016/j.jbiotec.2014.05.025.

Koller, Martin, Alexander Muhr, and Gerhart Braunegg. 2014. "Microalgae as Versatile Cellular Factories for Valued Products." *Algal Research* 6: 52–63. doi:10.1016/j.algal.2014.09.002.

Kothari, Richa, Arya Pandey, Shamshad Ahmad, and Ashwani Kumar. 2017. "Microalgal Cultivation for Value-Added Products: A Critical Enviro-Economical Assessment." *3 Biotech* 7 (4): 243. doi:10.1007/s13205-017-0812-8.

Kumar, Bikash, Komal Agrawal, Nisha Bhardwaj, Venkatesh Chaturvedi, and Pradeep Verma. 2019. "Techno-Economic Assessment of Microbe-Assisted Wastewater Treatment Strategies for Energy and Value-Added Product Recovery." In *Microbial Technology for the Welfare of Society*, pp. 147–181. doi: https://doi.org/10.1007/978-981-13-8844-6_7

Kumar, Bikash, and Pradeep Verma. 2021. "Techno-Economic Assessment of Biomass-Based Integrated Biorefinery for Energy and Value-Added Product." In *Biorefineries: A Step Towards Renewable and Clean Energy*, pp. 581–616. Springer doi: https://doi.org/10.1007/978-981-15-9593-6_23

Kuskoski, E. Marta, Agustìn G. Asuero, Ana M. Troncoso, Jorge Mancini-Filho, and Roseane Fett. 2005. "Aplicación de Diversos Métodos Químicos Para Determinar Actividad Antioxidante En Pulpa de Frutos." *Ciência e Tecnologia de Alimentos* 25 (4): 726–732. doi:10.1007/s10877-009-9165-0.

Meinerz, L. I., E. L. C. Ballester, L. J. Vaz, and W. Wasielesky, Jr. 2009. "Efeitos Da Temperatura Sobre a Absorção de Nutrientes e Crescimento Celular Da Microalga Thalassiosira Weissflogii." *Atlântica* 31 (2): 209–212. doi:10.5088/atl.2009.31.2.209.

Lafarga, Tomas, Francisco Gabriel Acién-Fernández, and Marco Garcia-Vaquero. 2020. "Bioactive Peptides and Carbohydrates from Seaweed for Food Applications: Natural Occurrence, Isolation, Purification, and Identification." *Algal Research* 48. 101909. doi:10.1016/j.algal.2020.101909.

Larson, Silva, Natalie Stoeckl, Mardiana E. Fachry, Muhammad Dalvi Mustafa, Imran Lapong, Agus Heri Purnomo, Michael A. Rimmer, and Nicholas A. Paul. 2021. "Women's Well-Being and Household Benefits from Seaweed Farming in Indonesia." *Aquaculture* 530: 735711. doi:10.1016/j.aquaculture.2020.735711.

Larsson, Susanna C., Leif Bergkvist, Ingmar Näslund, Jörgen Rutegård, and Alicja Wolk. 2007. "Vitamin A, Retinol, and Carotenoids and the Risk of Gastric Cancer: A Prospective Cohort Study." *American Journal of Clinical Nutrition* 85 (2): 497–503. doi:10.1093/ajcn/85.2.497.

Laurens, Lieve M. L., Madeline Lane, and Robert S. Nelson. 2020. "Sustainable Seaweed Biotechnology Solutions for Carbon Capture, Composition, and Deconstruction." *Trends in Biotechnology* 38 (11): 1232–1244. doi:10.1016/j.tibtech.2020.03.015.

Lauritano, Chiara, Jeanette H. Andersen, Espen Hansen, Marte Albrigtsen, Giovanna Romano, and Adrianna Ianora. 2016. "Bioactivity Screening of Microalgae for Antioxidant, Anti-Inflammatory, Antibacterial Activities." 3 (May): 1–12. doi:10.3389/fmars.2016.00068.

Lawrence, Karl P., Paul F. Long, and Antony R. Young. 2017. "Mycosporine-Like Amino Acids for Skin Photoprotection." *Current Medicinal Chemistry* 25 (40): 5512–5527. doi:10.2174/0929867324666170529124237.

Lee, I., J. I. Han. 2015. "Simultaneous Treatment (Cell Disruption and Lipid Extraction) of Wet Microalgae Using Hydrodynamic Cavitation for Enhancing the Lipid Yield." *Bioresource Technology* 186: 246–251.

Lee, Je Min, Hyungjae Lee, Seok Beom Kang, and Woo Jung Park. 2016. "Fatty Acid Desaturases, Polyunsaturated Fatty Acid Regulation, and Biotechnological Advances." *Nutrients* 8 (1): 1–13. doi:10.3390/nu8010023.

Lee, Seung-Hong, Joon-Baek Lee, Ki-Wan Lee, and You-Jin Jeon. 2010. "Antioxidant Properties of Tidal Pool Microalgae, Halochlorococcum Porphyrae and Oltamannsiellopsis Unicellularis from Jeju Island, Korea." *Algae* 25 (1): 45–56. doi:10.4490/algae.2010.25.1.045.

Lee, Xin Jiat, Hwai Chyuan Ong, Yong Yang Gan, Wei Hsin Chen, and Teuku Meurah Indra Mahlia. 2020. "State of Art Review on Conventional and Advanced Pyrolysis of Macroalgae and Microalgae for Biochar, Bio-Oil and Bio-Syngas Production." *Energy Conversion and Management* 210: 112707. doi:10.1016/j.enconman.2020.112707.

Di Lena, Gabriella, Irene Casini, Massimo Lucarini, and Ginevra Lombardi-Boccia. 2019. "Carotenoid Profiling of Five Microalgae Species from Large-Scale Production." *Food Research International* 120: 810–818. doi:10.1016/j.foodres.2018.11.043.

Leódido, Ana Carolina M., Luis E. C. Costa, Thiago S. L. Araújo, Douglas S. Costa, Nayara A. Sousa, Luan K. M. Souza, Francisca B. M. Sousa, et al. 2017. Anti-Diarrhoeal Therapeutic Potential and Safety Assessment of Sulphated Polysaccharide Fraction from Gracilaria Intermedia Seaweed in Mice. *International Journal of Biological Macromolecules* 97. 34–45. doi:10.1016/j.ijbiomac.2017.01.006.

Leston, Sara, Carolina Nebot, Margarida Nunes, Alberto Cepeda, Miguel Ângelo Pardal, and Fernando Ramos. 2015. "Sulfathiazole: Analytical Methods for Quantification in Seawater and Macroalgae." *Environmental Toxicology and Pharmacology* 39 (1): 77–84. doi:10.1016/j.etap.2014.11.007.

Li, Hua Bin, Ka Wing Cheng, Chi Chun Wong, King Wai Fan, Feng Chen, and Yue Jiang. 2007. "Evaluation of Antioxidant Capacity and Total Phenolic Content of Different Fractions of Selected Microalgae." *Food Chemistry* 102 (3): 771–776. doi:10.1016/j.foodchem.2006.06.022.

Liu, Hui, and Jilan Su. 2017. "Vulnerability of China's Nearshore Ecosystems under Intensive Mariculture Development." *Environmental Science and Pollution Research* 24 (10): 8957–8966. doi:10.1007/s11356-015-5239-3.

Liu, Jin, and Feng Chen. 2016. "Biology and Industrial Applications of Chlorella: Advances and Prospects." *Advances in Biochemical Engineering/Biotechnology* 153: 1–35. doi:10.1007/10_2014_286.

Llewellyn, Carole Anne, and Ruth Louise Airs. 2010. "Distribution and Abundance of MAAs in 33 Species of Microalgae across 13 Classes." *Marine Drugs* 8 (4): 1273–1291. doi:10.3390/md8041273.

Lourenço, S.O. 2006. *Cultivo de Microalgas Marinhas. Princípios e Aplicações.*

Lüning, Klaus. 1990. *Seaweeds: Their Environment, Biogeography, and Ecophysiology.*

Maadane, Amal, Nawal Merghoub, Tarik Ainane, Hicham El Arroussi, Redouane Benhima, Saaid Amzazi, Youssef Bakri, and Imane Wahby. 2015. "Antioxidant Activity of Some Moroccan Marine Microalgae: Pufa Profiles, Carotenoids and Phenolic Content." *Journal of Biotechnology* 215: 13–19. doi:10.1016/j.jbiotec.2015.06.400.

Magdouli, S., S. K. Brar, and J. F. Blais. 2016. "Co-Culture for Lipid Production: Advances and Challenges." *Biomass and Bioenergy* 92: 20–30. doi:10.1016/j.biombioe.2016.06.003.

Makkar, Harinder P. S., Gilles Tran, Valérie Heuzé, Sylvie Giger-Reverdin, Michel Lessire, François Lebas, and Philippe Ankers. 2016. "Seaweeds for Livestock Diets: A Review." *Animal Feed Science and Technology* 212: 1–17. doi:10.1016/j.anifeedsci.2015.09.018.

Mallik, Nirupama. 2002. "Biotechnological Potential of Immobilised Algae for Wastewater N, P and Metal Removal: A Review." *BioMetals* 15: 377–390.

Manolidi, Korina, Theodoros M. Triantis, Triantafyllos Kaloudis, and Anastasia Hiskia. 2018. "Neurotoxin BMAA and Its Isomeric Amino Acids in Cyanobacteria and Cyanobacteria-Based Food Supplements." *Journal of Hazardous Materials* 365: 346–365. doi:10.1016/j.jhazmat.2018.10.084.

Mantzouki, Evanthia, Miquel Lürling, Jutta Fastner, Lisette de Senerpont Domis, Elżbieta Wilk-Woźniak, Judita Koreivienė, Laura Seelen, et al. 2018. "Temperature Effects Explain Continental Scale Distribution of Cyanobacterial Toxins." *Toxins* 10 (4): 1–24. doi:10.3390/toxins10040156.

Manzoni Maroneze, Mariana, Eduardo Jacob-Lopes, Leila Queiroz Zepka, María Roca, and Antonio Pérez-Gálvez. 2019. "Esterified Carotenoids as New Food Components in Cyanobacteria." *Food Chemistry* 287: 295–302. doi:10.1016/j.foodchem.2019.02.102.

Manzoor, Mehnaza, Jagmohan Singh, Julie D. Bandral, Adil Gani, and Rafeeya Shams. 2020. "Food Hydrocolloids: Functional, Nutraceutical and Novel Applications for Delivery of Bioactive Compounds." *International Journal of Biological Macromolecules* 165: 554–567. doi:10.1016/j.ijbiomac.2020.09.182.

Markou, Giorgos, and Dimitris Georgakakis. 2011. "Cultivation of Filamentous Cyanobacteria (Blue-Green Algae) in Agro-Industrial Wastes and Wastewaters: A Review." *Applied Energy* 88 (10): 3389–3401. doi:10.1016/j.apenergy.2010.12.042.

Matos, Ângelo Paggi. 2017. "The Impact of Microalgae in Food Science and Technology." *Journal of the American Oil Chemists' Society* 94 (11): 1333–1350. doi:10.1007/s11746-017-3050-7.

Matos, Ângelo Paggi. 2019. "Microalgae as a Potential Source of Proteins." In *Proteins: Sustainable Source, Processing and Applications*, pp. 63–96. Academic Press doi:10.1016/b978-0-12-816695-6.00003-9.

Matsunaga, Tadashi, J. Grant Burgess, Noriko Yamada, Kazuo Komatsu, Seeichi Yoshida, and Youji Wachi. 1993. "An Ultraviolet (UV-A) Absorbing Biopterin Glucoside from the Marine Planktonic Cyanobacterium Oscillatoria Sp." *Applied Microbiology and Biotechnology* 39 (2): 250–253. doi:10.1007/BF00228614.

Matsuno, Takao. 2001. "Aquatic Animal Carotenoids." *Fisheries Science* 67 (5): 771–783. doi:10.1046/j.1444-2906.2001.00323.x.

Mehariya, Sanjeet, Rahul Kumar Goswami, Pradeep Verma, Roberto Lavecchia, and Antonio Zuorro. 2021a. "Integrated Approach for Wastewater Treatment and Biofuel Production in Microalgae Biorefineries." *Energies* 14 (8): 2282. doi:10.3390/en14082282.

Mehariya, Sanjeet, Rahul Kumar Goswami, Obulisamy Parthiba Karthikeysan, and Pradeep Verma. 2021b. "Microalgae for high-value products: A way towards green nutraceutical and pharmaceutical compounds." *Chemosphere* 280: 130553. doi:10.1016/j.chemosphere.2021.130553.

Mehariya, Sanjeet, Prasun Kumar, Tiziana Marino, Patrizia Casella, Antonio Lovine, Pradeep Verma, Dino Musmarra, and Antonio Molino. 2021c. "Aquatic Weeds: A Potential Pollutant Removing Agent from Wastewater and Polluted Soil and Valuable Biofuel Feedstock." In *Bioremediation Using Weeds*, pp. 59–77. Springer doi: https://doi.org/10.1007/978-981-33-6552-0_3

Méndez, A., G. Gascó, B. Ruiz, and E. Fuente. 2019. "Hydrochars from Industrial Macroalgae '*Gelidium Sesquipedale*' Biomass Wastes." *Bioresource Technology* 275: 386–393. doi:10.1016/j.biortech.2018.12.074.

Merel, Sylvain, David Walker, Ruth Chicana, Shane Snyder, Estelle Baurès, and Olivier Thomas. 2013. "State of Knowledge and Concerns on Cyanobacterial Blooms and Cyanotoxins." *Environment International* 59: 303–327. doi:10.1016/j.envint.2013.06.013.

Miki, Wataru. 1991. "Biological Functions and Activities of Animal Carotenoids." *Pure and Applied Chemistry* 63 (1): 141–146. doi:10.1351/pac199163010141.

Milledge, John J. 2011. "Commercial Application of Microalgae Other than as Biofuels: A Brief Review." *Reviews in Environmental Science and Biotechnology* 10 (1): 31–41. doi:10.1007/s11157-010-9214-7.

Mishra, Pragya, Vijay Pratap Singh, and Sheo Mohan Prasad. 2014. "Spirulina and Its Nutritional Importance: A Possible Approach for Development of Functional Food." *Biochemistry & Pharmacology: Open Access* 3 (6): 1–2. doi:10.4172/2167-0501.1000e171.

Mishra, Sanjiv K., Anupama Shrivastav, Imran Pancha, Deepti Jain, and Sandhya Mishra. 2010. "Effect of Preservatives for Food Grade C-Phycoerythrin, Isolated from Marine Cyanobacteria Pseudanabaena Sp." *International Journal of Biological Macromolecules* 47 (5): 597–602. doi:10.1016/j.ijbiomac.2010.08.005.

Miyake, Masato, Mayumi Erata, and Yasuo Asada. 1996. "A Thermophilic Cyanobacterium, Synechococcus Sp. MA19, Capable of Accumulating Poly-β-Hydroxybutyrate." *Journal of Fermentation and Bioengineering* 82 (5): 512–514. doi:10.1016/S0922-338X(97)86995-4.

Miyashita, Kazuo, Fumiaki Beppu, Masashi Hosokawa, Xiaoyong Liu, and Shuzhou Wang. 2020. "Nutraceutical Characteristics of the Brown Seaweed Carotenoid Fucoxanthin." *Archives of Biochemistry and Biophysics* 686: 108364. doi:10.1016/j.abb.2020.108364.

Mohamed, Zakaria A. 2008. "Polysaccharides as a Protective Response against Microcystin-Induced Oxidative Stress in Chlorella Vulgaris and Scenedesmus Quadricauda and Their Possible Significance in the Aquatic Ecosystem." *Ecotoxicology* 17 (6): 504–516. doi:10.1007/s10646-008-0204-2.

Molazadeh, Marziyeh, Hossein Ahmadzadeh, Hamid R. Pourianfar, Stephen Lyon, and Pabulo Henrique Rampelotto. 2019. "The Use of Microalgae for Coupling Wastewater Treatment With CO2 Biofixation." *Frontiers in Bioengineering and Biotechnology* 7: 42. doi:10.3389/fbioe.2019.00042.

Molfino, Alessio, Maria I. Amabile, Sara Mazzucco, Gianni Biolo, Alessio Farcomeni, Cesarina Ramaccini, Simonetta Antonaroli, Massimo Monti, and Maurizio Muscaritoli. 2017. "Effect of Oral Docosahexaenoic Acid (DHA) Supplementation on DHA Levels and Omega-3 Index in Red Blood Cell Membranes of Breast Cancer Patients." *Frontiers in Physiology* 8: 549. doi:10.3389/fphys.2017.00549.

Molino, Antonio, Angela Iovine, Patrizia Casella, Sanjeet Mehariya, Simeone Chianese, Antonietta Cerbone, Juri Rimauro, and Dino Musmarra. 2018. "Microalgae Characterization for Consolidated and New Application in Human Food, Animal Feed and Nutraceuticals." *International Journal of Environmental Research and Public Health* 15 (11): 1–21. doi:10.3390/ijerph15112436.

Moore, R. E. 1996. "Cyclic Peptides and Depsipeptides from Cyanobacteria: A Review." *Journal of Industrial Microbiology* 16 (2): 134–143. doi:10.1007/BF01570074.

Moraes, C. C., Luisa Sala, G. P. Cerveira, and S. J. Kalil. 2011. "C-Phycocyanin Extraction from Spirulina Platensis Wet Biomass." *Brazilian Journal of Chemical Engineering* 28 (1): 45–49. doi:10.1590/S0104-66322011000100006.

Mourelle, M. Lourdes, Carmen P. Gómez, and José L. Legido. 2017. "The Potential Use of Marine Microalgae and Cyanobacteria in Cosmetics and Thalassotherapy." *Cosmetics* 4 (4): 46. doi:10.3390/cosmetics4040046.

Msuya, Flower E. 2011. "The Impact of Seaweed Farming on the Socioeconomic Status of Coastal Communities in Zanzibar, Tanzania." *World Aquaculture* 42 (3): 45–48.

Mukund, S., M. Muthukumaran, and V Sivasubramanian. 2014. "In-Silico Studies on Cyanobacterial Metabolites against Lung Cancer EGFR Protein." *International Journal of Pharmacy and Pharmaceutical Sciences* 4: 89–98.

Mulbry, W., S. Kondrad, J. Buyer, D. L Luthria. 2009. "Optimization of an Oil Extraction Process for Algae from the Treatment of Manure Effluent." *Journal of the American Oil Chemists' Society* 86 (9): 909–915. doi:10.1007/s11746-009-1432-1.

Murthy, K. N. Chidambara, A. Vanitha, J. Rajesha, M. Mahadeva Swamy, P. R. Sowmya, and Gokare A. Ravishankar. 2005. "In Vivo Antioxidant Activity of Carotenoids from Dunaliella Salina - A Green Microalga." *Life Sciences* 76 (12): 1381–1390. doi:10.1016/j.lfs.2004.10.015.

Naguib, Yousry M. A. 2000. "Antioxidant Activities of Astaxanthin and Related Carotenoids." *Journal of Agricultural and Food Chemistry* 48 (4): 1150–1154. doi:10.1021/jf991106k.

Natrah, F. M. I., F. M. Yusoff, M. Shariff, F. Abas, and N. S. Mariana. 2007. "Screening of Malaysian Indigenous Microalgae for Antioxidant Properties and Nutritional Value." *Journal of Applied Phycology* 19 (6): 711–718. doi:10.1007/s10811-007-9192-5.

Nayak, Saswati, Radha Prasanna, Boddupalli M. Prasanna, and Dina B. Sahoo. 2007. "Analysing Diversity among Indian Isolates of Anabaena (Nostocales, Cyanophyta) Using Morphological, Physiological and Biochemical Characters." *World Journal of Microbiology and Biotechnology* 23 (11): 1575–1584. doi:10.1007/s11274-007-9403-x.

Neori, Amir, Michael D. Krom, Steve P. Ellner, Claude E. Boyd, Dan Popper, Ruth Rabinovitch, Patrick J. Davison, et al. 1996. "Seaweed Biofilters as Regulators of Water Quality in Integrated Fish-Seaweed Culture Units." *Aquaculture* 141 (3–4): 183–199. doi:10.1016/0044-8486(95)01223-0.

Neori, Amir, and Ana M. Nobre. 2012. "Relationship between Trophic Level and Economics in Aquaculture." *Aquaculture Economics and Management* 16 (1): 40–67. doi:10.1080/13657305.2012.649046.

Neori, Amir, Muki Shpigel, and David Ben-Ezra. 2000. "A Sustainable Integrated System for Culture of Fish, Seaweed and Abalone." *Aquaculture* 186 (3–4): 279–291. doi:10.1016/S0044-8486(99)00378-6.

Nowicka, Beatrycze, and Jerzy Kruk. 2010. "Occurrence, Biosynthesis and Function of Isoprenoid Quinones." *Biochimica et Biophysica Acta - Bioenergetics* 1797 (9): 1587–1605. doi:10.1016/j.bbabio.2010.06.007.

Nwoba, Emeka G., Jeremy M. Ayre, Navid R. Moheimani, Benjamin E. Ubi, and James C. Ogbonna. 2016. "Growth Comparison of Microalgae in Tubular Photobioreactor and Open Pond for Treating Anaerobic Digestion Piggery Effluent." *Algal Research* 17: 268–276. doi:10.1016/j.algal.2016.05.022.

Odjadjare, Ejovwokoghene C., Taurai Mutanda, and Ademola O. Olaniran. 2017. "Potential Biotechnological Application of Microalgae: A Critical Review." *Critical Reviews in Biotechnology* 37 (1): 37–52. doi:10.3109/07388551.2015.1108956.

Okimura, Takasi, Zedong Jiang, Hirofumi Komatsubara, Katsuya Hirasaka, and Tatsuya Oda. 2020. "Therapeutic Effects of an Orally Administered Edible Seaweed-Derived Polysaccharide Preparation, Ascophyllan HS, on a Streptococcus Pneumoniae Infection Mouse Model." *International Journal of Biological Macromolecules* 154: 1116–1122. doi:10.1016/j.ijbiomac.2019.11.053.

Pacheco, Diana, Ana Cristina Rocha, Leonel Pereira, and Tiago Verdelhos. 2020. "Microalgae Water Bioremediation: Trends and Hot Topics." *Applied Sciences* 10 (5): 1886. doi:10.3390/app10051886.

Pádua, D., E. Rocha, D. Gargiulo, and A. A. Ramos. 2015. "Bioactive Compounds from Brown Seaweeds: Phloroglucinol, Fucoxanthin and Fucoidan as Promising Therapeutic Agents against Breast Cancer." *Phytochemistry Letters* 14: 91–98. doi:10.1016/j.phytol.2015.09.007.

Panis, G., and J. Rosales Carreon. 2016. "Commercial Astaxanthin Production Derived by Green Alga Haematococcus Pluvialis: A Microalgae Process Model and a Techno-Economic Assessment All through Production Line." *Algal Research* 18: 175–190. doi:10.1016/j.algal.2016.06.007.

Park, Kyung Hyun, and Hyun Do Jeong. 1996. "Enhanced Resistance against Edwardsiella Tarda Infection in Tilapia (Oreochromis Niloticus) by Administration of Protein-Bound Polysaccharide." *Aquaculture* 143 (2): 135–143. doi:10.1016/0044-8486(95)01224-9.

Patel, Alok, Leonidas Matsakas, Ulrika Rova, and Paul Christakopoulos. 2019. "A Perspective on Biotechnological Applications of Thermophilic Microalgae and Cyanobacteria." *Bioresource Technology* 278: 424–434. doi:10.1016/j.biortech.2019.01.063.

Pellizzari, Franciane, and Renata Perpetuo Reis. 2011. "Seaweed Cultivation on the Southern and Southeastern Brazilian Coast." *Brazilian Journal of Pharmacognosy* 21 (2): 305–312. doi:10.1590/S0102-695X2011005000057.

Pereira, L. 2018. *Therapeutic and Nutritional Uses of Algae.* Edited by L. Pereira. Boca Raton, FL: CRC Press/Taylor & Francis Group.

Perez-Garcia, Octavio, Yoav Bashan, Yoav Bashan, and Yoav Bashan. 2015. "Microalgal Heterotrophic and Mixotrophic Culturing for Bio-Refining: From Metabolic Routes to Techno-Economics." In *Algal Biorefineries: Volume 2: Products and Refinery Design*, pp. 61–131. Springer doi:10.1007/978-3-319-20200-6_3.

Pérez-Lloréns, José Lucas, Ole G. Mouritsen, Prannie Rhatigan, M. Lynn Cornish, and Alan T. Critchley. 2020. "Seaweeds in Mythology, Folklore, Poetry, and Life." *Journal of Applied Phycology* 32 (5): 3157–3182. doi:10.1007/s10811-020-02133-0.

Peteiro, César, Noemí Sánchez, Clara Dueñas-Liaño, and Brezo Martínez. 2014. "Open-Sea Cultivation by Transplanting Young Fronds of the Kelp Saccharina Latissima." *Journal of Applied Phycology* 26 (1): 519–528. doi:10.1007/s10811-013-0096-2.

Peteiro, César, Noemí Sánchez, and Brezo Martínez. 2016. "Mariculture of the Asian Kelp Undaria Pinnatifida and the Native Kelp Saccharina Latissima along the Atlantic Coast of Southern Europe: An Overview." *Algal Research* 15: 9–23. doi:10.1016/j.algal.2016.01.012.

Di Pippo, Francesca, Neil T.W. Ellwood, Alessandra Gismondi, Laura Bruno, Federico Rossi, Paolo Magni, and Roberto de Philippis. 2013. "Characterization of Exopolysaccharides Produced by Seven Biofilm-Forming Cyanobacterial Strains for Biotechnological Applications." *Journal of Applied Phycology* 25 (6): 1697–1708. doi:10.1007/s10811-013-0028-1.

Plaza, Merichel, Miguel Herrero, Alejandro Cifuentes, and Elena Ibáñez. 2009. "Innovative Natural Functional Ingredients from Microalgae." *Journal of Agricultural and Food Chemistry* 57 (16): 7159–7170. doi:10.1021/jf901070g.

Posten, Clemens. 2009. "Design Principles of Photo-Bioreactors for Cultivation of Microalgae." *Engineering in Life Sciences* 9 (3): 165–177. doi:10.1002/elsc.200900003.

Poulose, Navya, Arya Sajayan, Amrudha Ravindran, T. V. Sreechithra, Vishnu Vardhan, Joseph Selvin, and George Seghal Kiran. 2020. "Photoprotective Effect of Nanomelanin-Seaweed Concentrate in Formulated Cosmetic Cream: With Improved Antioxidant and Wound Healing Properties." *Journal of Photochemistry and Photobiology B: Biology* 205: 111816. doi:10.1016/j.jphotobiol.2020.111816.

Pratt, Robertson, T. C. Daniels, John J. Eiler, J. B. Gunnison, W. D. Kumler, John F. Oneto, Louis A. Strait, et al. 1944. "Chlorellin, an Antibacterial Substance from Chlorella." *American Association for the Advancement of Science* 99 (2574): 351–352. doi:10.1126/science.99.2574.351.

Prieto, P., M. Pineda, and M. Aguilar. 1999. "Spectrophotometric Quantitation of Antioxidant Capacity through the Formation of a Phosphomolybdenum Complex: Specific Application to the Determination of Vitamin E." *Analytical Biochemistry* 269: 337–341. doi:10.1037/a0037168.

Pulz, Otto, and Wolfgang Gross. 2004. "Valuable Products from Biotechnology of Microalgae." *Applied Microbiology and Biotechnology* 65 (6): 635–648. doi:10.1007/s00253-004-1647-x.

Qiu, Baosheng, Jiyong Liu, Zhili Liu, and Shengxiang Liu. 2002. "Distribution and Ecology of the Edible Cyanobacterium Ge-Xian-Mi (Nostoc) in Rice Fields of Hefeng County in China." *Journal of Applied Phycology* 14 (5): 423–429. doi:10.1023/A:1022198605743.

Quijano, Guillermo, Juan S. Arcila, and Germán Buitrón. 2017. "Microalgal-Bacterial Aggregates: Applications and Perspectives for Wastewater Treatment." *Biotechnology Advances* 35 (6): 772–781. doi:10.1016/j.biotechadv.2017.07.003.

Radmann, E. M, and J.A.V. Costa. 2008. "Conteúdo Lipídico e Composição de Ácidos Graxos de Microalgas Expostas Aos Gases CO2, SO2 e NO." *Química Nova* 31 (7): 1609–1612.

Ramli, Norulhuda Mohamed, J. A. J. Verreth, Fatimah M. Yusoff, K. Nurulhuda, N. Nagao, and Marc C. J. Verdegem. 2020. "Integration of Algae to Improve Nitrogenous Waste Management in Recirculating Aquaculture Systems: A Review." *Frontiers in Bioengineering and Biotechnology* 8: 1004. doi:10.3389/fbioe.2020.01004.

Rangel-Yagui, Carlota De Oliveira, Eliane Dalva Godoy Danesi, João Carlos Monteiro De Carvalho, and Sunao Sato. 2004. "Chlorophyll Production from Spirulina Platensis: Cultivation with Urea Addition by Fed-Batch Process." *Bioresource Technology* 92 (2): 133–141. doi:10.1016/j.biortech.2003.09.002.

Rawat, I., R. Ranjith Kumar, T. Mutanda, and F. Bux. 2011. "Dual Role of Microalgae: Phycoremediation of Domestic Wastewater and Biomass Production for Sustainable Biofuels Production." *Applied Energy* 88 (10): 3411–3424. doi:10.1016/j.apenergy.2010.11.025.

Reid, Gregor K., Helen J. Gurney-Smith, Mark Flaherty, Amber F. Garber, Ian Forster, Kathy Brewer-Dalton, and Duncan Knowler, et al. 2019. "Climate Change and Aquaculture: Considering Adaptation Potential." *Aquaculture Environment Interactions* 11: 603–624. doi:10.3354/AEI00333.

Reid, Gregor K., Sébastien Lefebvre, Ramón Filgueira, Shawn M. C. Robinson, Ole J. Broch, Andre Dumas, and Thierry B. R. Chopin. 2020. "Performance Measures and Models for Open-Water Integrated Multi-Trophic Aquaculture." *Reviews in Aquaculture* 12 (1): 47–75. doi:10.1111/raq.12304.

Renugadevi, K., C. Valli Nachiyar, P. Sowmiya, and Swetha Sunkar. 2018. "Antioxidant Activity of Phycocyanin Pigment Extracted from Marine Filamentous Cyanobacteria Geitlerinema Sp TRV57." *Biocatalysis and Agricultural Biotechnology* 16: 237–242. doi:10.1016/j.bcab.2018.08.009.

Rizzello, Carlo G., Davide Tagliazucchi, Elena Babini, Giuseppina Sefora Rutella, Danielle L. Taneyo Saa, and Andrea Gianotti. 2016. "Bioactive Peptides from Vegetable Food Matrices: Research Trends and Novel Biotechnologies for Synthesis and Recovery." *Journal of Functional Foods* 27: 549–569. doi:10.1016/j.jff.2016.09.023.

Robles Medina, A., A. Gimenez Gimenez, F. Garcia Camacho, J. A. Sanchez Perez, E. Molina Grima, and A. Contreras Gomez. 1995. "Concentration and Purification of Stearidonic, Eicosapentaenoic, and Docosahexaenoic Acids from Cod Liver Oil and the Marine Microalga Isochrysis Galbana." *Journal of the American Oil Chemists' Society* 72 (5): 575–583. doi:10.1007/BF02638859.

Rodriguez-Garcia, Ignacio, and Jose Luis Guil-Guerrero. 2008. "Evaluation of the Antioxidant Activity of Three Microalgal Species for Use as Dietary Supplements and in the Preservation of Foods." *Food Chemistry* 108 (3): 1023–1026. doi:10.1016/j.foodchem.2007.11.059.

Ronga, Domenico, Elisa Biazzi, Katia Parati, Domenico Carminati, Elio Carminati, and Aldo Tava. 2019. "Microalgal Biostimulants and Biofertilisers in Crop Productions." *Agronomy* 9 (4): 1–22. doi:10.3390/agronomy9040192.

Rose, David P., and Jeanne M. Connolly. 1999. "Omega-3 Fatty Acids as Cancer Chemopreventive Agents." *Pharmacology and Therapeutics* 83 (3): 217–244. doi:10.1016/S0163-7258(99)00026-1.

Rudke, Adenilson Renato, Cristiano José de Andrade, and Sandra Regina Salvador Ferreira. 2020. "Kappaphycus Alvarezii Macroalgae: An Unexplored and Valuable Biomass for Green Biorefinery Conversion." *Trends in Food Science and Technology* 103: 214–224. doi:10.1016/j.tifs.2020.07.018.

Ruiz-Marin, Alejandro, Leopoldo G. Mendoza-Espinosa, and Tom Stephenson. 2010. "Growth and Nutrient Removal in Free and Immobilized Green Algae in Batch and Semi-Continuous Cultures Treating Real Wastewater." *Bioresource Technology* 101 (1): 58–64. doi:10.1016/j.biortech.2009.02.076.

Runnegar, Maria T., Chaoyu Xie, Barry B. Snider, Grier A. Wallace, Steven M. Weinreb, and John Kuhlenkamp. 2002. "In Vitro Hepatotoxicity of the Cyanobacterial Alkaloid Cylindrospermopsin and Related Synthetic Analogues." *Toxicological Sciences* 67 (1): 81–87. doi:10.1093/toxsci/67.1.81.

Safi, Carl, Bachar Zebib, Othmane Merah, Pierre Yves Pontalier, and Carlos Vaca-Garcia. 2014. "Morphology, Composition, Production, Processing and Applications of Chlorella Vulgaris: A Review." *Renewable and Sustainable Energy Reviews* 35: 265–278. doi:10.1016/j.rser.2014.04.007.

Saha, Sushanta Kumar, and Patrick Murray. 2018. "Exploitation of Microalgae Species for Nutraceutical Purposes: Cultivation Aspects." *Fermentation* 4 (2). doi:10.3390/fermentation4020046.

Saini, Dinesh Kumar, Sunil Pabbi, and Pratyoosh Shukla. 2018. "Cyanobacterial Pigments: Perspectives and Biotechnological Approaches." *Food and Chemical Toxicology* 120: 616–624. doi:10.1016/j.fct.2018.08.002.

Sakai, M. 1999. "Fish Immunostimulants: The Application for Aquaculture." *Faculty of Agriculture Miyazaki University*: 889.

Sakai, M., H. Kamiya, S. Ishii, S. Atsuta, and M. Kobayashi. 1992. "The Immunostimulating Effects of Chitin in Rainbow Trout, Oncorhynchus Mykiss." *Aquaculture* 1: 413–17.

Salama, El Sayed, Mayur B. Kurade, Reda A. I. Abou-Shanab, Marwa M. El-Dalatony, I. l. Seung Yang, Booki Min, and Byong Hun Jeon. 2017. "Recent Progress in Microalgal Biomass Production Coupled with Wastewater Treatment for Biofuel Generation." *Renewable and Sustainable Energy Reviews* 79: 1189–1211. doi:10.1016/j.rser.2017.05.091.

Salehi, Sharifi-Rad, Seca, Pinto, Michalak, Trincone, Mishra, Nigam, Zam, and Martins. 2019. "Current Trends on Seaweeds: Looking at Chemical Composition, Phytopharmacology, and Cosmetic Applications." *Molecules* 24 (22): 4182. doi:10.3390/molecules24224182.

Santhakumaran, Prasanthkumar, Sunil Meppath Ayyappan, and Joseph George Ray. 2020. "Nutraceutical Applications of Twenty-Five Species of Rapid-Growing Green-Microalgae as Indicated by Their Antibacterial, Antioxidant and Mineral Content." *Algal Research* 47: 101878. doi:10.1016/j.algal.2020.101878.

Santos, Thaisa Duarte, Bárbara Catarina Bastos de Freitas, Juliana Botelho Moreira, Kellen Zanfonato, and Jorge Alberto Vieira Costa. 2016. "Development of Powdered Food with the Addition of Spirulina for Food Supplementation of the Elderly Population." *Innovative Food Science and Emerging Technologies* 37: 216–220. doi:10.1016/j.ifset.2016.07.016.

Sathasivam, Ramaraj, Ramalingam Radhakrishnan, Abeer Hashem, and Elsayed F. Abd_Allah. 2019. "Microalgae Metabolites: A Rich Source for Food and Medicine." *Saudi Journal of Biological Sciences* 26 (4): 709–722. doi:10.1016/j.sjbs.2017.11.003.

Satyanarayana, K. G., A. B. Mariano, and J. V. C. Vargas. 2011. "A Review on Microalgae, a Versatile Source for Sustainable Energy and Materials." *International Journal of Energy Research* 35 (4): 291–311. doi:10.1002/er.1695.

Scherer, S., T. W. Chen, and P. Böger. 1988. "A New UV-A/B Protecting Pigment in the Terrestrial Cyanobacterium Nostoc Commune." *Plant Physiology* 88 (4): 1055–1057. doi:10.1104/pp.88.4.1055.

Seifert, Marc, Glenn McGregor, Geoff Eaglesham, Wasantha Wickramasinghe, and Glen Shaw. 2007. "First Evidence for the Production of Cylindrospermopsin and Deoxy-Cylindrospermopsin by the Freshwater Benthic Cyanobacterium, Lyngbya Wollei (Farlow Ex Gomont) Speziale and Dyck." *Harmful Algae* 6 (1): 73–80. doi:10.1016/j.hal.2006.07.001.

Shaaban, M. M. 2001. "Nutritional Status and Growth of Maize Plants as Affected by Green Microalgae as Soil Additives." *OnLine Journal of Biological Sciences* 6: 475–79.

Shanab, Sanaa M. M., Soha S. M. Mostafa, Emad A. Shalaby, and Ghada I. Mahmoud. 2012. "Aqueous Extracts of Microalgae Exhibit Antioxidant and Anticancer Activities." *Asian Pacific Journal of Tropical Biomedicine* 2 (8): 608–615. doi:10.1016/S2221-1691(12)60106-3.

Sharma, Naveen K., Sri Prakash Tiwari, Keshwanand Tripathi, and Ashwani K. Rai. 2011. "Sustainability and Cyanobacteria (Blue-Green Algae): Facts and Challenges." *Journal of Applied Phycology* 23 (6): 1059–1081. doi:10.1007/s10811-010-9626-3.

Sheih, I. Chuan, Tony J. Fang, and Tung Kung Wu. 2009. "Isolation and Characterisation of a Novel Angiotensin I-Converting Enzyme (ACE) Inhibitory Peptide from the Algae Protein Waste." *Food Chemistry* 115 (1): 279–284. doi:10.1016/j.foodchem.2008.12.019.

Shim, Jae Young, Hye Seoung Shin, Jae Gab Han, Hyeung Suk Park, Byung Lak Lim, Kyung Won Chung, and Ae Son Om. 2008. "Protective Effects of Chlorella Vulgaris on Liver Toxicity in Cadmium-Administered Rats." *Journal of Medicinal Food* 11 (3): 479–485. doi:10.1089/jmf.2007.0075.

Shimizu, Mineo, and Takashi Tomoo. 1994. "Anti-Inflammatory Constituents of Topically Applied Crude Drugs. V. Comstituents and Anti-Inflammatory Effect of Aoki, Aucuba Japonica THuNB." *Biological and Pharmaceutical Bulletin* 17 (5): 665–667.

Silva-Stenico, Maria Estela, Caroline Souza Pamplona Silva, Adriana Sturion Lorenzi, Tânia Keiko Shishido, Augusto Etchegaray, Simone Possedente Lira, Luiz Alberto Beraldo Moraes, and Marli Fátima Fiore. 2011. "Non-Ribosomal Peptides Produced by Brazilian Cyanobacterial Isolates with Antimicrobial Activity." *Microbiological Research* 166 (3): 161–175. doi:10.1016/j.micres.2010.04.002.

Silva, D. M., L. M. P. Valente, I. Sousa-Pinto, R. Pereira, M. A. Pires, F. Seixas, and P. Rema. 2015. "Evaluation of IMTA-Produced Seaweeds (Gracilaria, Porphyra, and Ulva) as Dietary Ingredients in Nile Tilapia, Oreochromis Niloticus L., Juveniles. Effects on Growth Performance and Gut Histology." *Journal of Applied Phycology* 27 (4): 1671–1680. doi:10.1007/s10811-014-0453-9.

Singh, Amit Kumar, Risha Ganguly, Shashank Kumar, and Abhay K. Pandey. 2017. "Microalgae: A Source of Nutraceutical and Industrial Products." In *Molecular Biology and Pharmacognosy of Beneficial Plants*, pp. 34–51. Delhi: Lenin Media.

Singh, P., B. Rani, A. K. Chauhan, and R. Maheshwari. 2012. "Lycopene's Antioxidant Activity in Cosmetics Meadow." *International Research Journal of Pharmacy* 3 (1): 46–47.

Singh, Rachana, Parul Parihar, Madhulika Singh, Andrzej Bajguz, Jitendra Kumar, Samiksha Singh, Vijay P. Singh, and Sheo M. Prasad. 2017. "Uncovering Potential Applications of Cyanobacteria and Algal Metabolites in Biology, Agriculture and Medicine: Current Status and Future Prospects." *Frontiers in Microbiology* 8: 515. doi:10.3389/fmicb.2017.00515.

Soletto, D., L. Binaghi, A. Lodi, J. C. M. Carvalho, and A. Converti. 2005. "Batch and Fed-Batch Cultivations of Spirulina Platensis Using Ammonium Sulphate and Urea as Nitrogen Sources." *Aquaculture* 243 (1–4): 217–24. doi:10.1016/j.aquaculture.2004.10.005.

Spolaore, Pauline, Claire Joannis-Cassan, Elie Duran, and Arsène Isambert. 2006. "Commercial Applications of Microalgae." *Journal of Bioscience and Bioengineering* 101 (2): 87–96. doi:10.1263/jbb.101.87.

Sudhakar, K., R. Mamat, M. Samykano, W. H. Azmi, W. F. W. Ishak, and Talal Yusaf. 2018. "An Overview of Marine Macroalgae as Bioresource." *Renewable and Sustainable Energy Reviews* 91: 165–179. doi:10.1016/j.rser.2018.03.100.

Suetsuna, Kunio, and Jiun Rong Chen. 2001. "Identification of Antihypertensive Peptides from Peptic Digest of Two Microalgae, Chlorella Vulgaris and Spirulina Platensis." *Marine Biotechnology* 3 (4): 305–309. doi:10.1007/s10126-001-0012-7.

Switzer, L. 1982. *Spirulina, the Whole Food Revolution. Bantam Books.*

Takaichi, Shinichi. 2011. "Carotenoids in Algae: Distributions, Biosyntheses and Functions." *Marine Drugs* 9 (6): 1101–1118. doi:10.3390/md9061101.

Takamatsu, Satoshi, Tyler W. Hodges, Ira Rajbhandari, William H. Gerwick, Mark T. Hamann, and Dale G. Nagle. 2003. "Marine Natural Products as Novel Antioxidant Prototypes." *Journal of Natural Products* 66 (5): 605–608. doi:10.1021/np0204038.

Tanaka, Kuniaki, Yoshifumi Tomita, Mari Tsuruta, Fumiko Konishi, Masao Okuda, Kunisuke Himeno, and Kikuo Nomoto. 1990. "Oral Administration of Chlorella Vulgaris Augments Concomitant Antitumor Immunity." *Immunopharmacology and Immunotoxicology* 12 (2): 277–291. doi:10.3109/08923979009019673.

Tandeau de Marsac, Nicole, and Jean Houmard. 1993. "Adaptation of Cyanobacteria to Environmental Stimuli: New Steps towards Molecular Mechanisms." *FEMS Microbiology Letters* 104 (1–2): 119–189. doi:10.1016/0378-1097(93)90506-W.

Thanigaivel, S., Natarajan Chandrasekaran, Amitava Mukherjee, and John Thomas. 2016. "Seaweeds as an Alternative Therapeutic Source for Aquatic Disease Management." *Aquaculture* 464: 529–536. doi:10.1016/j.aquaculture.2016.08.001.

Tibbetts, Sean M., Joyce E. Milley, and Santosh P. Lall. 2016. "Nutritional Quality of Some Wild and Cultivated Seaweeds: Nutrient Composition, Total Phenolic Content and in Vitro Digestibility." *Journal of Applied Phycology* 28 (6): 3575–3585. doi:10.1007/s10811-016-0863-y.

Torres, Fábio A. E., Thais G. Passalacqua, Angela M. A. Velásquez, Rodrigo A. de Souza, Pio Colepicolo, and Márcia A. S. Graminha. 2014. "New Drugs with Antiprotozoal Activity from Marine Algae: A Review." *Brazilian Journal of Pharmacognosy* 24 (3): 265–276. doi:10.1016/j.bjp.2014.07.001.

Torres, Maria Dolores, Noelia Flórez-Fernández, and Herminia Domínguez. 2019. "Integral Utilization of Red Seaweed for Bioactive Production." *Marine Drugs* 17 (6): 314. doi:10.3390/md17060314.

Troell, M., P. Rönnbäck, C. Halling, N. Kautsky, and A. Buschmann. 1999. "Ecological Engineering in Aquaculture: Use of Seaweeds for Removing Nutrients from Intensive Mariculture." In *Sixteenth International Seaweed Symposium*, pp. 603–611. Springer, Netherlands. doi:10.1007/978-94-011-4449-0_74.

Vairappan, Charles S., Chong Sim Chung, A. Q. Hurtado, Flower E. Soya, Genevieve Bleicher Lhonneur, and Alan Critchley. 2008. "Distribution and Symptoms of Epiphyte Infection in Major Carrageenophyte-Producing Farms." *Journal of Applied Phycology* 20 (5): 477–483. doi:10.1007/s10811-007-9299-8.

Valduga, E., P. O. Tatsch, L. Tiggemann, H. Treichel, G. Toniazzo, J. Zeni, M. Luccio. 2009. "Produção de Carotenoides: Microrganismos Como Fonte de Pigmentos Naturais." *Química Nova* 32 (9): 2429–2436.

Valduga, Eunice, Pihetra Oliveira Tatsch, Lídia Tiggemann, Helen Treichel, Geciane Toniazzo, Jamile Zeni, Marco Di Luccio, and Agenor Fúrigo. 2009. "Carotenoids Production: Microorganisms as Source of Natural Dyes." *Quimica Nova* 32 (9): 2429–36. doi:10.1590/S0100-40422009000900036.

Van Poppel, Geert, and R. Alexandra Goldbohm. 1995. "Epidemiologic Evidence for N-Carotene and Cancer." *The American Journal of Clinical Nutrition* 62 (6): 1393–1402.

Varshney, Prachi, Paulina Mikulic, Avigad Vonshak, John Beardall, and Pramod P. Wangikar. 2015. "Extremophilic Micro-Algae and Their Potential Contribution in Biotechnology." *Bioresource Technology* 184: 363–72. doi:10.1016/j.biortech.2014.11.040.

Vazquez-Villegas, Patricia, Mario A. Torres-Acosta, Sergio A. Garcia-Echauri, Jose M. Aguilar-Yanez, Marco Rito-Palomares, and Federico Ruiz-Ruiz. 2018. "Genetic Manipulation of Microalgae for the Production of Bioproducts." *Frontiers in Bioscience - Elite* 10 (2): 254–275. doi:10.2741/e821.

Velea, Sanda, Lucia Ilie, and Laurenţiu Filipescu. 2011. "Optimization of Porphyridium Purpureum Culture Growth Using Two Variables Experimental Design: Light and Sodium Bicarbonate." *UPB Scientific Bulletin, Series B: Chemistry and Materials Science* 73 (4): 81–94.

Venkatesan, Manigandan, Velusamy Arumugam, Rubanya Pugalendi, Karthik Ramachandran, Karthi Sengodan, Sri Ramkumar Vijayan, Umamaheswari Sundaresan, Saravanan Ramachandran, and Arivalagan Pugazhendhi. 2019. "Antioxidant, Anticoagulant and Mosquitocidal Properties of Water Soluble Polysaccharides (WSPs) from Indian Seaweeds." *Process Biochemistry* 84: 196–204. doi:10.1016/j.procbio.2019.05.029.

Vijayakumar, Subramaniyan, and Muniraj Menakha. 2015. "Pharmaceutical Applications of Cyanobacteria-A Review." *Journal of Acute Medicine* 5 (1): 15–23. doi:10.1016/j.jacme.2015.02.004.

Vo, Thanh Sang, Bo Mi Ryu, and Se Kwon Kim. 2013. "Purification of Novel Anti-Inflammatory Peptides from Enzymatic Hydrolysate of the Edible Microalgal Spirulina Maxima." *Journal of Functional Foods* 5 (3): 1336–1346. doi:10.1016/j.jff.2013.05.001.

Volkmann, Harriet, Ulisses Imianovsky, Jorge L. B. Oliveira, and Ernani S. Sant'Anna. 2008. "Cultivation of Arthrospira (Spirulina) Platensis in Desalinator Wastewater and Salinated Synthetic Medium: Protein Content and Amino-Acid Profile." *Brazilian Journal of Microbiology* 39 (1): 98–101. doi:10.1590/S1517-83822008000100022.

Wang, Shuang, Shuang Zhao, Benjamin Bernard Uzoejinwa, Anqing Zheng, Qingyuan Wang, Jin Huang, and Abd El Fatah Abomohra. 2020. "A State-of-the-Art Review on Dual Purpose Seaweeds Utilization for Wastewater Treatment and Crude Bio-Oil Production." *Energy Conversion and Management* 222: 113253. doi:10.1016/j.enconman.2020.113253.

Wei, Na, Josh Quarterman, and Yong Su Jin. 2013. "Marine Macroalgae: An Untapped Resource for Producing Fuels and Chemicals." *Trends in Biotechnology* 31 (2): 70–77. doi:10.1016/j.tibtech.2012.10.009.

Wei, Yuxi, Qi Liu, Jia Yu, Qiang Feng, Ling Zhao, Huiping Song, and Wenxiu Wang. 2015. "Antibacterial Mode of Action of 1,8-Dihydroxy-Anthraquinone from Porphyra Haitanensis against Staphylococcus Aureus." *Natural Product Research* 29 (10): 976–979. doi:10.1080/14786419.2014.964705.

Wells, Mark L., Philippe Potin, James S. Craigie, John A. Raven, Sabeeha S. Merchant, Katherine E. Helliwell, Alison G. Smith, Mary Ellen Camire, and Susan H. Brawley. 2017. "Algae as Nutritional and Functional Food Sources: Revisiting Our Understanding." *Journal of Applied Phycology* 29 (2): 949–982. doi:10.1007/s10811-016-0974-5.

Wu, G. F., Q. Y. Wu, and Z. Y. Shen. 2001. "Accumulation of Poly-β-Hydroxybutyrate in Cyanobacterium Synechocystis Sp. PCC6803." *Bioresource Technology* 76 (2): 85–90. doi:10.1016/S0960-8524(00)00099-7.

Wu, Hailong, Yuanzi Huo, Fang Han, Yuanyuan Liu, and Peimin He. 2015. "Bioremediation Using Gracilaria Chouae Co-Cultured with Sparus Macrocephalus to Manage the Nitrogen and Phosphorous Balance in an IMTA System in Xiangshan Bay, China." *Marine Pollution Bulletin* 91 (1): 272–279. doi:10.1016/j.marpolbul.2014.11.032.

Xu, Yi, and Steven Qian. 2014. "Anti-Cancer Activities of ω-6 Polyunsaturated Fatty Acids." *Biomedical Journal* 37 (3): 112–119. doi:10.4103/2319-4170.131378.

Yaakob, Zahira, Ehsan Ali, Afifi Zainal, Masita Mohamad, and Mohd S. Takriff. 2014. "An Overview: Biomolecules from Microalgae for Animal Feed and Aquaculture." *Journal of Biological Research (Greece)* 21 (1): 1–10. doi:10.1186/2241-5793-21-6.

Yan, Na, Chengming Fan, Yuhong Chen, and Zanmin Hu. 2016. "The Potential for Microalgae as Bioreactors to Produce Pharmaceuticals." *International Journal of Molecular Sciences* 17 (6): 1–24. doi:10.3390/ijms17060962.

Yang, Jia, Ming Xu, Xuezhi Zhang, Qiang Hu, Milton Sommerfeld, and Yongsheng Chen. 2011. "Life-Cycle Analysis on Biodiesel Production from Microalgae: Water Footprint and Nutrients Balance." *Bioresource Technology* 102 (1): 159–165. doi:10.1016/j.biortech.2010.07.017.

You, Tao, and Stanley M. Barnett. 2004. "Effect of Light Quality on Production of Extracellular Polysaccharides and Growth Rate of Porphyridium Cruentum." *Biochemical Engineering Journal* 19 (3): 251–258. doi:10.1016/j.bej.2004.02.004.

Zhang, Rui, Alexander K. L. Yuen, Rocky de Nys, Anthony F. Masters, and Thomas Maschmeyer. 2020. "Step by Step Extraction of Bio-Actives from the Brown Seaweeds, Carpophyllum Flexuosum, Carpophyllum Plumosum, Ecklonia Radiata and Undaria Pinnatifida." *Algal Research* 52: 102092. doi:10.1016/j.algal.2020.102092.

3 Identification, Cultivation and Potential Utilization of Microalgae in Domestic Wastewater Treatment

Debanjan Sanyal, Sneha Athalye, Shyam Prasad, Dishant Desai, Vinay Dwivedi and Santanu Dasgupta

CONTENTS

3.1 INTRODUCTION

Pollution-related problems are a significant concern for today's aware society. Information about environmental laws has become widespread and their implementation increasingly stricter. Domestic wastewater treatment and discharge is a major challenge in many developing countries (Scott et al. 2004; Moe and Rheingans 2006; Qadir et al. 2010). Lack of suitable wastewater treatment results in environmental pollution, contamination of scarce drinking water supplies and poses a threat to public health (Bwapwa 2018; Kumar et al. 2018, 2019a). Centralised wastewater treatment plants typically fail in these developing regions due to the complexity of the physical, chemical and biological treatment processes, high cost of operation, maintenance and lack of trained personnel (Diaz-Elsayed et al. 2019; Munasinghe 2019; Kumar et al. 2021).

DOI: 10.1201/9781003155713-3

Therefore, there is a need for sustainable, simple and effective wastewater treatment systems that can meet the wastewater treatment and sanitation needs in the developing regions of the world (Jayaswal et al. 2018; Agrawal and Verma 2020a, b; 2021a, b). Microalgae-based wastewater treatment systems are gaining attention in recent years as sustainable treatment processes that offer simple and cost-effective solutions for different types of wastewater (Almomani 2019; Li et al. 2019; Nagarajan et al. 2019; Kang et al. 2018; Almomani et al. 2017; Judd et al. 2017; Goswami et al. 2021a; Mehariya et al. 2021a). Algae are heterogeneous, largely eukaryotic, aquatic organisms that range from single cells to highly differentiated plants. The process of transforming light energy into chemical energy by means of sunlight is known as photosynthesis. Algae can use sunlight to utilize carbon from carbon dioxide and release oxygen into the atmosphere (Rehnstam Holm and Godhe 2003; Goswami et al. 2020).

Algae are a very diverse group of largely aquatic photosynthetic organisms that account for almost 50% of the photosynthesis that takes place on Earth (Day et al. 2017). Algae have a wide range of pigments to harvest light energy for photosynthesis, giving different types of algae their characteristic colour. Algae produce oxygen that is utilized to degrade organic substances and sequester carbon dioxide during growth. In the absence of sunlight, the speed of photosynthesis in algae is reduced, thus limiting their role in removing nutrients from organic sources. Algae have diversified applications in energy production such as algal biofuel or algal oil (ethanol, biogas, biobutanol, biogasoline, biodiesel, biohydrogen, etc.), which is a substitute to liquid fossil fuels (Darzins et al. 2010; Scott et al. 2010; Agrawal et al. 2020; Bhardwaj et al. 2020; Goswami et al. 2021b).

In 1960, the use of algae for wastewater treatment was described for the first time together with the system for energy generation from the algal biomass (Oswald and Golueke 1960). Algal cells were effectively used for wastewater treatment as they absorbed nutrients, metals and carbon-based compounds present in the wastewater (Laurens et al. 2017). Large-scale cultivation of algae for treatment of polluted air or water can result in the production of biomass that has a high calorific value. This biomass can be used as a substitute for coal and might be utilized for other applications like biochar and biofertilisers. Hydrothermal liquefaction (HTL) could be a very useful thermochemical method of manufacturing bio-oil from microalgae biomass in the coming years (Lopez Barreiro et al. 2013).

The advantages of using microalgae for wastewater treatment include the low cost of operation and the possibility of utilizing the generated algae biomass. Additionally, microalgae can produce oxygen which supports the aerobic biological treatment of wastewater through the heterotrophic bacteria (Barnharst et al. 2018). Algae can also be utilized as bioindicators for various contaminants, ensuring that the impurities stay within mandated limits (García-Seoane et al. 2018).

The ability to utilize algal capacity for wastewater treatment along with production of multiple by-products is largely limited due to the space needed for large-scale growth operations and the time needed to remove waste from site (Vassilev and Vassileva 2016; Mehariya et al. 2021b). Algae can also be grown in a photobioreactor (PBR), which is an alternate cultivation system. Almost any transparent container can be called a PBR; however, the term is more commonly used to define a closed system, as opposed to an open tank or pond.

Domestic and industrial wastewater can contain a complex mix of chemical and biological contaminants. Many sources of wastewater contain nitrogen (N) and phosphorus (P) based compounds in excess. If not removed, a build-up of these nutrients can cause eutrophication (extreme algal growth) and toxicity in aquatic environments. Consequently, there are limits on their release to prevent environmental damage. Allowable median concentrations in wastewater discharged to inland water bodies are total nitrogen <10 ppm and total phosphorus <0.5 ppm [EPA Victoria (1995)]. There are no conventional methods available that can separate N and P from wastewater. Critically, they become disproportionally more expensive to reach lower nutrient concentrations, making removal to very low concentrations uneconomical. Separate unit operations or enhanced designs are needed to enable both N and P removal. Advanced strategies and technologies are necessary in order to remove both N and P residuals. The concentrated form of the nutrients cannot be reused readily and the potential formation of harmful by-products creates a further need for safe disposal of any waste sludge generated (Christenson and Sims 2011).

Rather than viewing the nutrients N and P as a waste that has to be disposed of, wastewater treatment with algae facilitates their recovery using a cost-effective and environmentally friendly method. For example, algae biomass can be utilized as a bioavailable fertiliser, reducing the dependence on synthetic fertilisers. This can be especially important for added food security as there is a current reliance on mining P from finite reserves. Algae have the additional benefit of removing toxic contaminants from wastewater like heavy metals (Mallick 2002) and cyclic organic pollutants (Wu et al. 2014). Micropollutants are synthetic or natural compounds that occur in aquatic environment at low concentrations. A number of these micropollutants don't seem to be sufficiently removed by conventional treatment processes. The removal of micropollutants is vital as they will build up within the ecosystem and have negative health and environmental impacts (Metcalf and Eddy 2003). Some algal species are equally capable of removing N and P and of adapting to different wastewaters (Tam and Wong 1989; Park et al. 2012; Abou-Shanab et al. 2013; Ji et al. 2013).

3.2 ALGAE NATURALLY PRESENT IN DOMESTIC WASTEWATER TREATMENT

The history of use of algal cultures for wastewater treatment spans about 75 years. Algae-based systems for the removal of toxic minerals like lead, cadmium, mercury, scandium, tin, arsenic and bromine are being developed (Soeder et al. 1978; Kaplan et al. 1988; Gerhardt et al. 1991; Cai et al. 1995; Hammouda et al. 1995). Currently, significant interest is present in nations like Australia, Mexico, Thailand, Taiwan and U.S. for utilizing this algae-based technology (Borowitzka and Borowitzka 1988, 1989a, b; Moreno et al. 1990; Wong and Chan 1990; Renaud et al. 1994).

Bio-treatment with microalgae, this fascinating idea was launched in the U.S. by Oswald and Gotaas (1957) and has been intensively tested in many countries (Goldman 1979; Shelef and Soeder 1980; De Pauw and Van Vaerenbergh 1983; De la Noue and De Pauw 1988). Palmer (1974) surveyed microalgal genera from numerous waste stabilization ponds. The algae found in most abundance and frequency were

Chlorella, Ankistrodesmus, Scenedesmus, Euglena, Chlamydomonas, Oscillatoria, Micractinium and *Golenkinia*. Palmer (1969) listed the algae in the order of their tolerance to organic pollutants as reported by 165 authors. The list was compiled of 60 genera and 80 species. The foremost tolerant eight genera were found to be *Euglena, Oscillatoria, Chlamydomonas, Scenedesmus, Chlorella, Nitzschia, Navicula* and *Stigeoclonium*. One thousand algal taxa are reported one or more times as pollution tolerant which include 240 genera, 725 species and 125 varieties and forms. Pictures of some of the most commonly found algae in domestic wastewater are presented in Figure 3.1. The foremost tolerant genera include eight algae, five cyanobacteria, six flagellates and six diatoms. Periphyton are one of the most important

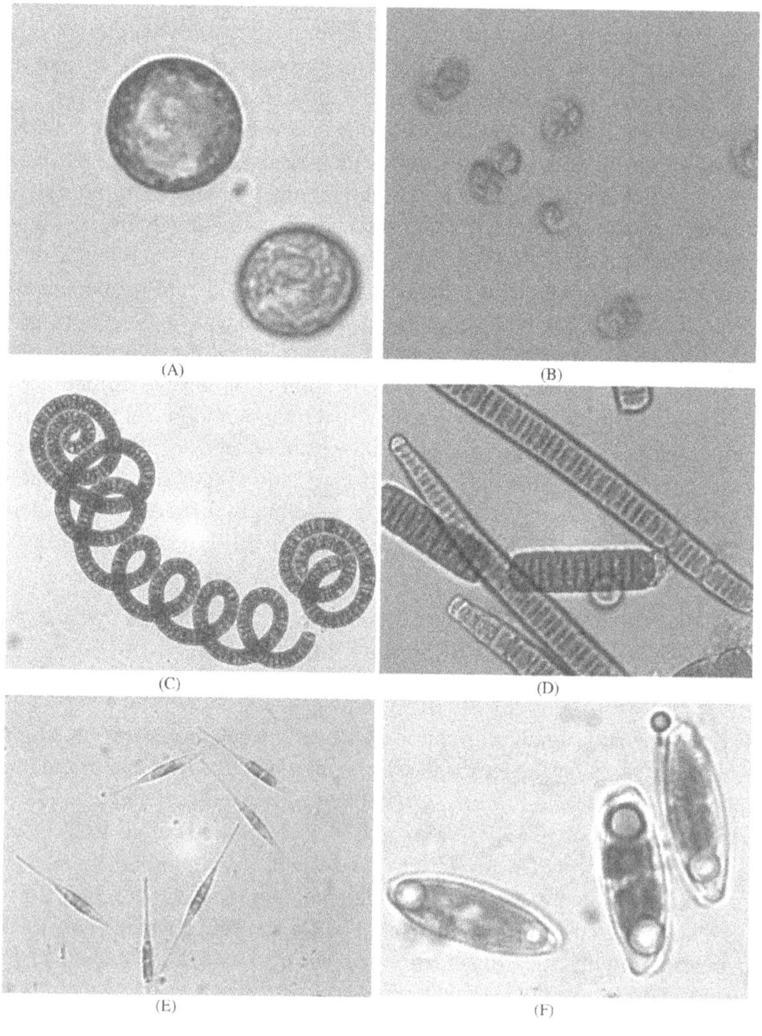

FIGURE 3.1 Examples of microalgae in domestic wastewater. (a) *Chlorella* sp., (b) *Chlamydomonas* sp., (c) *Spirulina* sp., (d) *Oscillatoria* sp., (e) *Nitzschia* sp., (f) *Navicula* sp.

TABLE 3.1
Microalgae Found in Wastewater

S. No.	Name of the Algae	Domestic Wastewater	References
1.	*Chlorella Vulgaris*	Primary settler	Gupta et al. (2016)
2.	Cyanobacteria	Secondarily treated domestic effluent	Pouliot et al. (1986)
3.	*Chlamydomonas reinhardtii*	Centrate wastewater	Kong et al. (2010)
4.	*Desmodesmus communis*		Samori et al. (2013)
5.	*Micractinium* sp.	Primary wastewater	Craggs et al. (2012)
	Desmodesmus sp.		
6.	*Botryococcus braunii*	Secondarily treated	Orpez et al. (2009)
7.	*Chlorella* sp., *Micractinium* sp., *Actinastrum* sp.	Primary treated	Woertz et al. (2009)

algae associated with substrates in aquatic habitats. Periphyton is a complex mixture of microalgae, bacteria, protozoa and invertebrate. Periphyton have been widely used as a tool for biologically monitoring water quality (Leland and Carter 1985; Newman et al. 1985; Cosgrovea et al. 2004). These organisms exhibit high diversity and also major component in energy flow and nutrient cycling in aquatic ecosystems (Table. 3.1). A large group of diatoms among the algae are present in highly polluted environments (Ghosh and Love 2011). Diatoms have the ability to decrease nutrient level and increases O_2 levels in wastewater (Bowler et al. 2009).

3.3 ALGAE AS AN INDICATOR FOR POLLUTION

Bioindicators are taxa or groups of organisms that show signs when they are affected by environmental pressure because of human activities or the destruction of biotic systems (McGeoch 1998; Shahabuddin 2003). The major groups of organisms that have been used as indicators of environmental pollution include bacteria, fungi, protozoa, algae, higher plants, macro invertebrates and fish. The presence or absence of the indicator organisms reflects the aquatic environmental conditions. Therefore, to conserve valuable resources from further deterioration, there is a need for regular monitoring of the aquatic systems.

Algae can act as an indicator of the drop-in water quality parameters like dissolved oxygen, pH, temperature, salinity and nutrients (N and P). They can be useful in identifying various types of water degradation, complementary to other environmental indicators. Algae are one of the most rapid bioindicators of water quality changes due to their short life spans, quick response to pollutants and easy determination of population numbers (Plafkin et al. 1989). The use of algae as biological indicators of pollution has been studied by rating pollution-tolerant algae in the rivers by Palmer (1969).

Due to constant population growth and technological and industrial progress, the nature of the aquatic environment has undergone numerous changes and deteriorated significantly. In urban areas, water pollution problems always influence the biological imbalance both qualitatively and quantitatively. When water pollution causes an algal bloom in a river, lake or any marine environment, the proliferation of newly introduced nutrients stimulates plant and algae growth, which in turn reduces oxygen levels in the water. These pollutants kill sea weeds, mollusca, marine birds, fishes, crustaceans and many other organisms that are a part of the food chain.

Algae are related to water pollution in a number of important ways. The enrichment of water sources with inorganic phosphorous and nitrogen via various contamination sources is responsible for the growth of algae in water bodies. Research in the freshwater ecology of algae related to water pollution is sparse and detailed study is necessary for identifying indicator species.

Algal accumulations are typically species-rich and algal species exhibit wider distributions among ecosystems and geographical regions. As primary producers, algae are most directly affected by physical and chemical factors. Algal assemblages are sensitive to some pollutants and readily accumulate pollutants and algal metabolism is also sensitive to the variation of environmental and natural factors. Algae are also used in laboratory bioassays to study water quality, using media for culturing indicator species from the field or defined media to which varying degrees or concentrations of the pollutant are added (Ho 1980; Guckert et al. 1992; Grimshaw et al. 1993; Knauer et al. 1997). Algae are easily cultured in the laboratory and sampling is easy, inexpensive and creates minimal impact on resident biota; relatively standard methods exist for the evaluation of functional and non-taxonomic structural characteristics of algal communities (Stevenson and Lowe 1986; Rott 1991; Round 1991; McCormick and Cairns 1994; Van Dam et al. 1994; Stevenson and Pan 1999). Alterations and shifts in the species composition and productivity of algal assemblages in response to anthropogenic stresses should be considered in order to predict the effects on food web interactions and other ecosystem components (McCormick and Cairns 1994).

The advantages that periphyton communities have over other organisms for monitoring purposes include the following: fixed habitats, so they cannot avoid pollution; relatively quick recolonization after perturbations in water quality or flow, the ability to enable a rapid resumption of monitoring; the ease of sample preparation for analysis; and widespread, common taxa, enabling their pollution tolerances to become well known (Biggs 1985). Many characteristics of periphyton community structure and function can be used to develop indicators of ecological conditions in the aquatic ecosystem (Hill et al. 1999). Periphyton are sensitive to many environmental conditions, which can be detected by changes in species composition, cell density, ash-free dry mass (AFDM), chlorophyll and enzyme activity (e.g., alkaline and acid phosphatase). Each of these characteristics may be used, singly or cumulatively, to assess conditions such as biological integrity and trophic condition.

Trophic state is an absolute scale that describes the biological condition of a body of water. The trophic status for any one wetland is a condition determined by the surrounding catchment, landform and geology. The Trophic State Index (TSI) is a grouping system designed to rate water bodies based on the amount of biological productivity they sustain.

Aquatic populations are impacted by anthropogenic stress, resulting in a variety of alterations in the biological integrity of aquatic systems. Kolkwitz and Marsson (1908) were the pioneers who classified algal species based on their tolerance to various kinds of pollution. They stated that the presence of certain species of algae could define various zones of degradation in a river. To define the effects of various types of river degradation, it is important to use a variety of algal parameters (Patrick 1973). Palmer (1969) published a composite rating of algal species that could be used to indicate clean and polluted waters.

Diatoms have also been used extensively in water quality monitoring (Round 1991). They exist in a wide range of ecological conditions, colonising almost all suitable habitats; they can thus provide multiple indicators of environmental change (Stevenson and Bahls 1999). Indices of water quality using diatoms gave the most precise data compared to chemical and zoological assessment (Leclercq 1988). Patrick (1949) developed community indices and provided information that demonstrated that a healthy community is made up of numerous species in several groups of organisms, including algae. Patrick (1971) proposed a numerical approach to study water quality using diatom flora attached to glass slides as artificial substrates. Dixit et al. (1992) discussed diatom flora as a powerful indicator of environmental change and its emergence as a preferred indicator in monitoring studies.

3.4 ROLE OF ALGAE IN WASTEWATER TREATMENT

Rapidly increasing worldwide urban population has led to an increase in consumption of fresh water and dumping of used water into aquatic bodies. According to a survey by Central Pollution Control Board of India, only 35% of sewage water was treated out of the total wastewater generated in urban areas (CPCB 2009). Furthermore, sewage wastewater discharge from urban areas and towns in India was 61,754 million litres per day (MLD) while the present sewage treatment limit of the nation is 22,963 MLD. In this manner, there is a gap of 38,791 MLD, which is released into adjacent lakes and waterways untreated, causing water contamination. Common strategies to treat wastewater are physical, chemical and biological treatments. Physical treatment is removal of the suspended particles, chemical treatment uses flocculants and biological treatment utilizes microorganisms to treat the wastewater (Kumar et al. 2019).

Interest in biological wastewater treatment with microalga has grown due to their photosynthetic ability to convert solar energy into useful biomass (Bhatnagar et al. 2011; Li et al. 2014). Microalgae play the dual role of phycoremediation along with the production of useful biomass, which can be used in many areas (Rawat et al. 2011). Auto-phototrophic microalgae cultivation can not only remove nutrients such as N and P but can also fix CO_2 by using the CO_2 from flue gases obtained from power plant or other emission sources as a carbon source. Generally, 1 kg of dry algal biomass consumes about 1.83 kg of CO_2 (Chisti 2007).

Sewage wastewater bioremediation by microalgae is an environmentally friendly process with efficient nutrient recycling (Mulbry et al. 2010). A wide range of microalgae such as *Chlorella, Scenedesmus, Phormidium, Botryococcus, Chlamydomonas* and *Spirulina* have been reported for treating domestic wastewater, and efficacy of this method is promising (Chinnasamy et al. 2010; Kong et al. 2010; Wang et al. 2010). Wastewater characteristics vary between locations and fluctuate with time for one source, with the concentration of nutrients likely to vary independently of each other for a selected wastewater stream (Beuckels et al. 2015). This is often an obstacle in simultaneously removing N and P. The complex interaction of wastewater characteristics and environmental conditions creates discrepancies in reported removal behaviours. Careful analysis is required to consider the compounding interaction and understand its influence on nutrient removal.

Kumar et al. (2019a) conducted an experiment by using two types of domestic wastewater (sewage wastewater and kitchen wastewater) for *Chlorella* sp. cultivation. Both showed high biomass productivity in mixotrophic cultivation mode. The average biomass productivity in sewage wastewater was 0.6 g/L, which was higher when compared to kitchen wastewater. Nitrate and phosphate removal efficiencies increased with the increase in cultivation days in both wastewaters. The total nitrate removal in sewage wastewater was 67% and kitchen wastewater was 37%, while phosphate removal efficiency observed was 75% and 88%, respectively.

Sengar et al. (2011) performed an experiment by isolating 22 algae species from sewage wastewater. Out of all species, three species, i.e., *Gloeocapsa gelatinosa*, *Euglena viridis* and *Synedra affinis*, were found to be the most prominent. Blue green algae (BGA) such as *Oscillatoria limosa* and *Nostoc commune* have also been used for wastewater treatment due to their fast growth, rapid nutrient consumption and photo autotrophic nature (Vijayakumar et al. 2012; Kotteswari et al. 2012). Hossein et al. (2014) reported that *Oscillatoria limosa* and *Nostoc commune* reduce toxic pollution from municipal wastewater. The results demonstrated that BGA were highly efficient in decreasing the biological demand, chemical oxygen demand, total dissolved solids and total hardness. The 62–63% increase in dissolved oxygen indicated improvement in the quality of wastewater. Experimental results showed that *Oscillatoria limosa* was superior to *Nostoc commune* due to the higher potential to tolerate pollution.

Renuka et al. (2013) used native filamentous microalgal strains, selected microalgal strains from germplasm and native unicellular microalgal strain for sewage wastewater treatment. Results of the experiment showed highest nitrate (90%) and phosphate (97.8%) removal by filamentous microalgae inoculated in sewage wastewater. It also decreased total dissolved solids from 1120 mg/L to 806 mg/L while dissolved oxygen level increased from 0.4 mg/L to 9 mg/L. Filamentous algae gave the highest biomass production, 1.07 gm/L, compared to the other algae. Therefore, this group of filamentous algae can be used to remove organic matter from wastewater, produce biomass and enhance water quality.

Jia Yang et al. (2011) checked the growth and lipid accumulation properties of a freshwater microalgae *Chlorella ellipsoidea* in sewage wastewater. *C. ellipsoidea* produced 425 mg/L biomass in domestic effluent water with a 43% w/w lipid content. Moreover, after cultivation of *C. ellipsoidea*, nitrogen and phosphate content decreased by 99% and 90%, respectively, in the secondary effluent water. *Botryococcus braunii* is a microalga found in freshwater and brackish ponds and lakes around the world in large floating masses. This alga stores long chain hydrocarbons and ether lipids, similar in many ways to crude oil, in its extracellular matrix (Brown et al. 1969; Wake and Hillen 1981). Sydney et al. (2011) tested twenty strains of microalgae for growth in secondary domestic wastewater. Out of the twenty strains, *Botryococcus braunii* gave the best growth performance with 1.88 g/L biomass while lipid accumulation was 36.14%. *B. braunii*, if cultivated with success in 20 cm ponds of treated wastewater, represent a potential yield of 3300 kg of lipids/ha/year. This is roughly five times as much oil as is achieved through soybean production (600 kg/ha/year), which represents around 90% of the Brazilian biodiesel production. In addition, *B. braunii* showed a considerable amount of CO_2 uptake, 144.91 mg CO_2/gm biomass/L/day.

Caporgno et al. (2015) cultivated *Chlorella kessleri*, *Chlorella vulgaris* and the marine microalgae species *Nannochloropsis oculata* in domestic wastewater. Both *Chlorella* species reached high biomass dry weights, 2.70±0.08 g/L and 2.91±0.02 g/L, respectively, accompanied by nitrogen concentration reduction around 96% and 95% and a phosphorous concentration reduction around 99% and 98%, respectively. *N. oculata* was able to uptake nutrients from wastewater to grow but with less efficiency, indicating the need of microalgae acclimatization or process optimisation to achieve high nutrient removal. In sewage wastewater treatment, *Chlorella vulgaris* performed the best in terms of nutrient removal efficiency. It consumed 95% nitrate, 95% ammonia and 99.3% phosphate (Chawla et al. 2020). Studies on freshwater cultivation of *Scenedesmus* sp. algal species indicate that lipid accumulation increased 11-fold when the intensity of light changes from 250 to 400 µmol photons/m^2/s (Hossein et al. 2019). *Chlorella sorokiniana* and *Scenedesmus obliquus* have been reported to tolerate high amounts of organic loading and physiological stress. Nutrient removal, pathogen removal and lipid accumulation with secondary or tertiary effluents have been demonstrated independently for these organisms. *S. obliquus* showed greater potential for removing organic carbon (76.13±1.59% Chemical Oxygen Demand (COD) removal), nutrients (98.54±3.30% N-removal, 97.99±3.59% P-removal) and comparable pathogens removal (99.93±0.12% total coliforms removal, 100% faecal coliform removal) in comparison to *C. sorokiniana* (69.38±1.81% COD removal, 86.93±3.49% N-removal, 68.24±11.69% P-removal, 99.78±0.12% total coliforms removal, 100% faecal coliform removal) with 15 days of cultivation (Gupta et al. 2016).

3.5 CULTIVATION METHODOLOGY FOR VARIOUS ALGAL SPECIES

Open and closed systems are the two main forms of microalgae cultivation. Open ponds are the oldest and simplest systems to grow microalgae on a large scale. CO_2 fixation in open systems ranges from 10–30% (Li et al. 2013). The main advantage of open pond cultivation is its simplicity, resulting in low operating costs. Open systems are classified into two categories: non-stirred and stirred ponds, also known as raceway ponds (Tredici 2004). Some limited number of microalgae species like *Dunaliella Chlorella* and *Spirulina* have been grown on a large scale using non-stirred open ponds (Borowitzka and Borowitzka 1990). Wu et al. (2013) cultivated *Scenedesmus* sp. in an open pond fed with secondary wastewater, and the highest rate of biomass production achieved was 20 gm m^{-2} d^{-1}. Normally, raceway pond systems have an oval-shaped pond with a continuous or alienated form containing a series of plates, where culture is moved with the help of paddle impeller to stimulate uniform mixing of nutrients and algae (Cai et al. 2013, Richardson et al. 2012,

Ketheesan and Nirmalakhandan 2011). Siham et al. (2016) had cultivated *Chlorella pyrenoidosa* in secondary sewage wastewater in open raceway ponds and reached a biomass productivity of 1.71±0.04 g/L.

The closed system (PBRs) was developed for microalgae cultivation under controlled conditions and prevents direct contact between the culture media and the atmosphere. In closed systems, light, CO_2, temperature and nutrients are provided under controlled conditions (Vasumathi et al. 2012; Wang et al. 2014). In closed systems, high rates of CO_2 mass transfer and very minimal amount of CO_2 and water losses due to evaporation can be achieved (Singh and Sharma 2012). Different types of PBRs have been developed like tubular, vertical column and flat panel. Multiple forms of PBRs have their own characteristics and are appropriate for different locations, conditions, microalgae species, economic requirements and purpose (Sierra et al. 2008).

Tubular PBR systems typically are composed of a series of transparent tubes of glass, Perspex or PVC, with diameters ranging from about 24 mm to 24 cm (Wang et al. 2012). The air and CO_2 are added through a gas exchange system, while the culture medium is moved from the tubes to a reservoir with a mechanical pump. This provides both mixing and prevents the accumulation of biomass (Razzak et al. 2013). With help of these PBRs, a wide variety of algae like *Spirulina, Porphyridium, Chlorella, Dunaliella, Haematococcus, Tetraselmis and Phaeodactylum* were grown on a pilot scale (Borowitzka and Borowitzka 1989b). Molinuevo-Salces et al. (2010) used two types of PBRs, an open type and a tubular type, to check the comparative degradation of swine slurry used as the cultivation medium. Significant results were seen in the open configuration with a 38–45% ammonia removal by the algae biomass.

Vertical column PBRs and bubble column PBRs are cylindrical containers with a height up to 4 m. This system has a low cost, high surface area in relation to volume and adequate heat transfer. Culture mixing is achieved by injecting the gas mixture from a spray system and an efficient release of O_2 (Wang et al. 2013). Flat panel PBRs have the highest incident light radiation within all the closed cultivation systems. Flat panels have a very short light path which allows light to easily penetrate the culture liquid (Reyna-Velarde et al. 2010). Flat panel reactors are made up of transparent materials to increase the consumption of solar energy and have the highest photosynthetic efficiency.

Algal treatment systems are often categorised as suspended or immobilised. The low cost and ease of suspended systems are their main advantage; however, they typically have low cell loadings, resulting in long hydraulic retention times (HRTs) (Pires et al. 2013). Harvesting of suspended cells is often difficult, causing inefficient resource recovery and risk of contaminating the discharged wastewater (Prajapati et al. 2013). More understanding is required on the impact that selection of reactor type and algal species has on remediation performance. Freshwater macroalgae have large cells and can be an alternative to avoid the restrictions in harvesting suspended microalgal systems (Neveux et al. 2016).

Passive techniques of immobilisation, referred to as attached growth systems or biofilms, involve the algal cells accumulating on and attached to a surface of a substrate through the excretion of extracellular polymeric substances (Kesaano and

Sims 2014; Cohen 2001; Schnurr and Allen 2015; Hoh et al. 2016). Active techniques of immobilisation involve physically separating the cells from the liquid phase and include covalent coupling, adsorption, liquid-liquid emulsion, membrane separation, entrapment and encapsulation (Cohen 2001; Mallick 2002; Moreno-Garrido 2008; Whitton et al. 2015). In general, immobilisation allows for higher cell loading than suspended systems, giving a better rate of nutrient removal. The resultant lower HRT and reduced reactor volume together with the convenience of harvesting are most advantageous for wastewater treatment (Whitton et al. 2015; Hoffmann 1998). These benefits must be balanced against the extra cost of the immobilisation process and materials, and it should be noted that there are few full-scale examples of immobilised algae wastewater treatment systems.

Many of the reported algal wastewater treatments use pure monocultures to scale back interferences caused by interactions between different algal species in mixed cultures (Muñoz and Guieysse 2006; Goncalves et al. 2017). A biofilm involves any syntrophic (cells jointly dependent upon each another with reference to food supply) consortium of microorganisms in which cells stick to each other (Hall-Stoodley et al. 2004; Lopez et al. 2010). Microalgae–bacteria consortia are promising for advanced nutrient removal from wastewater. Suspended algal-bacterial systems can easily wash out unless the HRT is long; attached microalgae–bacteria consortium is more realistic. Additionally, algae and bacteria form a synergistic relationship which may improve growth and nutrient removal, because the bacteria consume O_2 and produce CO_2, while the algae consume CO_2 and produce O_2 (Pires et al. 2013). Immobilisation can prevent contamination of the inoculated algae, which enables more control over the choice of species within the culture (Cohen 2001 and Covarrubias et al. 2012). This permits the foremost appropriate algal species to treat the wastewater to be selected and offers a pathway to decouple the role of algae from that of bacteria. Compared to algal monocultures, co-cultures of selected algal and bacterial species, or of multiple algal species, can improve the speed and robustness of remediation (Pires et al. 2013). Such co-cultures may be controlled by immobilisation, either by keeping each species in separate matrices or mixed within the matrix (Cohen 2001; de-Bashan et al. 2004). A carefully managed co-culture of bacteria and algae may additionally be the simplest way to enhance nutrient removal during dark periods. If the growth and preservation of two different microorganisms are compatible in a co-culture system, heterotrophic metabolism of aerobic bacteria and the abilities of nutrient uptake and photosynthetic oxygenation of algae can be equally symbiotic (Lee et al. 2015).

There are species-dependent responses to environmental conditions (Teoh et al. 2013; Singh and Singh 2015), which mean in an open system, seasonal differences can cause fluctuations within the species present and alter remediation rates (Mehrabadi et al. 2016). One approach to selecting species that are suitable for the treatment conditions is to isolate them from the treatment plant or a neighbouring aquatic system (de-Bashan et al. 2008; Ruiz-Martinez et al. 2012). Also in some cases, upon treating wastewater, the amount of N and P removed can differ between algal species (Rasoul-Amini et al. 2014; Mennaa et al. 2015; Shaker et al. 2015).

Starvation of algal cells before exposure to wastewater may increase the speed of nutrient removal (Solovchenko et al. 2016; Zhang et al. 2008). Hernandez et al.

(2006) observed better P removal after starving alginate-immobilised *Chlorella sorokiniana* but not with *C. vulgaris*, indicating caution must be used as the positive effect due to starvation is species-dependent.

The present review is an outcome of an exhaustive literature survey on the utility of algal-based technology in wastewater remediation, few real instances wherein the technique is applied, the technological restrictions, the challenges enforced and therefore the scope of future research in this area.

3.6 UTILIZATION OF HARVESTED ALGAE BIOMASS FOR VARIOUS PRODUCTS

Algae cultures have been developed as an important source of many products such as aquaculture feed, human food supplements and pharmaceuticals (Pulz and Gross 2004; Apt et al. 1999; Goswami et al. 2021c; Mehariya et al. 2021c; Saini et al. 2021), and they have been suggested as very good candidates for fuel production (Shenk et al. 2008). Both macroalgae and microalgae contain numerous compounds to promote germination, leaf or stem growth, flowering, and can also be used as biological protection agents against plant diseases After the recovery of oil and carbohydrates from macroalgae and microalgae, many nutrients are left in the spent biomass. One potential application for this leftover biomass is to use it as a biofertiliser. This will increase the economic potential of algae. Nitrogen is the second most important component for growth of plants in fields, and its requirement is fulfilled by fertilisers (Malik et. Al. 2011). BGA have the ability to improve various physical, chemical properties of farming land by increasing the yield and saving on the fertiliser nitrogen used. Most of the cyanobacteria can fix atmospheric nitrogen and can be used as effective biofertilisers. BGA like *Nostoc, Anabaena* and *Tolypothrix* can fix atmospheric nitrogen and are used as inoculants for growing paddy crops in both upland and low land conditions (Priyadarshani and Rath 2012). Cyanobacteria play a major role in keeping the soil fertile for increasing rice growth and yield (Song et al. 2005).

Currently, various algae feeds are used to culture different fish growth stages like larvae, juvenile fish and finfish. Hundreds of microalgae have been examined as food over last few decades, but still, a fewer than twenty have gained importance in the context of aquaculture (Chen 2003; Da Silva and Barbosa 2008). The most commonly used algae for aquaculture feed are *Chlorella, Tetraselmis, Pavlova, Phaeodactylum, Nannochloropsis, Skeletonema* and *Thalassiosira*. Many companies are making aquaculture feed by using *Chlorella* and *Spirulina*, or their mixture. *Hypneacervi-cornis* and *Cryptonemia crenulata* microalgae, rich in protein, have been tested as a shrimp diet (Chen 2003; Da Silva and Barbosa 2008).

After aquafeed, the next most prevalent use for algal biomass is in biofuel production via various processes. The process of conversion of raw material lipids mainly triacyleglycerols/free fatty acids into non-toxic and ecofriendly biodiesel is called transesterification. In recent years, biodiesel production from oil seed crops like soybean oil, palm oil and rapeseed oil has gained prominence. The production cost of biodiesel mainly depends on the type of raw material used, accounting for 50–85% of final fuel price. For production of cost-effective biodiesel, assessment of feedstock is important in terms of productivity, quality and exploitation of by-products (Nikolaison et al. 2012). Fuel conversion from algae is broadly based on the feedstock's high concentrations of lipids: fatty, oil-containing acid molecules that can be extracted to create biofuels. Crude algal oil having high viscosity can be converted into low molecular weight compounds in the form of fatty acid alkyl esters.

3.7 CHALLENGES AND FUTURE PROSPECTS

Apart from biofuel production, microalgae may serve as a potential renewable source for other commercial applications (Milledge 2011) such as wastewater treatment, CO_2 mitigation, human nutrition, feed, cosmetics production, high-value compounds, pigments, biofertiliser and drug synthesis. To date, the value-added production by microalgae still faces high production costs compared to other sources (Wijffels and Barbosa 2010). Despite the high nutrient levels, the utilization of waste water as a medium for microalgae at industrial scale is still challenging. First, the high organic compounds, consisting of tannins, lignin and phenolic compounds, could negatively affect growth (Habib et al. 2003; Neoh et al. 2013; Nur et al. 2016). The dark coloration due to high concentrations of suspended solids could inhibit light penetration, which is a critical factor for photosynthetic growth (Pacheco et al. 2015; Ding et al. 2016; Nur et al. 2017). The presence of heterotrophic bacteria may affect biomass productivity (Cho et al. 2011; Li et al. 2011). The pH and salinity of the wastewater needs to be adjusted before it can be used as growth medium for microalgae. The heavy metals in wastewater could prevent utilization for pharmaceutical or human consumption (Ahmad et al. 2017). However, the conditions as mentioned above could be prevented by employing some pre-treatment process to lower COD, color and heavy metals. To increase the salinity, the cultivation might be relocated to seashore areas or blending can be done with hypersaline wastewater from chemical manufacturing and oil production industries (Woolard and Irvine 1995) to make the cultivation feasible for marine microalgae. Several researchers reported that a high turbidity from the wastewater may induce heterotrophic or mixotrophic growth that benefits biomass production, lipid accumulation and nutrient uptake (Abreu et al. 2012; Gupta et al. 2016). Furthermore, to provide CO_2 for mixotrophic cultivation and lowering COD and BOD content, wastewater could be processed in an anaerobic fermenter, resulting in methane that could be used directly as an energy source.

Another proposed strategy is the cultivation of microalgae–bacteria consortia for wastewater treatment to enhance production of the bioactive compounds. Wastewater provides ideal media for bacterial growth and decomposition of organic matter in the presence of oxygen (Pacheco et al. 2015). During this process, microalgae provide oxygen for aerobic bacteria to degrade organic matter and consume the carbon dioxide produced by bacterial respiration. Thus, these systems are an economical alternative to conventional aerobic treatments of wastewater (Prajapati et al. 2013; Delgadillo-Mirquez et al. 2016). However, bacteria may compete with microalgae for the available nutrients such as N and P, by producing metabolites that are inhibitory to microalgae growth (Joint et al. 2002).

To optimise value-added product accumulation in microalgae, a two-step cultivation process seems promising. In the first step, microalgae could be cultivated under mixotrophic conditions to increase biomass productivity and then culture could be transferred to heterotrophic conditions. The type of cultivation systems such as open pond cultivation or closed cultivation using PBRs could also influence the growth, biomass and products. More research must be done to further optimise wastewater utilization for high-value product generation by microalgae. However, it has become clear from the research executed over the past decade that optimisation should be done based on a combination of approaches before valorisation could become economically feasible (Kumar and Verma 2021a,b,c).

3.8 CONCLUSION

This chapter has shown the role of algae in wastewater treatment, different strategies and cultivation methodologies for various algal species and utilization of harvested algae biomass. These strategies can be used to carry out more studies in wastewater treatment. It is also important to remember that the effectiveness of the strategies and their results will depend on the species, the experimental facilities available and the economic resources accessible to the development of the project. Therefore, to allow the application of microalgae for bioremediation purposes, it is essential that integrated processes, combining cultivation in wastewater and use of biomass for the production of commercially valuable by-products such as fertilisers and biofuels, be explored. Furthermore, there is a need to carefully select a robust algal strain, which can be cultivated with a high growth rate. An effective strain along with development of a large-scale cultivation system for biomass production and treatment of large volumes of wastewater will decrease the overall cost of algal wastewater treatment.

ABBREVIATIONS

HTL, hydrothermal liquefaction; AFDM, ash-free dry mass; TSI, trophic state index; MLD, million litres per day

CONFLICT OF INTEREST STATEMENT

The authors declare no conflict of interest.

REFERENCES

Abou-Shanab, R.A.I., Kim, S.H., Ji, M.K., Lee, S.H., Roh, H.S., Jeon, B.H. (2013) Municipal wastewater utilization for biomass and biodiesel production by *Scenedesmus obliquus* HM103382 and *Micractinium reisseri* JN169781. J Renew Sustain Energy. 2013;5(5):052006. doi: 10.1063/1.4821504.

Abreu, A.P., Fernandes, B., Vicente, A.A., Teixeira, J., Dragone, G. (2012) Mixotrophic cultivation of *Chlorella vulgaris* using industrial dairy waste as organic carbon source. Bioresour. Technol. 118, 61–66.

Agrawal, K., Bhatt, A., Bhardwaj, N., Kumar, B., Verma, P. (2020) Algal biomass: potential renewable feedstock for biofuels production–part I. In: Biofuel production technologies: Critical analysis for sustainability. Springer, Singapore. pp. 203–237.

Agrawal, K., Verma, P. (2020a) Advanced oxidative processes: An overview of their role in treating various wastewaters. In: Advanced oxidation processes for effluent treatment plants. Springer. pp. 87–102.

Agrawal, K., Verma, P. (2020b) Degradation and detoxification of waste via bioremediation: a step toward sustainable environment. In: Emerging technologies in environmental bioremediation. Elsevier. pp. 67–83.

Agrawal, K., Verma, P. (2021a) "Omics"—A step toward understanding of complex diversity of the microbial community. In: Wastewater treatment cutting edge molecular tools, techniques and applied aspects. Elsevier. pp. 471–487.

Agrawal, K., Verma, P. (2021b) Metagenomics: A possible solution for uncovering the "Mystery Box" of microbial communities involved in the treatment of wastewater. In: Wastewater treatment cutting edge molecular tools, techniques and applied Aspects. Elsevier. pp. 41–53.

Ahmad, A., Bhat, A.H., Buang, A. (2017) Immobilized *Chlorella vulgaris* for efficient palm oil mill effluent treatment and heavy metals removal. Desalin Water Treat. 81, 105–117.

Almomani, F., Judd, S., Shurair, M., Bhosale, R., Kumar, A., Khreisheh, M. (2017) Potential use of mixed indigenous microalgae for carbon dioxide bio-fixation and advanced wastewater treatment. In: Paper Presented at the Environmental Division 2017 – Core Programming Area at the 2017 AIChE Spring Meeting and 13th Global Congress on Process Safety; San Antonio, United States.

Almomani, F.A. (2019) Assessment and modeling of microalgae growth consideringthe effects of CO_2, nutrients, dissolved organic carbon and solar irradiation. J. Environ. Manage. 247, 738–748.

Apt, K.E., Behrens, P.W. (1999) Commercial developments in microalgal biotechnology. J Appl Phycol 35:215–226.

Barnharst, T., Rajendran, A., Hu, B. (2018) Bioremediation of synthetic intensive aqua-culture wastewater by a novel feed-grade composite biofilm. Int. Biodeterior. Biodegrad. 126, 131–142.

Beuckels, A., Smolders, E. and Muylaert, K. (2015) Nitrogen availability influences phosphorus removal in microalgae-based wastewater treatment. Water Res. 77, 98–106.

Bhardwaj, N., Agrawal, K., Verma, P. (2020) Algal biofuels: An economic and effective alternative of fossil fuels. In: Microbial strategies for techno-economic biofuel production. Springer. pp. 59–83.

Bhatnagar, A., Chinnasamy, S., Singh, M., Das, K.C. (2011) Renewable biomass production by mixotrophic algae in the presence of various carbon sources and wastewaters. Appl. Energy. 88, 3425–3431.

Biggs, B.J.F. (1985) The use of periphyton in the monitoring of water quality. In: R.D. Pridmore, A.B. Cooper (Eds.), Biological monitoring in freshwaters: Proceedings of a seminar. Water and soil miscellaneous publication. Ministry of Works and Development, Wellington, New Zealand. Vol. 82. pp. 117–142.

Borowitzka, L.J., Borowitzka, M.A. (1990) Commercial production of b-carotene by Dunaliella salina in open ponds. Bull. Mar. Sci. 47, 244–252.

Borowitzka, L.J., Borowitzka, M.A. (1989a) Carotene (Provitamin A) production with algae. In: E.J. Vandamme (Ed.), Biotechnology of vitamins, pigments and growth factors. Elsevier Applied Science, London. pp. 15–26.

Borowitzka, L.J., Borowitzka, M.A. (1989b) Industrial production: methods and economics. In: R.C. Cresswell, T.A.V. Rees, N. Shah (Eds.), Algal and cyanobacterial biotechnology. Longman Scientific, London.

Borowitzka, L.J., Borowitzka, M.A. (1989b) Industrial production: methods and economics. In: R.C. Cresswell, T.A.V. Rees, N. Shah (Eds.), Algal and cyanobacterial biotechnology. Longman Scientific, London. pp. 244–316.

Borowitzka, M.A., Borowitzka, L.J. (1988) Microalgal biotechnology. Cambridge University Press, Cambridge.

Bowler, C., Karl, D.M., Colwell, R.R. (2009) Microbial oceanography in a sea of opportunity. Nature. 459, 180–184.

Brown, A.C., Knights, B.A., Conway, E. (1969) Hydrocarbon content and its relationship to physiological state in the green alga *Botryococcus braunii*. *Phytochemistry*. 8, 543–547.

Bwapwa, J.K. (2018) Review on main issues causing deterioration of water quality and water scarcity: case study of South Africa. Environ. Manag. Sustain. Dev. 7 (3), 14.

Cai, T., Park, S.Y., Li, Y. (2013) Nutrient recovery from wastewater streams by microalgae: status and prospects. Renew. Sustain. Energy Rev. 19, 360–369.

Cai X.H., Traina, S.J., Logan, T.J., Gustafson, T., Sayre, R.T. (1995) Applications of eukaryotic algae for the removal of heavy metals from water. Mol. Mar. Biol. Biotechnol. 4 (4), 338–344.

Caporgno MP, Taleb A, Olkiewicz M, Font J, Pruvost J, Legrand J, Bengoa C (2015) Microalgae cultivation in urban wastewater: nutrient removal and biomass production for biodiesel and methane. Algal Res 10:232–239.

Central Pollution Control Board (2009). Annual Report, Ministry of Environment & Forests, India.

Chawla, P., Malik, A., Sreekrishnan, T.R., Dalvi, V., Gola, D. (2020). Selection of optimum combination via comprehensive comparison of multiple algal cultures for treatment of diverse wastewaters. Environmental Technology & Innovation. 18: 100758

Chen, Y.C. (2003) Immobilized *Isochrysis galbana* (Haptophyta) for long-term storage and applications for feed and water quality control in clam (*Meretrix lusoria*) cultures. J. Appl. Phycol. 15, 439–444.

Chinnasamy S, Bhatnagar A, Hunt RW, Das KC (2010) Microalgae cultivation in a wastewater dominated by carpet mill effluents for biofuel applications. Bioresour Technol 101:3097–3105

Chisti, Y., 2007. Biodiesel from microalgae. Biotechnology advances, 25(3): 294–306.

Cho, S., Lee, D., Luong, T.T., Park, S., Oh, Y.K., Lee, T. (2011) Effects of carbon and nitrogen sources on fatty acid contents and composition in the green microalga *Chlorella* sp. 227. J. Microbiol. Biotechnol. 21, 1073–1080.

Christenson, L., Sims, R. (2011) Production and harvesting of microalgae for wastewater treatment, biofuels, and bioproducts. Biotechnol. Adv. 29 (6), 686–702.

Cohen, Y. (2001) Biofiltration – The treatment of fluids by microorganisms immobilized into the filter bedding material: a review. Biores. Technol. 77 (3), 257–274.

Cosgrovea J.D.W., Morrison P., Hillman K. (2004) Periphyton indicate effects of wastewater discharge in the near-coastal zone, Perth, Western Australia. Estuar. Coast. Shelf. Sci. 61 (2), 331–338.

Covarrubias, S.A., De-Bashan, L.E., Moreno, M., Bashan, Y. (2012) Alginate beads provide a beneficial physical barrier against native microorganisms in wastewater treated with immobilized bacteria and microalgae. Appl. Microbiol. Biotechnol. 93 (6), 2669–2680.

Craggs R, Sutherland D, Campbell H (2012) Hectare-scale demonstration of high rate algal ponds for enhanced wastewater treatment and biofuel production. J Appl Phycol 24:329–337.

Da Silva, R.L., Barbosa, J.M. (2008) Seaweed meal as a protein source for the white shrimp *Lipopenaeus vannamei*. J. Appl. Phycol. 21, 193–197.

Darzins, A., Pienkos, P., Edye, L. (2010) Current status and potential for algal biofuels production (PDF). IEA Bioenergy Task 39 pp 1–131.

Day J.G., Gong Y., Hu Q. (2017) Microzooplanktonic grazers–A potentially devastating threat to the commercial success of microalgal mass culture. Algal Res. 27, 356–365.

De la Noue, J., De Pauw, N. (1988) The potential of microalgal biotechnology. A review of production and uses of microalgae. Biotechnol. Adv. 6, 725–770.

De Pauw, N., Van Vaerenbergh, E. (1983) Microalgal wastewater treatment systems: Potentials and limits. In: P.F. Ghette (Ed.), Phytodepuration and the Employment of the Biomass Produced. Centro Ric. Produz, Animali, Reggio Emilia, Italy. pp. 211–287.

De-Bashan, L.E., Hernandez, J.P., Morey, T., Bashan, Y. (2004) Microalgae growth promoting bacteria as "helpers" for microalgae: A novel approach for removing ammonium and phosphorus from municipal wastewater. Water Res. 38 (2), 466–474.

De-Bashan, L.E., Trejo, A., Huss, V.A.R., Hernandez, J.P., Bashan, Y. (2008) Chlorella sorokiniana UTEX 2805, a heat and intense, sunlight-tolerant microalga with potential for removing ammoniumfromwastewater. Biores. Technol. 99 (11), 4980–4989.

Delgadillo-Mirquez, L., Filipa, L., Behnam, T., Dominique, P. (2016) Nitrogen and phosphate removal from wastewater with a mixed microalgae and bacteria culture. Biotechnol. Rep. 1, 118–126.

Diaz-Elsayed, N., Rezaei, N., Guo, T., Mohebbi, S., Zhang, Q. (2019) Wastewater-based resource recovery technologies across scale: a review. Resour. Conserv. Recycl. 145, 94–112.

Ding, G.T., Yaakob, Z., Takriff, M.S., Jailani, S., Muhammad, S.A.R. (2016) Biomass production and nutrients removal by a newly isolated microalgal strain *Chlamydomonas* sp. In palm oil mill effluent (POME). Int. J. Hydrogen Energ. 41, 4888–4895.

Dixit S.S., Smol J.P., Kingston J.C., Charles D.F. (1992) Diatoms: Powerful indicators of environmental change. Environ. Sci. Technol. 26, 23–33.

du Plessis, A. (2019) Current and future water scarcity and stress. In: Water as an Inescapable Risk. Springer. pp. 13–25.

EPA Victoria (1995) Managing sewage discharges to inland waters, Publication 473, Victoria.

García-Seoane, R., Fernández, J., Villares, R., Aboal, J. (2018) Use of macroalgae to biomonitor pollutants in coastal waters: Optimization of the methodology. Ecol. Indic. 84, 710–726.

Gerhardt, M.B., Green, F.B., Newman, R.D., Lundquist, T.J., Tresan, R.B., Oswald, W.J. (1991) Removal of selenium using a novel algal bacteria process. Res. J. Water Pollut. Contr. Fed. 63, 779–805.

Ghosh, S., Love, N.G. (2011) Application of RbcL based molecular diversity analysis to algae in wastewater treatment plants. Bioresour. Technol. 102, 3619–3622.

Goldman, J. (1979) Outdoor algal mass Cultures-I. Appl. Water Res. 13, 1–19.

Goncalves, A.L., Pires, J.C.M., Simões, M. (2017) A review on the use of microalgal consortia for wastewater treatment. Algal. Res. 24, 403–415.

Goswami, R.K., Agrawal, K., Mehariya, S., Molino, A., Musmarra, D., Verma, P. (2020) Microalgae-based biorefinery for utilization of carbon dioxide for production of valuable bioproducts. In: Chemo-biological systems for CO2 utilization. Taylor & Francis. pp. 199–224.

Goswami, R.K., Mehariya, S., Verma, P, Lavecchia, R., Zuorro, A. (2021a) Microalgae-based biorefineries for sustainable resource recovery from wastewater. J. Water Process Eng. 40, 101747.

Goswami, R.K., Mehariya, S., Karthikeyan, O.P.K., Verma, P. (2021b) Advanced microalgae-based renewable biohydrogen production systems: A review. Biores. Technol. 320 (A), 124301.

Goswami, R.K., Agrawal, K., Verma, P. (2021c) An overview of microalgal carotenoids: Advances in the production and its impact on sustainable development bioenergy research: Evaluating strategies for commercialization and sustainability. John Wiley & Sons. pp 105–128.

Grimshaw, H.J, Rosen M., Swift D.R., Rodberg K., Noel J. (1993) Marsh phosphorous concentrations, phosphorous content and species composition in Everglades periphyton communities. Archiv für Hydrobiol. 128 (3), 257–276.

Guckert, J.B., Nold, S.C., Boston, H.L., White, D.C. (1992) Periphyton response in an industrial receiving stream: Lipid-based physiological stress analysis and pattern recognition of microbial community structure. Can. J. Fish. Aquat. Sci. 49 (12), 2579–2587.

Gupta, P.L., Choi, H.J., Lee, S. (2016) Enhanced nutrient removal from municipal wastewater assisted by mixotrophic microalgal cultivation using glycerol. Environ. Sci. Pollut. Res. 23, 10114–10123.

Gupta, S.K., Ansari, F.A., Shriwastav, A., Sahoo, N.K., Rawat, I. and Bux, F. (2016) Dual role of *Chlorella sorokiniana and Scenedesmus obliquus* for comprehensive wastewater treatment and biomass production for bio-fuelsz. Journal of Cleaner Production. 115: 255–264.

Habib, M.A.B., Yusoff, F.M., Phang, S.M., Kamarudin, M.S., Mohamed, S. (2003) Growth and nutritional values of Molina micrura fed on *Chlorella vulgaris* grown in digested palm oil mill effluent. Asian Fish Sci. 16, 107–119.

Hall-Stoodley, L., Costerton, J.W., Stoodley, P. (2004). Bacterial biofilms: from the natural environment to infectious diseases. Nat. Rev. Microbiol. 2 (2), 95–108.

Hammouda, O., Gaber, A., Abdel-Raouf, N. (1995) Microalgae and wastewater treatment. Ecotoxicol. Environ. Saf. 31 (3), 205–210.

Hernandez, J.P., de-Bashan, L.E. and Bashan, Y. (2006) Starvation enhances phosphorus removal from wastewater by the microalga Chlorella spp. Co-immobilized with Azospirillum brasilense. Enzyme Microb Technol. 38 (1–2), 190–198.

Hill, M.O., Mountford, J.O., Roy, D.B., Bunce, R.G.H. (1999) Ellenberg's indicator values for British plants. Vol. 2. Institute for Terrestrial Ecology, Huntingdon, UK. ECOFACT, Technical Annex.

Ho, S.C. (1980) On the chemical and algal growth potential of the surface water of the Muda river irrigation system, West Malaysia. In: J.I. Furtado (Ed.), Tropical ecology and development; 5th Proceeding of the International Tropical Ecology Symposium. Kuala Lumpur, Malaysia. pp. 989–998.

Hoffmann, J.P. (1998) Wastewater treatment with suspended and non-suspended algae. J. Phycol. 34 (5), 757–763.

Hoh, D., Watson, S., Kan, E. (2016) Algal biofilm reactors for integrated wastewater treatment and biofuel production: A review. Chem. Eng. J. 287, 466–473.

Hossein, A., Kondiram, D., Gorakh, P. (2014) Application of Phycoremediation Technology in the Treatment of Sewage Water to Reduce Pollution Load. Advances in Environmental Biology. 8(7): 2419–2423.

Aratboni, H.A., Rafei, N., Garcia-Granados, R., Alemzadeh, A., Morones-Ramirez, J.R. (2019) Biomass and lipid induction strategies in microalgae for biofuel production and other applications. Microb. Cell. Fact. 18, 178.

Jayaswal, K., Sahu, V., Gurjar, B. (2018) Water pollution, human health and ivièr-ation. In: Water remediation. Springer. pp. 11–27.

Ji, M.K., Abou-Shanab, R.A.I., Kim, S.H., Salama, E.S., Lee, S.H., Kabra, A.N., Lee, Y.S., Hong, S., Jeon, B.H. (2013) Cultivation of microalgae species in tertiary municipal wastewater supplemented with CO_2 for nutrient removal and biomass production. Ecol. Eng. 58, 142–148.

Joint, I., Henriksen, P., Fonnes, G.A., Bourne, D., Thingstad, T.F., Riemann, B. (2002) Competition for inorganic nutrients between phytoplankton and bacterioplankton in nutrient manipulated mesocosms. Aquat. Microb. Ecol. 29, 145–159.

Judd, S.J., Al Momani, F.A.O., Znad, H., Al Ketife, A.M.D. (2017) The cost benefit of algal technology for combined CO_2 mitigation and nutrient abatement. Renew. Sustain. Energy Rev. 71, 379–387.

Kang, D., Kim, K., Jang, Y., Moon, H., Ju, D., Kwon, G., Jaehnig, D. (2018) Enhancement of wastewater treatment efficiency through modulation of aeration and blue light on wastewater-borne algal-bacterial consortia. Int. Biodeter. Biodegr. 135, 9–18.

Kaplan, D., Christiaen, D., Arad, S. (1988) Binding of heavy metals by algal polysaccharides. In: T. Stadler, J. Mollion, M.C. Verdus, Y. Karamanos, H. Morvan, D. Christiaen (Eds.), Algal biotechnology. Elsevier Applied Science, London. pp. 179–187.

Ketheesan, B., Nirmalakhandan, N. (2011) Development of a new airlift- driven raceway reactor for algal cultivation. Appl. Energ. 88, 3370–3376.

Kesaano, M., Sims, R.C. (2014) Algal Biofilm Based Technology for Wastewater Treatment. Algal Research, 5, 231–240.

Knauer, K., Behra, R., Sigg, L. (1997) Effects of free Cu^{2+} and Zn^{2+} ions on growth and metal accumulation in freshwater algae. Environ. Toxicol. Chem. 16 (2), 220–229.

Kolkwitz, R., dan Marsson, M. (1908) Oekologie der pflanzlichen Saprobien. Berichte der Deutschen Botanischen Gesellschaft. 26, 505–519.

Kong, M., Chen, X.G., Xing, Ke., Park, H.J. 2010. Antimicrobial properties of chitosan and mode of action: A state of the art review. International Journal of Food Microbiology. 144 (1): 51–63.

Kotteswari M, Murugesan S, Ranjith Kumar R. (2012) Phycoremediation of dairy effluent by using the microalgae *Nostoc* sp. Int J Environ Res Dev.2:35–43.

Kumar, B., Agrawal, K., Bhardwaj, N., Chaturvedi, V., Verma, P. (2018) Advances in concurrent bioelectricity generation and bioremediation through microbial fuel cells. In: Microbial fuel cell technology for bioelectricity. Springer, Cham. pp. 211–239.

Kumar, B., Agrawal, K., Bhardwaj, N., Chaturvedi, V., Verma, P. (2019a) Techno-economic assessment of microbe-assisted wastewater treatment strategies for energy and value-added product recovery. In: Microbial technology for the welfare of society. pp 147–181.

Kiran Kumar, P., Krishna, S.V., Naidu, S.S., Verma, K., Bhagawan, D., Himabindu, V. (2019a) Biomass production from microalgae *Chlorella* grown in sewage, kitchen wastewater using industrial CO2 emissions: Comparative study. Carbon Resources Conversionz. 2 (2): 126–133.

Kumar R.S., Kumar, R., Yadav, S. (2019b) Environmental Issues in Textiles. pp 129–151.

Kumar, B., Agrawal, K., Verma, P. (2021) Current perspective and advances of microbe assisted electrochemical system as a sustainable approach for mitigating toxic dyes and heavy metals from Wastewater. J. Hazard. Toxic. Radioact. Waste. 25 (2), 04020082.

Kumar, B., and Verma, P. (2021a) Techno-economic assessment of biomass-based integrated biorefinery for energy and value-added product. In: Biorefineries: A step towards renewable and clean energy. Springer. pp. 581–616.

Kumar, B., Verma, P. (2021b) Life cycle assessment: blazing a trail for bioresources management. Energy Conversion and Management X, Elsevier. 100063

Laurens, L.M., Chen-Glasser, M., McMillan, J.D. (2017) A perspective on renewable bioenergy from photosynthetic algae as feedstock for biofuels and bioproducts. Algal Res. 24, 261–264.

Leclercq L. (1988) Utilization de trio's indices, chimique, diatomique et biocénotique, pour l'évaluation de la qualité de l'eau de la Joncquiere, rivière calcaire polluée par le village de Doische (Belgique, Prov. Namur) Mém. Soc. Roy. Bot. Belg. 10, 26–34.

Lee, C.S., Lee, S.A., Ko, S.R., Oh, H.M., Ahn, C.Y. (2015) Effects of photoperiod on nutrient removal, biomass production, and algal-bacterial population dynamics in lab-scale photobioreactors treating municipal wastewater. Water Res. 68, 680–691.

Leland, H.V., Carter, J.L. (1985) Effects of copper on production of periphyton, nitrogen fixation and processing of leaf litter in a Sierra Nevada, California stream. Freshw. Biol. 15 (2), 155–173.

Li, K., Liu, Q., Fang, F., Luo, R. et al. (2019) Microalgae-based wastewater treatment for nutrients recovery: A review. Bioresource Technology. 291: 121934.

Li, S.; Luo, S.; Guo, R. (2013) Efficiency of CO_2 fixation by microalgae in a closed raceway pond. Biores. Technol. 136, 267–272.

Li, T.; Zheng, Y.; Yu, L.; Chen, S. (2014). Mixotrophic cultivation of a Chlorella sorokiniana strain for enhanced biomass and lipid production. Biomass Bioenergy. 66, 204–213.

Li, Y., Chen, Y., Chen, P., Min, M., Zhou, W., Martinez, B., Zhu, J., Ruan, R. (2011) Characterization of a microalgae Chlorella sp. Well adapted to highly concentrated municipal wastewater for nutrient removal and biodiesel production. Bioresour. Technol. 102 (8), 5138–5144.

Lopez Barreiro, D., Prins, W., Ronsse, F., Brilman, W. (2013) Hydrothermal liquefaction (HTL) of microalgae for biofuel production: State of the art review and future prospects. Biomass Bioenergy. 53, 113–127.

Mallick, N. (2002) Biotechnological potential of immobilized algae for wastewater N, P and metal removal: A review. BioMetals. 15 (4), 377–390.

Malik, A.H., Suryapani, S., Ahmad, J. (2011) Chemical Vs Organic Cultivation of Medicinal and Aromatic Plants: the choice is clear. Int. J. Med. Arom. Plants. 1 (1): 5–13.

McCormick, P.V., Cairns, J. (1994) Jr Algae as indicators of environmental change. J. Appl. Phycol. 6, 509–526.

McGeoch, M.A. (1998) The selection, testing & application of terrestrial insects as bioindicators. Biol. Rev. 73, 181–201.

Mehariya, S., Kumar, P., Marino, T., Casella, P., Lovine, A., Verma, P., Musuma, D., Molino, A. (2021a) Aquatic weeds: a potential pollutant removing agent from wastewater and polluted soil and valuable biofuel feedstock, bioremediation using weeds. Springer. pp. 59–77.

Mehariya, S., Goswami, R., Verma, P., Lavecchia, R., Zuorro, A. (2021b) Integrated approach for wastewater treatment and biofuel production in microalgae biorefineries. Energies. 14 (8), 2282.

Mehariya, S., Goswami, R.K., Karthikeysan, O.P., Verma, P. (2021c) Microalgae for high-value products: A way towards green nutraceutical and pharmaceutical compounds. Chemosphere. 280, 130553.

Mehrabadi, A., Farid, M.M., Craggs, R. (2016) Variation of biomass energy yield in wastewater treatment high rate algal ponds. Algal Res. 15, 143–151.

Mennaa, F.Z., Arbib, Z., Perales, J.A. (2015) Urban wastewater treatment by seven species of microalgae and analgal bloom: Biomass production, N and P removal kinetics and harvestability. Water Res. 83, 42–51.

Metcalf and Eddy (2003) Water engineering: treatment and reuse, 4th ed., revised by G. Tchobanoglous, F. Burtun, H. Stensel. McGraw-Hill, New York.

Milledge, J.J. (2011) Commercial application of microalgae other than as biofuels: a brief review. Rev. Environ. Sci. Biotechnol. 10 (1), 31–41.

Moe, C.L., Rheingans, R.D., 2006. Global challenges in water, sanitation and health. J. Water Health. 4, 41–57.

Molinuevo-Salces, B., García-Gonzalez, M.C., Gonzalez-Fernandez, C. (2010) Performance comparison of two photobioreactors configurations (open and closed to the atmosphere) treating anaerobically degraded swine slurry. Bioresour. Technol. 101, 5144–5149.

Moreno, A., Rueda, O., Cabrera, E., Luna-del-Castillo, J.D. (1990) Standarization in wastewater biomass growth. Ig. Mod. 94 (1), 24– 32.

Moreno-Garrido, I. (2008) Microalgae immobilization: Current techniques and uses. Biores. Technol. 99 (10), 3949–3964.

Mulbry, W., Kangas, P., Kondrad, S. (2010) Toward scrubbing the bay: Nutrient removal using small algal turf scrubbers on Chesapeake Bay tributaries. Ecol. Eng. 36, 536–541.

Munasinghe, M. (2019) Water supply and environmental management. Routledge.

Muñoz, R., Guieysse, B. (2006) Algal–bacterial processes for the treatment of hazardous contaminants: A review. Water Res. 40 (15), 2799–2815.

Nagarajan, D., Kusmayadi, A., Yen, H.-W., Dong, C.-D., Lee, D.-J., Chang, J.-S. (2019) Current advances in biological swine wastewater treatment using microalgae-based processes. Bioresour. Technol. 289 (6), 121718.

Neoh, C.H., Yahya, A., Adnan, R., Majid, A., Ibrahim, Z.Z. (2013) Optimization of decolorization of palm oil mill effluent (POME) by growing cultures of *Aspergillus fumigatus* using response surface methodology. Environ. Sci. Pollut. Res. Int. 20 (5), 2912–2923.

Neveux, N., Magnusson, M., Mata, L., Whelan, A., de Nys, R., Paul, N.A. (2016) The treatment of municipal wastewater by the macroalga Oedogonium sp. and its potential for the production of biocrude. Algal Res. 13, 284–292.

Newman, M.C., Alberts, J.J., Greenhut, V.A. (1985) Geochemical factors complicating the use of aufwuchs to monitor b10-accumulation of arsenic, cadmium, chromium, copper, zinc. Water Res. 19(9), 1157–1165.

Nikolaison, L., Dahl, J., Bech, K.S., Bruhn, A., Rasmussen, M.B., Bjerre, A.B., Nielsen, H.B., Ambus, P., Rost, K.A., Kadar, Z., et al. Energy production fom macroalgae. In Proceedings of the 20th European Biomass Conference, Milan, Italy, 18–22 June 2012.

Nur, M.M.A., Kristanto, D., Setyoningrum, T.M., Budiaman, I.G.S. (2016) Utilization of microalgae cultivated in palm oil mill wastewater to produce lipid and carbohydrate by employing microwave- assisted irradiation. Recent. Innov. Chem. Eng. 9 (2), 107–116.

Nur, M.M.A., Setyoningrum, T.M., Budiaman, I.G.S. (2017) Potency of *Botryococcus braunii* cultivated on palm oil mill effluent wastewater as a source of biofuel. Enviro. Eng. Res. 22 (4), 417–425.

Orpez, R., Martinez, M.E., Hodaifa, G., Yousfi, F.E., Jbari, N., Sanchez, S. (2009) Growth of the microalga *Botryococcus braunii* in secondarily treated sewage. Desalination. 246, 625–30.

Oswald, W.J., Golueke, C.G. (1960) Biological transformation of solar energy. Adv. Appl. Microbiol. 2, 223–262.

Oswald, W.J., Gotaas, H.B. (1957) Photosynthesis in sewage treatment. Trans. Am. Soc. Civil. Eng. 122, 73–105.

Pacheco, M.M., Hoeltz, M., Moraes, M.S., Schneider, R.C. (2015) Microalgae: cultivation techniques and wastewater phycoremediation. J. Environ. Sci. Health A Tox. Subst. Environ. Eng. 50 (6), 585–601.

Palmer, C.M. (1969): A composite rating of algae tolerating organic pollution. J. Phycol. 5, 78–82.

Palmer, C.M. (1974) Algae in American sewage stabilization's ponds. Rev. Microbiol. (S-Paulo). 5, 75–80.

Park, K.C., Whitney, C., McNichol, J.C., Dickinson, K.E., MacQuarrie, S., Skrupski, B.P., Zou, J., Wilson, K.E., O'Leary, S.J.B., McGinn, P.J. (2012) Mixotrophic and photoautotrophic cultivation of 14 microalgae isolates from Saskatchewan, Canada: Potential applications for wastewater remediation for biofuel production. J. Appl. Phycol. 24 (3), 339–348.

Patrick, R. (1949) A proposed biological measure of stream conditions, based on a survey of the Conestoga Basin, Lancaster County, Pennsylvania. Proc. Acad. Nat. Sci. Philadelphia. 101, 277–341.

Patrick, R. (1971) Diatom communities. In: J. Cairns (Ed.), The structure and function of freshwater microbial communities. Virginia Polytechnic Institute and State University, Blacksburg, Virginia, USA.

Patrick, R. (1973) Use of algae, especially diatoms, in the assessment of water quality. In: J. Cairns Jr, K.L. Dickson (Eds.), Biological methods for the assessment of water quality, Special Technical Publication. American Society for Testing and Materials, Philadelphia, Pennsylvania. 528: 76–95.

Pires, J.C.M., Alvim-Ferraz, M.C.M., Martins, F.G., Simões, M. (2013) Wastewater treatment to enhance the economic viability of microalgae culture. Environ. Sci. Poll. Res. 20 (8): 5096–5105.

Plafkin, J.L., Barbour, M.T., Porter, K.D., Gross., S.K., Hughes, R.M. (1989) Rapid assessment protocols for use in streams & rivers: Benthic Macro invertebrates & Fish. EPA: Washington, D.C. In: D.M. Rosenberg, V.H. Resh (Eds.), Freshwater Biomonitoring & Benthic Macroinvertebrates. Chapman & Hall, New York, NY.

Pouliot, Y., Talbot, P., De la Noue, J. (1986) Biotraitement du purin de pore par production de biomass. Entropie. 130 (131), 73–77.

Prajapati, S.K., Kaushik, P., Malik, A., Vijay, V.K. (2013) Phycoremediation coupled production of algal biomass, harvesting and anaerobic digestion: Possibilities and challenges. Biotechnol. Adv. 31 (8), 1408–1425.

Prajapati, S.K., Kaushik, P., Malik, A., Vijay, V.K. (2013) Phycoremediation coupled production of algal biomass, harvesting and anaerobic digestion: possibilities and challenges. Appl. Environ. Microbiol. 31, 1408–1425.

Priyadarshani, I., and Rath, B. 2012. Commercial and industrial applications of micro algae - A review J. Algal Biomass Utln. 3 (4): 89–100.

Pulz, O., Gross, W. (2004) Valuable products from biotechnology of microalgae. Appl Microbiol Biotechnol **65,** 635–648.

Qadir, M., Wichelns, D., Raschid-Sally, L., McCornick, P.G., Drechsel, P., Bahri, A., Minhas, P. (2010) The challenges of wastewater irrigation in developing countries agricultural. Water Manag. 97, 561–568.

Rasoul-Amini, S., Montazeri-Najafabady, N., Shaker, S., Safari, A., Kazemi, A., Mousavi, Ray, N.E., Terlizzi, D.E., Kangas, P.C. (2014) Nitrogen and phosphorus removal by the Algal Turf Scrubber at an oyster aquaculture facility. Ecol. Eng. 78.

Rawat I, Ranjith Kumar R, Mutanda T, Bux F (2011) Dual role of microalgae: phycoremediation of domestic wastewater and biomass production for sustainable biofuels production. Appl Energy 88:3411–3424.

Razzak S A, Hossain M M, Lucky R A, et al (2013) Integrated CO_2, capture, wastewater treatment and biofuel production by microalgae culturing—A review[J]. Renewable & Sustainable Energy Reviews 27(6):622–653.

Rehnstam Holm, A.S., Godhe A. (2003) Genetic engineering of algal species. Eolss Publishers, Oxford, UK.

Renaud, S.M., Parry, D.L., Thinh, L.V. (1994) Microalgae for use in tropical aquaculture. 1. Gross chemical and fatty acid composition of twelve species of microalgae from the Northern Territory, Australia. J. Appl. Phycol. 6 (3), 337–345.

Renuka.N., Sood, A., Rath, S.K., Prasamma, R., Ahluwalia, A.S. (2013) Evaluation of microalgal consortia for treatment of primary treated sewage effluent and biomass production. Journal of applied phycology 25 (5), 1529–1537.

Reyna-Velarde, R., Cristiani-Urbina, E., Hernandez-Melchor, D.J., Thalasso, F., Canizares-Villanueva, R.O. (2010) Hydrodynamic and mass transfer characterization of a flat-panel airlift photobioreactor with high light path. Chem. Eng. Proc. 49, 97–103.

Richardson, J.W., Johnson, M.D., Outlaw, J.L. (2012) Economic comparison of open pond raceways to photo bioreactors for profitable production of algae for transportation fuels in the Southwest. Algal Res. 1, 93–100.

Rott, E. (1991) Methodological aspects and perspectives in the use of periphyton for monitoring and protecting rivers. In: B.A. Whitton, E. Rott, G. Friedrich (Eds.), Use of algae for monitoring rivers. Institut für Botanik, Universitat Innsbruck, Innsbruck, Austria.

Round, F.E. (1991) Diatoms in river water-monitoring studies. J. Appl. Phycol. 3, 129–145.

Ruiz-Martinez, A., Martin Garcia, N., Romero, I., Seco, A., Ferrer, J. (2012) Microalgae cultivation in wastewater: Nutrient removal from anaerobic membrane bioreactor effluent. Biores. Technol. 126, 247–253.

Saini, K.C., Yadav, D.S., Mehariya, S., Rathore, P., Kumar, B., Marino, T., Leone, G.P., Verma, P., Musmarra, D., Molino, A. (2021) Overview of extraction of astaxanthin from *Haematococcus pluvialis* using CO2 supercritical fluid extraction technology vis-a-vis

quality demands. In: Global perspectives on astaxanthin from industrial production to food, health, and pharmaceutical applications. Academic Press, Elsevier. pp: 341–354.
Samori G., Samori C., Guerrini F., Pistocchi R., 2013 Growth and nitrogen removal capacity of Desmodesmus communis and of a natural microalgae consortium in a batch culture system in view of urban wastewater treatment: part I. Water Research 47:791–801.
Schenk, P.M., Thomas-Hall, S.R., Stephens, E. *et al.* (2008). Second Generation Biofuels: High-Efficiency Microalgae for Biodiesel Production. Bioenerg. Res. 1, 20–43.
Schnurr, P.J., Allen, D.G. (2015) Factors affecting algae biofilm growth and lipid production: A review. Renew. Sustain. Energy Rev. 52, 418–429.
Scott, C.A., Faruqui, N.I., Raschid-Sally, L. (2004) Wastewater use in irrigated agri-culture: Management challenges in developing countries wastewater use in irrigated agriculture: Confronting the livelihood and environmental realities. CABI Publishing, Wallingford, UK. pp. 1–10.
Scott, S.A., Davey, M.P., Dennis, J.S., Horst, I., Howe, C.J., Lea-Smith, D.J., Smith, A.G. (2010) Biodiesel from algae: Challenges and prospects. Curr. Opin. Biotechnol. 21 (3), 277–286.
Sengar, R.M.S., Singh, K.K., Surendra, S. (2011). Application of Phycoremediation Technology in the Treatment of Sewage water to reduce pollution load. Indian Journal of Scientific research. 2 (4): 33–39.
Shaker, S., Nemati, A., Montazeri-Najafabady, N., Mobasher, M.A., Morowvat, M.H., Ghasemi, Y. (2015) Treating urban wastewater: Nutrient removal by using immobilized green algae in batch cultures. Int. J. Phytoremed. 17 (12), 1177–1182.
Shelef, G., Soeder, C.J. (1980) Algal biomass: production and use. Elsevier/North Holland Biomedical Press, Amsterdam. p. 852.
Sierra, E., Acien, F.G., Fernandez, J.M., García, J.L., Gonzalez, C., Molina, E. (2008) Characterization of a flat plate photobioreactor for the production of microalgae. Chem. Eng. J. 138, 136–147.
Dahmani, S., Zerrouki, D., Ramana, L., Rawal, I., Bux, F. (2016) Cultivation of *Chlorella pyrenoidosa* in outdoor open raceway pond using domestic wastewater as medium in arid desert region. Biores. Technol. 219, 749–752.
Singh, R.N., Sharma, S. (2012). Development of suitable photobioreactor for algae production – A review. Renew. Sust. Energ. Rev. 16, 2347–2353.
Singh, S.P., Singh, P. (2015) Effect of temperature and light on the growth of algae species: A review. Renew. Sust. Energ. Rev. 50, 431–444.
Soeder, C.J., Payer, H.D., Runkel, K.H., Beine, J., Briele, E. (1978) Sorption and concentration of toxic minerals by mass cultures of Chlorococcales. Mitt. Internat. Verein. Limnol. 21, 575–584.
Solovchenko, A., Verschoor, A.M., Jablonowski, N.D., Nedbal, L. (2016) Phosphorus from wastewater to crops: An alternative path involving microalgae. Biotechnol. Adv. 34 (5), 550–564.
Song, T., Martensson, L., Eriksson, T., Zheng, W.W. and Rasmussen, U. (2005) Biodiversity and Seasonal Variation of the Cyanobacterial Assemblage in a Rice Paddy Field in Fujian, China. FEMS Microbiology Ecology, 54: 131–140.
Stevenson, R.J., Bahls, L.L. (1999) Periphyton protocols. In: M.T. Barbour, J. Gerritsen, B.D. Snyder, J.B. Stribling (Eds.), Rapid bioassessment protocols for use in streams and wadeable rivers: Periphyton, benthic macroinvertebrates, and fish. 2nd ed. US Environmental Protection Agency, Office of Water, Washington D.C. pp. 6.1–6.22.
Stevenson, R.J., Lowe, R.L. (1986) Sampling and interpretation of algal patterns for water quality assessment. In: R.G. Isom (Ed.), Rationale for sampling and interpretation of ecological data in the assessment of freshwater ecosystems. American Society for Testing and Materials, Philadelphia, USA. pp. 118–149.
Stevenson, R.J., Pan, Y. (1999) Assessing ecological conditions in rivers and streams with diatoms. In: E.F. Stoemer, J.P. Smol (Eds.), The diatom: Applications to the environmental and earth science. Cambridge University Press, Cambridge, UK. pp. 11–40.

Sydney, E.B., da Silva, T.E.; Tokarski, A., Novak, A.C., de Carvalho, J.C., Woiciecohwski, A.L., Larroche, C., Soccol, C.R. (2011). Screening of microalgae with potential for biodiesel production and nutrient removal from treated domestic sewage. Appl. Energy 88, 3291–3294.

Tam, N.F.Y., Wong, Y.S. (1989) Wastewater nutrient removal by *Chlorella pyrenoidosa* and *Scenedesmus sp*. Environ. Poll. 58 (1), 19–34.

Teoh, M.L., Phang, S.M., Chu, W.L. (2013) Response of Antarctic, temperate, and tropical microalgae to temperature stress. J. Appl. Phycol. 25 (1), 285–297.

Tredici, M.R. (2004) Mass production of microalgae: photobioreactors. In: Handbook of Microalgal Culture. Biotechnology and Applied Phycology. 178–214.

Van Dam, H., Mertens, A., Sinkeldam, J. (1994) A coded checklist and ecological indicator values of freshwater diatoms from the Netherlands. Aquat. Ecol. 28 (1), 117–133.

Vassilev, S.V., Vassileva, C.G. (2016) Composition, properties and challenges of algae biomass for biofuel application: an overview. Fuel. 181, 1–33.

Vasumathi, K.K., Premalatha, M., Subramanian, P. (2012) Parameters influencing the design of photobioreactor for the growth of microalgae. Renew. Sustain. Energy Rev. 16 (7), 5443–5550.

Vijayakumar, S. (2012). Potential applications of cyanobacteria in industrial effluents- a review. J. Bioremed. Biodeg. 3, 1–6.

Wake, L.V., Hillen, L.W. (1981) Nature and hydrocarbon content of blooms of the alga *Botryococcus braunii* occurring in Australian freshwater lakes. Aust. J. Mar. Freshwat. Res. 32, 353–367.

Wang J, Koo Y, Alexander A, Yang Y, Westerhof S, Zhang Q, Schnoor JL, Colvin VL, Braam J, Alvarez PJJ. (2013). Phytostimulation of poplars and Arabidopsis exposed to sliver nanoparticles and Ag^+ at sublethal concentrations. Environ Sci Technol. 47:5442–5449.

Wang, S.K., Stiles, A.R., Guo, C., Liu, C.Z. (2014) Microalgae cultivation in photobioreactors: An overview of light characteristics. Eng. Life Sci. 14 (6), 550–559.

Wang, B., Lan, C.Q., Horsman, M. (2012) Closed photobioreactors for production of microalgal biomasses. Biotechnol. Adv. 30, 904–912.

Wang, W., Luo, Y., Qiao, W. (2010) Possible solutions for sludge dewatering in China. 4 (1) 102–107.

Whitton, R., Ometto, F., Pidou, M., Jarvis, P., Villa, R., Jefferson, B. (2015) Microalgae for municipal wastewater nutrient remediation: mechanisms, reactors and outlook for tertiary treatment. Environ. Technol. Rev. 4 (1), 133–148.

Wijffels, R.H., Barbosa, M.J. (2010) An outlook on microalgal biofuels. Science. 329 (5993), 796–799.

Woertz, I., Feffer, A., Lundquist, T., Nelson, Y. (2009) Algae grown on dairy and municipal wastewater for simultaneous nutrient removal and lipid production for biofuel feedstock. J. Environ. Eng. 135, 1115–1122.

Wong, P.K., Chan, K.Y. (1990) Growth and value of *Chlorella salina* grown on highly saline sewage effluent. Agric. Ecosyst. Environ. 30 (3–4), 334–350.

Woolard, C.R., Irvine, R.L. (1995) Treatment of hypersaline wastewater in the sequencing batch reactor. Wat. Res. 29 (4), 1159–1168.

Wu, Y.H., Hu, H.Y., Yu, Y., Zhang, T.Y., Zhu, S.F., Zhuang, L.L., Zhang, X., Lu, Y. (2014) Microalgal species for sustainable biomass/lipid production using wastewater as resource: A review. Renew. Sustain. Energy Rev. 33, 675–688.

Wu, Y.-H., Li, X., Yu, Y., Hu, H.-Y., Zhang, T.-Y., Li, F.-M. (2013) An integrated microalgal growth model and its application to optimize the biomass production of Scenedesmus sp. LX1 in open pond under the nutrient level of domestic secondary effluent. Biores. Technol. 144, 445–451.

Yang J, Li X, Hu H-Y, Zhang X, Yu Y, Chen Y-S. (2011). Growth and lipid accumulation properties of a freshwater microalga, Chlorella ellipsoidea YJ1, in domestic secondary effluents. Appl. Energy. 88: 3295–3299.

Zhang, E., Wang, B., Wang, Q., Zhang, S., Zhao, B. (2008) Ammonia-nitrogen and orthophosphate removal by immobilized Scenedesmus sp. isolated from municipal wastewater for potential use in tertiary treatment. Biores. Technol. 99 (9), 3787–3793.

4 Phycoremediation: A Promising Solution for Heavy Metal Contaminants in Industrial Effluents

Chandrashekaraiah P.S., Santosh Kodgire, Ayushi Bisht, Debanjan Sanyal, Santanu Dasgupta

CONTENTS

DOI: 10.1201/9781003155713-4

4.1 INTRODUCTION

Increasing modern practices, industrialization, urbanization, and anthropogenic activities contaminate water bodies with various heavy metals (Kumar et al. 2019; Yang et al. 2008). The metals with an atomic density greater than 5 g/cm^3 are termed heavy metals. Tannery, textile, electroplating, and petrochemical industries are considered as major sources of heavy metals such as lead, cadmium, chromium, and nickel (Figure 4.1). Heavy metals released from these industries affect the aquatic ecosystem and are the reason for the loss of biodiversity (He et al. 1998). Due to the non-biodegradable, bioaccumulation, and toxic nature of heavy metals, these are considered as potential hazards to humans and environments (Kumar et al. 2018). Heavy metals enter the human food chain through aquaculture and agriculture products (Leung et al. 2014; Gupta et al. 2015a). Human poisoning of heavy metals is associated with diarrhea, gastrointestinal disorders, paralysis, pneumonia, and depression (Kumar et al. 2021). Earlier studies showed that heavy metals are carcinogenic, neurotoxic, and teratogenic (Jarup, 2003; Oskarsson et al. 2004). It was also reported that each metal exhibits specific toxicity on the human body, e.g., higher accumulation of lead causes anemia, multiple sclerosis, neurological disorders, and impaired kidney function.

The conventional approaches used for heavy metal removal from water bodies are physical and chemical methods, e.g., chemical oxidation or reduction, chemical precipitation, filtration, ion exchange, reverse osmosis (RO), electrochemical treatment, evaporation recovery, and membrane technologies (Kumar et al. 2021). These methods are inefficient, more expensive, and work well for very low concentrations of heavy metals (Kumar et al. 2015). Chemical methods require more energy, generate toxic chemical sludge, are affected by media pH, and require more amount of reagents (Kumar et al. 2015). Thus, there is an increase in demand for eco-friendly, renewable, and cost-effective remediation technologies (Agrawal et al. 2020a; Agrawal and Verma, 2020; Kanchana et al. 2014). Water is a major input in many industries; from the point of economics and water crisis,

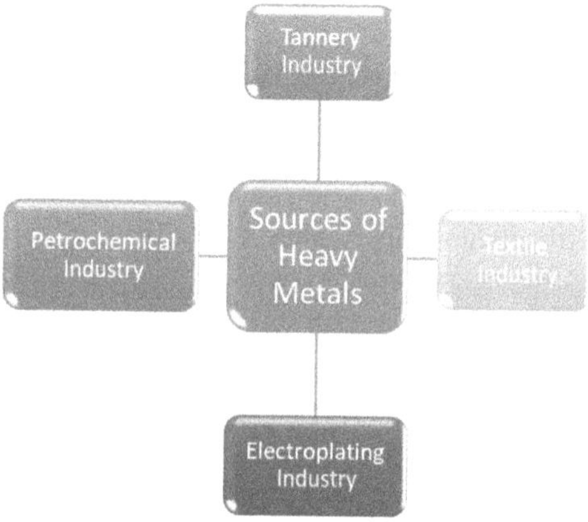

FIGURE 4.1 Sources of heavy metals.

water recycling is considered as probable option. The use of microbial biomass for the remediation of wastewater has been broadly studied (Chaturvedi et al. 2013a). In this context, macroalgae or microalgae-based phycoremediation of heavy metals is considered as green and economic approach (Kumar et al. 2015; Mehariya et al. 2021a,b; Olguin, 2003). Algae are photosynthetic organisms, fix atmospheric CO_2, using sunlight and low-cost fertilizers. Unlike plants, they do not compete with arable land and are able to grow in various types of wastewaters (Latiffi et al. 2015; Rawat et al. 2011). Algae can grow autotrophically and heterotrophically, have a high tolerance to heavy metals, and possess large surface area/volume ratios (Elumalai et al. 2013). Algae possess various mechanisms for heavy metal removal. The cell wall of algae contains various functional groups such as carboxylates, amines, and hydroxyls, etc. (Lesmana et al. 2009; Song et al. 2014); these functional groups are involved in the biosorption of metals. Algae are also known to accumulate and detoxify heavy metals. For effective removal of heavy metals, live and dead (dry) algae in free or immobilized form can be used in batch or continuous cultivation systems. These characteristics make algae a suitable candidate for heavy metal remediation from various industrial wastes (Gupta et al. 2015b). Algae used for heavy metal remediation include *Chlorella*, *Spirogyra*, *Cladophora*, *Botryococcus*, *Chaetoceros*, *Synechocystis*, and cyanobacteria (*Microcystis* and *Anabaena*) (Cetinkaya Donmez et al. 1999; Chaturvedi et al. 2013b; Mallick and Rai, 1992). After heavy metal removal from wastewater by algae, the water can be recycled and algal biomass can be harvested for various purposes (Agrawal et al. 2020b; Goswami et al. 2021; Kumar and Verma, 2021). This chapter gives an overview of various algal technologies used for the removal of heavy metals from industrial wastewater, with special emphasis given to recycling strategies.

4.2 CONVENTIONAL METHODS OF HEAVY METAL REMOVAL

There are various conventional approaches for treating industrial wastewater before it is discharged into the environment (Agrawal et al. 2020c). The pros and cons of conventional methods are discussed.

4.2.1 Ion Exchange Method

The ion exchange method is generally used in industry for the removal of heavy metals from wastewater. It uses anion or cation exchangers made of insoluble synthetic resin materials. The ions from resins are exchanged with either anions or cations of metals and get removed (Ajmal et al. 2003; Suemitsu et al. 1986). The most commonly used resins are sodium silicates, zeolites, acrylic, and metha-acrylic. Resins work in wide pH ranges; weak resins exchange ions in a small pH range, whereas strong resins work well in a wide pH range (Khan et al. 2003). Heavy metals such as nickel, copper, zinc, mercury, gold, silver, and chromium were recovered by using the ion-exchange method. Besides high selectivity, this method has limitations such as high cost, low life, and does not effectively work in dilute concentrations of metal solution (Selvi et al. 2001).

4.2.2 Adsorption Method

Adsorption is an effective method for the removal of heavy metals from wastewater (Gottipati and Mishra, 2012). The various adsorbents used are activated carbons, carbon nanotubes (CNTs), zeolite, chitosan, and biosorbents (Ince and Ince, 2019). The efficiency of the adsorption process depends on many factors, including metal concentration, adsorption particle size and type, temperature, pH, and contact time. This method has limitations with the column life and management of exhausted adsorbent.

4.2.3 Membrane Filtration

It is another method used to remove heavy metals from wastewater. Based on the size of the particle to be separated, various membrane filtration methods are applied, including RO and electrodialysis (ED) (Agarwal et al. 2015). In RO, water is forced from a highly saline solution to a low saline solution and the semi-permeable membrane is used for the removal of heavy metals from wastewater. Algureiri and Abdulmajeed (2016) showed the RO membrane method is the most significant in the removal of nickel, lead, and copper from industrial wastewater. They found that the RO method could have the ability to remove more than 95% of nickel, lead, and copper. In ED, the electrical potential was applied to separate metal ions from wastewater, passed through the ion exchange membrane. Chen 2004 used this method for the ED of nickel, cobalt, cadmium, chromium, and mercury. The main limitation associated with this methods is the high operational cost, energy consumption, and production of sludge. In the electrolysis method, the soluble metal ions are separated from the wastewater and deposited on the surface of the electrodes (Ince and Ince, 2019). The various electrodes and anodes used in this process are lead oxide electrode, boron-doped diamond electrode, iron electrode, and graphite electrode.

4.2.4 Chemical Precipitation

It is a commonly used method for the removal of heavy metals in parts per million (ppm) levels from industrial wastewater. In this method, the solubility of metal is reduced and converted into precipitate, which is separated easily from the water surface. It can be applied to treat industrial wastewater containing chromium and nickel. The conventional precipitation methods include hydroxide (sodium hydroxide and calcium hydroxide) and sulfide (sulfide ions) precipitation. Brboot et al. 2011 studied the chemical precipitation method for the removal of heavy metals using a jar tester. For their study, they used magnesia (MgO) as a precipitator in the range of 1–5 $g.L^{-1}$ concentration to check the treatment efficiency. MgO at 1.5–3.0 $g.L^{-1}$ concentration was found optimum at a pH range of 9.5–10 for 99% metal removal.

4.2.5 Coagulation and Clotting Method

In this method, many chemicals or organic coagulants were used (Abu-Eishah, 2008); due to economic constraints and the negative impact on the environment, the chemical coagulation process is replaced with electrical coagulation and floatation (Ahluwalia and Goyal, 2006). The aluminum and iron-positive charges were produced through the electrical breakdown of aluminum and iron electrodes. Further, these charges were used in the process of clotting with metal ions (Ahmady-Asbchin et al. 2008; Selvi et al. 2001). The size and dimension of bubbles formed in this electrical floatation process depend on the pH value (Alinnor, 2007). Electrical coagulation and floatation have higher efficiency for metal removal than chemical coagulation (Amana et al. 2008). Also, it significantly decreases the concentration of chemicals but adversely affects the biological system. The major limitation associated with this method is the initial investment (Wang et al. 2008). All these conventional methods cited above require a huge investment for their utilization and construction; hence, they are costly. Also, the disposal of toxic residual sludge generated after treatment is not eco-friendly. Therefore, for the effective removal of heavy metals from industrial wastewater, there is an urgent need for the development of an economic, eco-friendly, and sustainable method.

4.3 PHYCOREMEDIATION

4.3.1 Phycoremediation by Live Algal Cultures

Biosorption occurs when a biological entity, for instance, algae, is used to adsorb any substance (Veglio and Beolchini, 1997). Biosorption results in a decrease of the sorbate in the solution and its deposition on the sorbent (Michalak et al. 2013). Algae species are used for this purpose in both live and dead forms, with each having its pros and cons (Brinza et al. 2007). When dead algae are used, the process is passive and only surface-level interaction takes place between the biomass and heavy metal ions. This process does not rely on the metabolism of algae and takes place generally within 10 min of interaction. When live algae are used, the process is both active and passive. It depends on the metabolism as well as surface interaction of the cell wall with the solution containing heavy metal ions (Salam, 2019). Overall, some general ways through which biosorption and bioaccumulation take place include active transport, ion exchange, precipitation of metal ions, and some electrostatic interactions (Rao and Prabhakar, 2011). Active transport occurs when the algal cell spends energy in the form of adenosine triphosphate (ATP) to carry the heavy metal ion inside the cell (this is exclusive to live algae cultures) (Wilde and Benemann, 1993). In ion exchange, non-toxic ions that are similar to heavy metal ions are exchanged with them (Dahman, 2017). Precipitation occurs when heavy metal ions interact and bind with certain groups of algal biomasses and form non-toxic precipitates (Muñoz et al. 2006). In 1955, it was proved that algae possess a negative charge on their surface. This charge attracts positively charged metal ions and forms complexes on the surface through electrostatic interactions (Ives, 1959). All these interactions are made possible due to the diverse functional groups naturally present on the algal cell wall. The cell wall and plasma membrane of the algal cells act as the first line of defense against heavy metals. The negative charge on the surface is the consequence of the negatively charged internal composition of the cell wall. Particularly, the presence of proteins results in the existence of –NH2 group, which in turn attracts metal

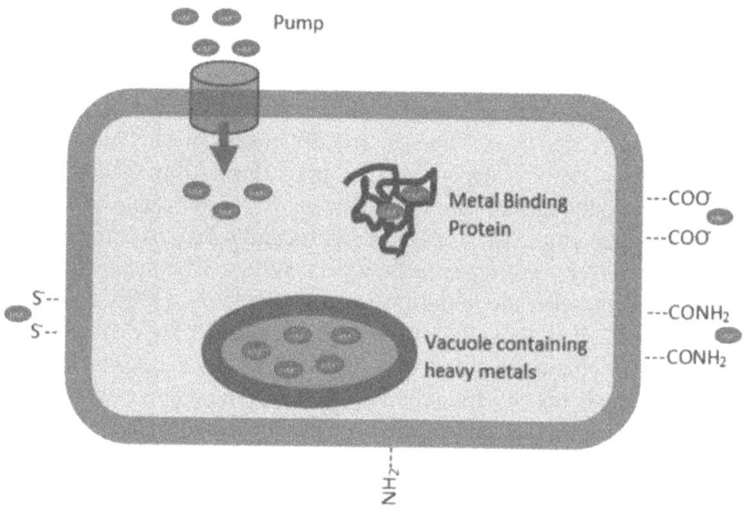

FIGURE 4.2 Biosorption and bioaccumulation of heavy metals by living algal cell.

ions (Figure 4.2). Group A category of transporters, viz. natural resistance associated macrophage protein (NRAMP) and Fe transporter (FTR), present in the plasma membrane of the cell facilitates the inward movement of metal ions (Kumar et al. 2015).

Factors like pH and temperature drastically influence the process of biosorption, particularly ion exchange (Negm et al. 2018). Any increase or decrease in the pH can alter the surface charge, which can change the affinity for the metal ion (Jin-fen et al. 2000). A study on *Palmaria palmata* (red algae) was conducted to check the pH effect. The pH was made similar to the wastewater from industries, i.e., 2–6, and it was noted that an increase in pH increased the uptake of copper and lead ions. The lower biosorption at pH 2 could be due to deprotonation, i.e., when pH is decreased, the surface charge of the algal cell becomes more positive (H^+ ion addition), thus blocking the metal ions from interacting with the functional groups (Prasher et al. 2004). Temperature has a substantial effect on live algae culture as it plays an important role in the biochemical reactions and photosynthesis of the algae (Du et al. 2013). Dead algal biomass is least affected by temperature. There are several other factors like the concentration of a metal ion in solution, amount of biomass, duration of contact, and presence of other ions in the solution that can affect the interaction (Salam, 2019). When the initial concentration of heavy metal ions is more, the uptake capacity reduces since the sites for binding are saturated at the beginning of the process (Anastopoulos and Kyzas, 2015). This was proved when the adsorption isotherms were studied for *Chlorella vulgaris* (Ting et al. 1991). The initial concentration of cadmium and zinc ions were maintained 2×10^{-5} M and the linear relationship was noted for the algal cells. The more the algal biomass, the more the uptake of metal ions; however, equilibrium has to be maintained during this. *Durvillaea potatorum* and *Ecklonia radiata* (Australian marine algae) were tested for lead and copper uptake, and the biomass dose of 200 mg was maintained to achieve an output of around 1.1–1.5 (mmol.g^{-1}) for both the metal ions (Matheickal and Yu, 1999). The contact time is directly proportional to the adsorption capacity unless the equilibrium is attained. Cell metabolism, physiology, type of metal, and temperature are also important factors while determining the contact time. Ye et al. 2015 worked with *Porphyra leucosticta* (red algae) and mentioned contact time to be a critical factor in determining the efficiency of adsorption. They gave 120 minutes for contact and observed a very high uptake for cadmium and lead. It is quite evident that the wastewater from industries always has other ions along with heavy metal ions (Ghafoor et al. 1994). If these ions are similar in structure, they can bind with the functional groups and create competition for free sites (Salam, 2019). Many brown algae species like *Undaria pinnatifida* and *Sargassum fulvellum* have been studied and observed to have many non-specific binding sites, which is extremely beneficial for industries because of the mixed composition of the water (Lee et al. 2002).

The phase of algal cell also contributes majorly in determining the uptake of heavy metal. When the cells are in the exponential growth phase, the metabolic activities are the most active and thus the active and passive uptake in live cultures is at peak levels. To investigate this, live *Coscinodiscus granii* cultures at the growth phase and stationary phase were examined for metal uptake in upwelling water. The concentration of metal ions in the solution with cells in the growth phase was considerably lower than the one in the stationary phase (Rabsch and Elbrachter, 1980). Live algal cultures are known for their better specificity over dead algal biomass (Salam, 2019). Metal accumulation inside the cell via active and passive transport is a characteristic feature of live cultures. Bioaccumulation of heavy metals inside the cells diminishes the photosynthetic ability of the algae, and at times to overcome this, these ions are stored in cellular compartments like vacuoles (Salama et al. 2019). To tolerate the presence of heavy metals inside the cell, algae have evolved to have an extraordinary defense system. Apart from compartmentalization in vacuoles, algae can biosynthesize specialized proteins like phytochelatins that bind to the metal ions and avoid inhibition of algal enzymes (Chekroun and Baghour, 2013). Making such non-toxic complexes and exclusion processes help algae to combat heavy metal ions naturally (Figure 4.2). Inside the cell, oxidation, vaporization, and methylation of the heavy metal change them into a form that is easier for the cell to excrete (Marella et al. 2020). The capability of phycoremediation of live algal cultures is shown by work on *Ulva lactuca* (sea lettuce). It successfully removed a copious number of heavy metals (cadmium, lead, copper, chromium, mercury, and nickel) from the solution in saline conditions (Henriques et al. 2019). Flocculation of living microalgae has been seen as an economic and effective method (Pugazhendhi et al. 2019). *C. vulgaris* JSC-7 was tested for zinc and cadmium uptake via flocculation (Alam et al. 2015). The flocculating strain showed better photosynthetic activity and metabolism along with 20% more uptake capacity than non-flocculating strain. In an investigation, immobilized *C. vulgaris* has proclaimed extraordinary results by removing almost all the heavy metal ions at a biomass concentration of 100 mg and metal concentration of less than 10 mg (Mehta and Gaur, 2001). The various live algal cultures with their heavy metal removal efficiency have been listed in Table 4.1.

Cyanobacteria (blue-green algae) makes one of the best choices for phycoremediation for two major reasons: i) unique structural organization of cell wall involving myeloid sheath and peptidoglycan layer and ii) presence of class-II metallothionein (MT-2), a

TABLE 4.1
Heavy Metal Removal by Live Algal Cultures

Sr. No.	Algal Cultures	Adsorbent Concentration on Dry Weight Basis	Heavy Metal	Process Description	Removal Efficiency	References
1	*Cladophora fracta*	8 g.L^{-1}	$Zn2^{+}$, $Cu2^{+}$, $Cd2^{+}$, $Hg2^{+}$	pH 5, 18 °C, 24 hours	0.228–2.338 mg.g^{-1}	Ji et al. 2012
2	*Chondrus crispus* (red algae)	0.5 g.L^{-1}	Cd (II), Ni (II), Zn (II), Cu (II), Pb (II)	pH 6, 120 min contact time	37.2–204 mg.g^{-1}	Romera et al. 2007
3	*Fucus vesiculosus* (brown algae)	0.5 g.L^{-1}	Cd (II), Cu (II), Pb (II)	pH 6, 23 °C, 150 min contact time	105.4–211.3 mg.g^{-1}	Mata et al. 2008
4	*Chlamydomonas reinhardtii* (green algae)	0.2 g.L^{-1}	$Pb2^{+}$ and $Cu2^{+}$	pH 6, ~30 °C, 60 min contact time	0.057 mg.g^{-1}	Flouty and Estephane, 2012
5	*Chlorella minutissima* (green algae)	4 g.L^{-1}	$Zn2^{+}$, $Mn2^{+}$, $Cd2^{+}$, and $Cu2^{+}$	Simulation of actual wastewater concentration pH 2–10, 10–37 °C, 180 min contact time	3.3–35.4 mg.g^{-1}	Yang et al. 2015

metal-binding protein with nine cysteine residues (Singh et al. 2012). Metallothionein shows a very high affinity for almost all heavy metals, and MT-2 particularly can form stable aggregates with metals (Sakulsak, 2012). An experiment on a commonly found cyanobacteria species, *Nostoc muscorum*, was conducted for lead and cadmium uptake. The algae adsorbed and accumulated the metal ions fairly well, with uptake of about 85%. It was also noted in the study that the number of ions entering the cell was considerably less than the amount attached to the surface (Dixit and Singh, 2014). Diatoms are undoubtedly one of the most versatile and dominant algal groups with incredible features, including permeable silica cell wall, exploitable nanoarchitecture, high-capacity vacuole, great potential for genetic engineering, and better photosynthetic activity than other algae (Marella et al. 2020). A detailed study was conducted on an immobilized mixture of *Nitzschia palea* and *Navicula incerta* species of diatoms under optimal conditions (Quraishi and Abbas, 2019). The study reported achieving the highest uptake for cadmium, nickel, and lead.

Chmielewska and Medved, 2001 studied the bioaccumulation of cadmium, lead, chromium, and nickel by green algae *Cladophora glomerata* from refinery sewage lagoon. The *Cladophora* sp. has shown higher bioaccumulation abilities for select metals. Pandi et al. 2007 measured the bioaccumulation potential of *Spirulina fusiformis* using chrome liquor from tannery industries. *S. fusiformis* effectively removed chromium up to 93–99% at 300 mg.L^{-1} concentration. Baumann et al. 2009 evaluated the cadmium and lead accumulation by seven algal species. They showed that macroalgae *P. palmata* and *Ulva intestinalis* accumulated the highest concentration of cadmium and lead, respectively. However, lead was accumulated by almost all tested species of macroalgae. Tonon et al. 2011 assessed the copper and cadmium bioaccumulation capacity of three *Gracilaria* sp., *G. tenuistipitata*, *G. domingensis*, and *G. birdiae*. The results of the study showed that *G. tenuistipitata* showed higher bioaccumulation of copper than of cadmium. Ajayan et al. 2011 evaluated the growth of *Oscillatoria quadripunctulata* and *Scenedesmus bijuga* in petroleum effluent contaminated with lead, cobalt, copper, and zinc. *O. quadripunctulata* showed removal efficiency of 100%, 100%, 50%, and 33.3% for lead, zinc, copper, and cobalt, respectively, whereas *S. bijuga* showed 25%, 50%, 50%, and 66% removal of these metals, respectively. Sbihi et al. 2012 studied the bioaccumulation capacity of diatom *Planothidium lanceolatum* for the removal of cadmium, copper, and zinc. The results showed that the algae were able to accumulate all these metals and more resistant to these metals than *Nitzschia* and *Chlorella*. Ajayan et al. 2015 studied the removal of lead, copper, chromium, and zinc from tannery wastewater using *Scenedesmus* sp. In the bioremoval study, 98% removal of lead, zinc, and copper and 96% removal of chromium was reported. Iye, 2015 studied the removal of chromium, cadmium, and arsenic from wastewater using *Botryococcus* sp. In the study, 94% chromium, 45% copper, and 9% arsenic removal were observed. Ballen-Segura et al. 2016 assessed the removal of chromium from tannery wastewater in Bogota, Colombia, using the microalga *Scenedesmus* sp. The algae were found to remove ~98% of chromium from the tannery wastewater. Subashini and Rajiv, 2018 investigated the removal of copper, zinc, iron, chromium, and nickel using *C. vulgaris* from tannery wastewater. The efficiency of heavy metal removal by algae was in the following order: copper (71%), zinc (50%), iron (45%), chromium (40%), and nickel (20%). Al Ketife et al. 2020 evaluated the mathematical bioaccumulation model for the removal of heavy metal from wastewater. They used Box–Behnken Methodology (BBM) and Response Surface in combination with best-fit simulation to study the bioaccumulation capacity. The results found that 99% of cadmium bioaccumulation was achieved with these models. Mubashar et al. 2020 evaluated the removal efficiency of lead, chromium, cadmium, and copper from textile industry wastewater using a consortium of microalgae and bacteria (*C. vulgaris* with *Enterobacter* sp MN17). The maximum removal was observed for cadmium (93%), followed by chromium and lead (79%), and copper (72%).

One of the limitations of using live algal cultures is that if the concentration of metal ions exceeds a certain limit, it experiences inhibitory effects. For instance, when green algae species *Scenedesmus obliquus* and *Scenedesmus quadricauda* were analyzed for the effect of heavy metals on their metabolism, values higher than 4.5 and 1.5 ppm, respectively, were considered detrimental for each strain (Omar, 2002). However, live algae cultures are widely used in all tanning, pharmaceutical, and petrochemical industry effluents. Species including *P. lanceolatum*, *Tetraselmis chuii*, and *Anabaena cylindrica* have shown incredibly high metal uptake when used in the live form (Kumar et al. 2015).

4.3.2 Phycoremediation by Dry/Dead Algal Biomass

Dead algal culture removes heavy metals only by biosorption on the cell wall with the aid of several functional groups (Figure 4.3). It does not indulge in accumulating the metal ions within the cell. Unlike live cultures, the metal ions do not damage and inhibit the cell activity, and thus, they can be used in wastewater with a very high concentration of heavy metals and other toxicants. Dead biomass is also easier to store and does not require any nutrient media (Wang et al. 2019). Since the removal takes place mainly via surface, the process is very rapid. For instance, when Herrero et al. 2006 used the dead culture of three strains of *Fucus* sp. for cadmium uptake, 90% of the metal was removed from the solution within 25 minutes of contact (Herrero et al. 2006). The metal ions are adsorbed by physio-chemical interactions and chelation via amino, carboxyl, sulfide, and imidazole groups on the surface of the cell (Salam, 2019; Figure 4.3). These functional groups exist because the biomolecules (proteins, lipids, and carbohydrates) are located in the mosaic cell wall (Rangabhashiyama and Balasubramanian, 2019). Some polyelectrolytes like teichoic acid linked to peptidoglycans have charged groups, which gives the cell wall an amphoteric property (Blumreisinger and Loos, 1983).

These functional groups provide affinity for heavy metal ions by making the cell wall protonated or deprotonated. Each type of algae has its specialized constituent responsible for the overall negative charge and biosorption. As most heavy metals have a positive charge, the attraction is high, and binding is convenient. In particular, brown algae have alginic acid and fucoidan, while red algae have galactan (Arad and Ontman, 2010; Mabeau et al. 1990). Green algae have a sulphated cell wall, which provides great affinity for the metal ion to attach (Synytsya et al. 2015). The variety of functional groups on the algal surface provide binding sites for a variety of metal ions. The bond formation occurs with a displacement of protons and sometimes by electrostatic interactions too. The type of bond formation depends on the composition of the cell wall. Generally, carbonyl groups will undergo covalent bonding while sulphate groups will indulge in ionic bonding (Crist et al. 1981). pH plays a critical role in determining the binding and uptake capacity because of its ability to alter the surface charge. Moreover, the fundamental processes followed by dead algae are complexation, precipitation, and ion exchange. An analysis of dead cyanobacteria

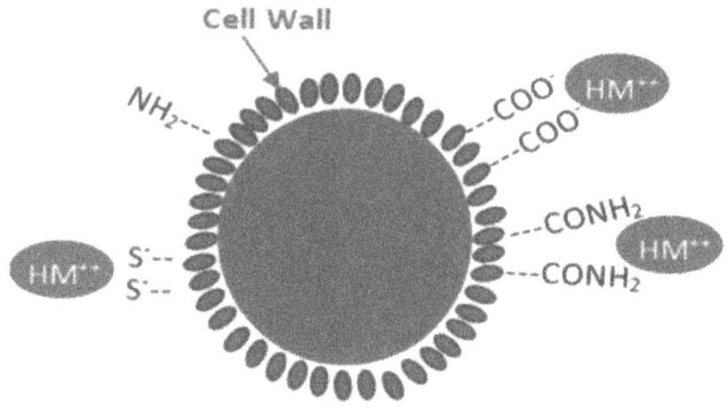

FIGURE 4.3 Biosorption by dry or dead algal cell.

TABLE 4.2
Heavy Metal Removal by Dry (Dead) Algal Biomass as Biosorbing Agents

Sl. No.	Dry Algal Biomass	Adsorbent Concentration	Heavy Metal	Process Description	Removal Efficiency	References
1	*Chlorella minutissima* (green algae)	4 g.L^{-1}	$Zn2^+$, $Mn2^+$, $Cd2^+$, and $Cu2^+$	Simulation of actual wastewater concentration pH 2–10, 10–37°C, 180 min contact time	16–303 mg.g^{-1}	Yang et al. 2015
2	*Scenedesmus abundans*	0.0344 g.L^{-1}	Cd	pH 7.8–8, 20°C, 36 hour contact time	280 mg.g^{-1}	Terry and Stone, 2002
3	*Chlorella vulgaris*	0.00018–0.018 g.L^{-1}	$Cd2^+$	$Cd2^+$ solutions prepared from 1 g.L^{-1} stock solutions of 3CdSO4 · 8H2O in distilled water 25°C, 5–105 min contact time	~10.9 mg.g^{-1}	Cheng et al. 2017
4	*Chlamydomonas reinhardtii* (green algae)	0.2 g.L^{-1}	$Pb2^+$ and $Cu2^+$	pH 6, ~30°C, 60 min contact time	0.109–0.286 mg.g^{-1}	Flouty and Estephane, 2012

concluded that the cells produced organic compounds, which enhanced the precipitation of heavy metal ions (Philippis et al. 2011). The various dry algal biomass with their heavy metal removal efficiency have been listed in Table 4.2.

El-Sikaily, 2007 evaluated the chromium removal capacity of dried biomass of green algae *U. lactuca* and found 92% of biosorption capacity. Kumar et al. 2009 evaluated five marine macroalgae for lead and cadmium biosorption capacity at a concentration of 20–80 mg.L^{-1}. *Chaetomorpha* sp. and *Valoniopsis pachynema* showed the highest adsorption for cadmium and lead, respectively. Da Kleinübing et al. 2011 used a dried marine algae *Sargassum filipendula* as biosorbent for the removal of copper and nickel metal ions, and this algae showed 1.324 and 1.070 mmol.g^{-1} of biosorption capacity for these metal ions, respectively. The biosorption was best fitted with the Langmuir isotherm model. Ibrahim, 2011 tested the dried red algal biomass *Gelidiella acerosa*, *Laurencia obtusa*, and *Hypnea* sp. for the removal of zinc, copper, nickel, and manganese from industrial effluent using a fixed-bed column. The biomass of *L. obtusa* showed the highest removal of metal (95%) followed by *G. acerosa* (85%), and *Hypnea* sp. (71%). Mirghaffari et al. 2015 examined the biosorption capacity of dry biomass of *S. quadricauda* for lead and cadmium from an aqueous solution. *S. quadricauda* showed maximum adsorption of 82% for lead, followed by 66% for cadmium. Latinwo et al. 2015 assessed the biosorption potential of green seaweed biomass for chromium, iron, silver, calcium, potassium, and magnesium from textile wastewater. The maximum biosorption was observed for silver (100%), followed by calcium (99.9%), iron (87.5%), chromium (86.8%), magnesium (59.7%), and potassium (57.2%). Palaniswamy and Veluchamy, 2017 measured the biosorbent capacity of dried biomass of microalgae *Spirulina platensis* for zinc from electroplating industry effluent. The maximum biosorption of zinc was found at pH 8, 35°C temperature, and 0.1 $g.L^{-1}$ of dry biomass with a contact time of 60 minutes. Mofeed, 2017 evaluated the biosorption capacity of dried biomass of two green algae, *Dunaliella salina* and *U. lactuca*, for lead, zinc, cadmium, iron, and copper from petrochemical industry effluent. *D. salina* was found to remove 22.34 $mg.g^{-1}$, 12.1 mg.g-1, and 6.32–6.73 $mg.g^{-1}$ lead, zinc, and cadmium, respectively. *U. lactuca* showed the maximum biosorption capacity of 15.86 $mg.g^{-1}$ for iron and 11.49 $mg.g^{-1}$ for copper from petrochemical industry effluent. El-Naggar et al. 2018 studied the biosorption ability of dried biomass of marine alga *Gelidium amansii* for the removal of lead (200 $mg.L^{-1}$) from aqueous solution. The maximum biosorption of lead was observed at 45°C temperature, pH 4.5, and the contact time of 60 minutes, with 1 $g.L^{-1}$ of *G. amansii* biomass dose.

4.4 CULTIVATION SYSTEMS FOR REMOVAL OF HEAVY METALS

4.4.1 Phycoremediation of Heavy Metals by Immobilized Algal Cultures

The small particle size and low density of algal biomass limit the selection of appropriate reactors for the treatment of effluent (Tsezos, 1986). Algal immobilization techniques are used by several researchers to treat wastewater effectively by fixing the cells on to inert media. In this method, cells are attached or entrapped in a polymer matrix (Lopez et al. 1997) and wastewater is passed through the matrix. The advantage of this method is that there is less clogging of the reactor even at the higher flow rate (Tsezos, 1986). The different techniques used in the immobilization of biomass include crosslinking of cells, covalent binding to carriers, encapsulation in a polymer gel, flocculation, adsorption on surfaces, and entrapment in a polymeric matrix. The supporting material used for immobilization is either natural such as alginate, agar, and carrageenan, or synthetic, including polyacrylamide, silica gel, and polyurethanes (Rangsayatorn et al. 2002). Natural supporting materials are better than synthetic materials due to the toxicity of synthetic materials to biomass. Alginate has been most widely utilized for immobilization of algal cells. However, the use of alginate for algae immobilization is not good for wastewater treatment due to the high content of sodium or orthophosphate and is associated with biomass leakage (Darnall et al. 1986; Robinson and Wilkinson, 1994).

The immobilization of algal biomass also has advantages and disadvantages. Generally, immobilization of algae tends to increase the metal accumulation by enhancing photosynthetic capability and increasing cell wall permeability (Brouers et al. 1989). It also protects living cells from metal toxicity (Bozeman et al. 1989). However, few researchers showed that free cells have higher metal absorption efficiency than immobilized cells (Wong and Pak, 1992). Free cells of *S. platensis* accumulated 98 $mg.g^{-1}$ while alginate- and silica-immobilized cells accumulated 71 $mg.g^{-1}$ and 37 $mg.g^{-1}$ of cadmium, respectively (Rangsayatorn et al. 2002). The reason behind decreased metal absorption capability is that the structure of the cell wall is changed during the immobilization process. In addition, immobilization of biomass is a costly process. Immobilized and granulated algae have great potential in removing toxic metal ions from wastewaters; hence, there is a need to develop an economically feasible method for producing and using them.

Akhtar et al. 2003 developed new biosorbents by immobilizing *Chlorella sorokiniana* within luffa sponge discs and showed that the microalgal-luffa sponge disc system has good biosorption properties for nickel. Microalgal-luffa sponge-immobilized discs removed nickel very rapidly, and achieved 97% equilibrium with in 5 minutes. Bayramoglu et al. 2006 evaluated the biosorption capacity of *C. reinhardtii* immobilized in alginate beads for three metals, cadmium, mercury, and lead, from an aqueous system. *C. reinhardtii* was able to maximally adsorb 253.6 $mg.g^{-1}$ of lead followed by 89.5 $mg.g^{-1}$ and 66.5 $mg.g^{-1}$ of mercury and cadmium, respectively. Kurniasih et al. 2013 assessed immobilized algal biomass for chromium removal and found a maximum of 11.494 $mg.g^{-1}$ of metal uptake. Rao et al. 2013 tested carrageenan and calcium alginate-immobilized biomass of *Chlorella pyrenoidosa* for chromium removal. The results of the study showed that calcium alginate-immobilized algae were found to remove 75 $mg.L^{-1}$ of chromium concentration at pH 3. Iye, 2015 studied the immobilized *Botryococcus* sp. for the removal of chromium, cadmium, and arsenic from wastewater. The study showed that the highest removal efficiency was observed for cadmium (76%) followed by arsenic (68%), and chromium (67%). Onyancha et al. 2008 evaluated the removal efficiency of chromium from tannery effluent using *Rhizoclonium hieroglyphicum* and *Spirogyra condensata*. The algae were immobilized in a glass column by Amberlite XAD-8. The pH played a major role during chromium removal; *R. hieroglyphicum* removed about 11.81 mg of chromium at pH 4, and *S. condensate* removed 14 mg of chromium at pH 5.

Immobilized algae with a cell density of 0.1 g dry weight/L were the most proficient for supplement and metal elimination in a pH scope of 6–8, and chitosan could be a promising algal support for wastewater detoxification. Wilkinson et al. 1990 studied the accumulation of mercury by free and immobilized algal systems of *Chlorella emersonii*. About 90% recovery was recorded after 12 days (initial concentration of mercury 1 $mg.L^{-1}$ with an initial cell loading of 106 cells/bead) and immobilized cell systems were found to accumulate more mercury than free cell systems. *Chlorella homosphaera* cells immobilized in alginate offer a good system to remove cadmium, zinc, and gold from wastewater (da Costa and Leite, 1991). Aggregation of cobalt, zinc, and manganese was recorded for *Chlorella salina* cells immobilized in alginate (Granham et al. 1992). There is greater potential of immobilized *Chlorella* and *Anabaena* in accumulating heavy metal ions, such as copper, nickel, and iron (Mallick and Rai, 1993; Mallick and Rai, 1992). The adsorption ability of immobilized *Chlorella* sp. was greater than the mobile biomass of *Chlorella* sp., reaching a removal capacity for Cu^{2+} ions of up to 24.91 $mg.g^{-1}$. Immobilized biosorbents have a better ability to adsorb Cu^{2+} than mobile ones. The primary concentration of 200 mg biosorbent has a higher adsorption ability than the initial concentration of 500 mg biosorbent (Wan Maznah et al. 2012). The adsorption capacity of immobilized biosorbent had the range of 0.69–7.43 $mg.g^{-1}$ on optimization of the initial concentration of Cu^{2+} waste, with variations of 2.99 $mg.L^{-1}$, 6.58 $mg.L^{-1}$, 9.92 $mg.L^{-1}$, 17.62 $mg.L^{-1}$, and 19.52 $mg.L^{-1}$. The most elevated adsorption capacity of 7.43 $mg.g^{-1}$ was reported for the immobilized mixed culture biosorbent with an initial Cu^{2+} concentration of 17.62 $mg.L^{-1}$. The optimum Cu^{2+} adsorption efficiency of 83.4% was observed at 30°C, pH 4, an initial biosorbent concentration of 200 mg, and 120 minutes contact time (Siwi et al. 2018). The algae *Chlorella* and *Spirulina* have the potential for the removal of copper and chromium ions from textile wastewater. The 1:3 ratio of beads to wastewater was the best ratio to remove metals from wastewater. *Spirulina* sp. showed the highest removal of both metals with 90% efficiency. *Chlorella* sp. removed chromium with 90% and copper with 89% efficiency (Hadiyanto et al. 2014). When the adsorption efficiency of algae is checked for live cells with dead cells conjugated on alginate matrix, the adsorption increased several folds for dead cells (Al-Rub et al. 2004). In spite of metabolic inactiveness, the dead algal cells did not show any decreased metal uptake. Dead *C. vulgaris* culture showed 95% of cadmium uptake in 5 minutes with better efficiency than live culture. (Cheng et al. 2017). Initial metal ion concentration, biomass dose, and competitive ions have the same effect on dead algae as well as on live ones. One additional advantage provided by dead culture is that the desorption is easy and the same batch of algae can be reused several times since no accumulation takes place inside the cell (Salam, 2019). Commercially, a few companies have started using immobilized algal biomass as biosorbents for metal removal from industrial wastewater. AlgaSorb by Biorecovery Inc. used silica gel-immobilized algal biomass to successfully remove mercury (Kanamarlapudi et al. 2018).

4.4.2 Phycoremediation of Heavy Metals by Batch Mode Cultivation of Algae

Algae are cultivated in batch mode for biomass generation and heavy metal removal. For biosorption mechanism of heavy metal removal in batch mode, four mathematical models are proposed, Freundlich model, Langmuir model, Brunauer-Emmett-Teller model, and an equilibrium model through thermodynamic approach (Gin et al. 2002; Ozer et al. 1994; Volesky and Holan, 1995). Metal removal in the batch system depends on the available binding sites of biosorbing agent. Several cations have varied binding affinity for available sites of biosorbing agents (Aksu and Kutsal, 1991; Guibal et al. 1995).

Napan, 2014 studied the bioaccumulation capacity of algae *S. obliquus* in batch mode for nickel, cadmium, mercury, lead, and copper of flue gas from coal-fired power plants. The results of the study showed that algae was able to accumulate all types of metal tested for the study. Sibi, 2016 evaluated the batch cultivation system using *C. vulgaris* for the biosorption of chromium from the electroplating industry waste. The results of the study showed that 81.3 $mg.L^{-1}$ chromium removal through biosorption reported an electrical conductivity of 2.893 mS/cm and it was best fitted with the Freundlich isotherm model. Isam et al. 2019 assessed the biosorption potential of red macroalgae *Gracilaria changii* grown under batch equilibrium method for lead and copper. Response surface methodology (RSM) with central composite design (CCD) was used to optimize the parameters of the study. The maximum removal of lead (96.3%) and copper (44.77%) were observed at 1 $g.L^{-1}$ biosorbent from 40 $mg.L^{-1}$ of initial concentrations, pH values of 4.5 and 5, and contact times of 115 minutes and 45 minutes, respectively. Wilan et al. 2020 studied the copper removal efficiency of mixed culture of *S. obliquus*, *Chlorococcum* sp., and *C. vulgaris* grown in batch mode. Biosorbent beads were made by mixing the dried biomass of these three algal cultures with sodium alginate polymer. The 98.56% removal of copper was reported at 25 $mg.L^{-1}$ of initial metal concentration at pH 5 and 25°C with a contact time of 180 minutes.

Sheoran and Bhandari, 2005 assessed the removal efficiency of heavy metals of acid mine drainage from mining industries using a consortium of cyanobacteria in a batch system. The removal was in the following decreasing order for copper (97%), followed by iron (95%), lead (88%), zinc (86%), cobalt (83%), nickel (62%), and manganese (45%). Jahan et al. 2014 investigated the removal efficacy of *Nostoc* in the batch system for iron, sodium, zinc, chromium, lead, and calcium from tannery wastewater. All the metals were effectively removed by *Nostoc*, and the highest removal efficiency was achieved for iron. Soeprobowati and Hariyati, 2017 evaluated the metal removal capacity of *Chaetoceros calcitrans*, *Arthrospira platensis*, and *C. pyrenoidosa* in batch mode for chromium, copper, lead, and cadmium from textile wastewater. The results of the study showed that *C. pyrenoidosa* was able to remove more than 80% of these metals as compared to *C. calcitrans* and *A. platensis*. Liehr et al. 1994 studied the metal sorption capacity of *C. glomerata* biofilm in a batch reactor developed using 0.4 μm polycarbonate membrane filters. They reported that algal biofilm was able to take up to 2200 μg of copper/g of its dry weight and anticipated that this was due to the higher pH inside the biofilm.

4.4.3 Phycoremediation of Heavy Metals by Continuous Mode Cultivation of Algae

Biomass separation from algal culture is relatively tough in a batch system for the removal of metals; hence, many researchers used a continuous system for algae cultivation. In continuous systems, algal cells usually found as immobilized resins include continuous stirred tank reactors, fluidized bed columns, packed bed columns, and moving bed columns. However, continuous removal of metal ions by adsorption column was considered the most efficient method at both pilots as well as at industrial scales (Ali, 2014). In a continuous system, several approaches were used to improve the absorption efficiency of heavy metals. The various parameters optimized during continuous algal cultivation include metal concentration, matrix type, initial pH, bed height, biomass loading, flow rates, and biosorbents particle size, etc. (Vijayaraghavan et al. 2005). For a continuous algal cultivation system, the mathematical model used for metal removal was the distributed parameter model in which the parameter selected was used for reactor designing and optimising operational conditions (Aksu, 1998).

In earlier studies, it was reported that the continuous cultivation system was better than a batch system for metal removal. This was because in the continuous cultivation system, the actively growing algal cells sequester metal ions more efficiently than a batch system where most of the algal cells were found in a stationary state (Canizares-Villanueva et al. 2000). However, for bivalent metal ions, the highest removal percentage was reported in the batch cultivation system. Using continuous algae cultivation system, around 99% of metal removal efficiency could be achieved. However, metal accumulation in the continuous cultivation system was less than that in the batch system (Aksu, 1998; Wilkinson et al. 1989). Lead and nickel were effectively removed by using fixed-bed continuous reactors (Olgun and Atar, 2012). Industrial wastewater was passed through the fixed-bed reactor column where metals were adsorbed by algal biomass. Barquilha et al. 2017 evaluated the nickel and copper removal efficiency of free and immobilized marine algae *Sargassum* sp. in batch and continuous cultivation systems. Calcium alginate matrix was used to immobilize the algae. The immobilized algae showed a higher affinity for copper ions than nickel ions, and fixed-bed columns displayed higher biosorption ability than the batch system. The results of the study were best fitted to the Langmuir model.

4.5 COMMERCIAL ADSORBENTS VERSUS ALGAL ADSORBENTS

Adsorption is one of the effluent treatment techniques that involves the accumulation of heavy metal on the adsorbent surface in the form of a film-like deposit (Lakherwal, 2014). It is one of the most preferred methods owing to the low costs, high reproducibility, efficiency, and simplicity of the process (Arora, 2019). There are different types of adsorbent materials, which perform well in depositing heavy metals from effluents. An adsorbent can either be commercial, viz. graphene, or non-commercial, e.g., algae-derived biosorbents (Agarwal and Singh, 2017). Commercially derived adsorbents include CNTs, zeolites, graphene, activated carbon, other polymers, and plant-based biosorbents (Duan et al. 2020).

Activated carbon is a frequently used adsorbent because of its tailor-made structure with high porosity and vast surface area covered with reactive functional groups (Shafeeyan et al. 2010). However, before use, it has to undergo several steps of heat treatment followed by surface modifications (Karnib et al. 2014). It is a well-known fact that industries prefer materials that are ready to use and do not require constant monitoring; therefore, a better alternative is needed. Most algal biosorbents require only $CaCl_2$ pre-treatment to enhance the uptake capacity (Mehta and Gaur, 2005). Even after a range of structural modifications, activated carbon showed 86% of lead uptake while *Sargassum hystrix* (brown algae) showed 95% uptake (Jalali et al. 2002; Karnib et al. 2014).

Carbon-based materials can prove to be excellent adsorbents owing to their extraordinary mechanical properties and configurable morphology (Duan et al. 2020). Graphene's unique two-dimensional structure acts as a non-absorbent, offering a low-cost and stable removal process (Kyzas et al. 2013). The only drawback lies in synthesizing a compatible oxide form of graphene for adsorption. Su et al. 2017 worked on making an iron oxide-graphene oxide nano-adsorbent for arsenic uptake.

The process was cumbersome as the pristine form did not give desired results and had to undergo the Hummers method before amorphous iron oxide-graphene oxide nanocomposite synthesis (Su et al. 2017). However, when *Mougeotia genuflexa* (algal species) was used to adsorb arsenic from the solution, no treatment was required. The biosorbent capacity turned out to be 96–98%, and the recovery of algal biomass was convenient (Özgür et al. 2011).

Another widely used carbon material for heavy metal remediation is CNTs. These are cylindrical layer(s) of graphite with unique chemical, physical, and electrical properties. Both multi-walled and single-walled CNTs are used for adsorption purposes after functionalization by a variety of methods, e.g., oxidation by acids (Fiyadh et al. 2019). The biggest disadvantage of CNT use is the interactive forces between the nanostructure, which resulted in its tendency to bundle together and aggregate (Porwal et al. 2017). Despite these limitations, CNTs are still widely used for heavy metal treatment; for instance, a study showed an uptake capacity of 116.3 $mg.g^{-1}$ for lead when CNTs and magnetic nanoparticles were used together (Wang et al. 2015). Algal biomass offers an advantage over CNTs as it does not aggregate. Also, the natural presence of amino, hydroxyl, and sulphate functional groups spares the time of functionalization unlike CNTs and other carbon-based nanostructures. The uptake capacity was also comparable to the commercial ones, viz. *Spirogyra* species gave efficient adsorption of 140.84 mg lead at only 0.5 $g.L^{-1}$ of algae dose (Gupta and Rastogi, 2008). One important issue about carbon-based adsorption materials that is often not addressed is their low recovery from water post adsorption. It cannot be overlooked that these materials are harmful to all life forms depending on water for their survival. Contrastingly, algae are easily desorbed and recovered simply by desorption eluents like ethylenediaminetetraacetic acid (EDTA) and NaOH (Brinza et al. 2007). Algae are not as harmful as carbon materials to the aquatic life forms and are a much greener approach.

Zeolites tend to have a three-dimensional, highly porous crystalline aluminosilicate network of chambers with a high heavy metal uptake capacity (Jiang et al. 2018). Zeolites occur naturally, but most industries prefer manufacturing them according to the task. To produce good quality zeolites, a source of alumina and silica is required along with a template. The complete process requires several cycles of heating and cooling with constant monitoring; this is time-consuming and detrimental for the industry. Zeolites tend to change into amorphous compounds by dealuminating if not checked timely (Pfenninger, 1999). When zeolite was used to remove mercury from water, it was successful in removing it according to the pollution directive guidelines, although some optimization was needed because of the variable levels of mercury (Chojnacki et al. 2004). When *Euglena gracilis* was tested for mercury uptake in both light and dark conditions, significant adsorption was noted by volatilization (Devars et al. 2000). Algae can be subjected to a vast range of conditions (pH and temperature) and can be genetically modified to perform efficient adsorption in the presence of other elements that hinder reactions in the case of commercial adsorbents (Schiewer and Volesky, 2000). The various commercial and algal biosorbents with their heavy metal removal efficiency have been listed in Table 4.3.

Today, many plant-based materials like rice husk, coconut waste, modified eggshells, etc. are being used for the biosorption of heavy metals (Agarwal and Singh, 2017).

TABLE 4.3
Heavy Metal Removal by Commercial and Algal Biosorbents

Sr. No.	Type of Adsorbents	Adsorbent Concentration	Heavy Metal	Process Description	Removal Efficiency	References
			Commercial Adsorbent			
1	Zeolite (clinoptilolite)	0.35 g.L^{-1}	Hg^{2+}	Industrial effluents from copper smelter and refinery (S, pH 6.28) (M, pH 8.48), 21°C), 15 min contact time	0.021 mg.g^{-1}	Chojnacki et al. 2004
2	Zeolite (synthetic from coal fly ash)	4 g.L^{-1} (desulfurization wastewater) and 15 g.L^{-1} (metal plating wastewater)	Pb^{2+}, Cd^{2+}, Cu^{2+}, Ni^{2+}, and Mn^{2+}	Desulfurization wastewater from wet limestone-gypsum system and metal plating wastewater from local metal plating factory, pH 2.7–4.2, 30°C, 240 min contact time	30.8–65.7 mg.g^{-1}	He et al. 2016
3	Surface modified activated carbon (SiO2/AC)	0.02 g	Ni^{2+}, Cd^{2+}, Zn^{2+}, Pb^{2+}, and Cr^{2+}	Batch mode adsorption pH 2, 24 hours contact time	60.9 mg.g^{-1}	Karnib et al. 2014
4	Carbon nanotube	0.01 g	Pb^{2+}	Industrial effluents pH 5, 25°C, 80 min contact time	102.04 mg.g^{-1}	Kabbashi et al. 2009
5	Single walled CNT	0.02 g	Pb^{2+}, Cd^{2+}, and Cu^{2+}	pH 5, 24°C, 120 min contact time	33.5 mg.g^{-1}	Moradi, 2011
6	Graphene manganese ferrite (MnFe2O4-G) composite	0.025 g.L^{-1}	Pb^{2+} and Cd^{2+}	Batch mode adsorption, pH 5–7, 37°C, 120–180 min contact time	76–100 mg.g^{-1}	Chella et al. 2015

Sr. No.	Type of Adsorbents	Adsorbent Concentration	Heavy Metal	Process Description	Removal Efficiency	References
7	Rice husk carbon	1–10 $g.L^{-1}$	Cr (CrO4 2- and Cr2O7 2- form)	pH 2, 21°C, 5–180 min contact time	47.61 $mg.g^{-1}$	Khan et al. 2016
8	Green bean husk (*Vigna radiata*)	0.25 and 3.0 $g.L^{-1}$	$Sb(3^+)$	Experimental Sb conditions by dissolving potassium antimony (III) tartrate trihydrate in deionized water, pH 4, 24.8°C, ~120 min contact time	20.14 $mg.g^{-1}$	Iqbal et al. 2013
			Algal adsorbents			
9	*Fucus spiralis* (Brown algae)	0.5 $g.L^{-1}$	Cd, Ni, Zn, Cu, Pb	pH 3–6	204–50 $mg.g^{-1}$	Romera et al. 2007
10	*Ascophyllum nodosum* (Brown algae)	0.5 $g.L^{-1}$	Cd, Ni, Zn, Cu, Pb	pH 3–6	178.6–42 $mg.g^{-1}$	Romera et al. 2007
11	*Sargassum* sp.	1 $g.L^{-1}$	Cu	pH 5, 60 min contact time	62.91 $mg.g^{-1}$	Sheng et al. 2008
12	*Spirulina platensis*	1–4 $g.L^{-1}$	Cu	pH 6.7, 20°C (room temperature), 60–180 min contact time	92.6–96.8 $mg.g^{-1}$	Solisio et al. 2006
13	*Sargassum Muticum*	5 $g.L^{-1}$	Cu	pH 4.5, 240 min contact time	71 $mg.g^{-1}$	Herrero et al. 2011

Various parts of a plant like leaves, roots, peels have been experimented and have given eminent results. For instance, when orange peel was functionalized by copolymerization, it gave an uptake capacity of 293.3 $mg.g^{-1}$ for cadmium (Feng et al. 2011; Jain et al. 2016). At optimum pH, when sunflower (*Helianthus annuus*) residues were used to treat water polluted by lead and cadmium, adsorption capacities of 182 $mg.g^{-1}$ and 70 $mg.g^{-1}$, respectively, were observed (Jalali and Aboulghazi, 2013). Although using plant residues for remediation is an economic and green approach, it comes with its disadvantages. When left untreated, the plant wastes tend to increase the biological and chemical oxygen demand of the water. It can also be responsible for increasing the total organic carbon levels, which can harm aquatic animals (Tripathi and Ranjan, 2015).

Algal biomass is abundant in nature, and since their natural habitat is water, they provide better results with no or minimum modifications, unlike commercial adsorbents. Macroalgae or seaweeds have given consistently higher uptake than other adsorbents without the need for any immobilization because of their considerable size (Schiewer and Volesky, 2000). They have the potential to offer useful by-products like lipids and proteins along with efficient bioremediation (Ansari et al. 2019). They provide opportunities for genetic engineering and making an ideal strain with high stability in different conditions. These complex machines of nature work with utmost cleanliness and do not leave behind any sludge or compound that can be harmful to any life form. Some microalgae also bioconcentrate the toxicants and subsequently convert them into a compound, which is not detrimental for life forms (Phang et al. 2015). When a comparative study between different adsorbents was conducted by Richards et al. 2019, it was found that *Fucus vesiculosus* (seaweed algae) was the most effective choice. It did not show signs of saturation even at the maximum concentration of the metal and did not require any modification to enhance its activity.

4.6 RECYCLING AND REGENERATION OF ALGAE

To minimize the cost and for recovery of metals, desorption, recycling, and regeneration of biosorbents are crucial steps after the removal of heavy metals. Algae can be regenerated by a proper desorption method and regenerated biomass may be recycled for reuse (Kanamarlapudi et al. 2018). The main objective of desorption is to retain the adsorption capacity of algal biomass and to recover the metals. In the desorption process, the metal was recovered in concentrated form and algal biomass was regenerated in its original state. The ideal eluent for desorption should be low-cost, non-damaging to the algal biomass, and environment-friendly. Several types of desorption eluents are used, such as dilute mineral acids (H_2SO_4, HCl, and HNO_3), organic acids (acetic, citric, and lactic acids), bases such as NaOH, $NaHCO_3$, Na_2CO_3, and KOH, complexing agents (thiosulphate, NaCl, EDTA, etc.) for the recovery of metals simultaneously with the regeneration of biomass (Lata et al. 2015). The desorption percent is determined by the following equation:

$$\text{Desorption percentage } (\%) = (\text{Amount of species desorbed/Amount of species loaded}) \times 100\% \quad (6.1)$$

The maximum desorption percentage implies higher regeneration of algal biomass. The 98% desorption of lead, cadmium, and mercury from *C. reinhardtii* was achieved by 0.1 M HCl (Tuzun et al. 2005). The same concentration of HCl was able to desorb around 90% of lead from *Spirogyra neglecta*, *Nostoc* sp., and *Oedogonium* sp. biomass (Gupta and Rastogi, 2008; Singh et al. 2012). Regenerated microalgal-luffa sponge-immobilized disc retained 92.9% of the initial binding capacity for nickel up to five cycles of reuse in a fixed bed column reactor (Akhtar et al. 2003).

A survey of literature reveals that HCl is the most commonly used desorbing agent; with it, highest metal recovery was observed, but at the same time the biosorption efficiency of biomass was reduced. Gupta and Rastogi, 2008a studied the desorption of metal ion cadmium using 0.1 M HCl from *Oedogonium* sp. In the study, 84.8% cadmium recovery was achieved. Recycling and regeneration of algal biomass were done for up to five cycles. Gupta et al. 2010 evaluated nickel recovery using 0.1 M NaOH from *Oedogonium* sp. biomass. They have observed that ~70% of nickel recovery and biomass regeneration up to four cycles showed the same efficiency. Gupta and Rastogi, 2009 examined 0.1 M NaOH as desorbing agent for chromium recovery using *Oedogonium* sp. biomass. The results of the study revealed that ~75% of chromium was recovered, and it was repeated up to four cycles. Gupta and Rastogi, 2008b measured the recovery of lead using 0.1 $mol.L^{-1}$ HCL from *Nostoc* sp. and *Oedogonium* sp. biomass. The study revealed that ~90% of lead recovery was observed in five cycles. Gupta and Rastogi, 2008c evaluated the chromium recovery using *N. muscorum* biomass using 0.1 M EDTA as a desorbing agent. The study showed ~90% recovery of chromium after five times recycling of *N. muscorum* biomass.

Martins et al. 2006 evaluated disodium EDTA as a desorbing agent for lead recovery from *Sargassum* sp. biomass and observed 95% recovery of lead. The recovery of cadmium was studied by Deng et al. (2008) with various desorbing agents such as HNO_3, H_2O_2, EDTA, and $Ca(NO_3)_2$ using *Cladophora fascicularis* biomass. In the study, EDTA was observed as the most preferable desorbent with 83% of cadmium recovery. Jalali et al. 2002 performed 10 adsorption-desorption cycles and obtained 98% of lead recovery from *Sargassum* sp. without damaging biomass during the regeneration process. Lewis, 1996 used a column made of non-living immobilized biomass of microalgae for testing the quantitative regeneration efficiency with 10 ppm copper heavy metal solution. Upon elution with 0.012 N HCl for 23 cycles, >97% of biomass-binding efficiency for copper was achieved. Mahan and Holcombe, 1992 tested the silica gel-immobilized dead biomass of microalgae *Stichococcus bacillaris* for quantitative elution of lead with 25, 50, and 100 ppm solution. The results revealed that the biomass was found effective for lead elution with 0.012 N HCl, and its elution capacity declined by 15% after 20 cycles. The recovery of metals using desorption medium depends on the metals to be recovered and was independent of the biosorbents used. The limitations of reusing algae biomass for metal recovery are its additional cost and loss of biomass after regeneration.

4.7 CONCLUSION

Heavy metal pollution is a serious concern at the global level. Due to strict environmental regulations, wastewater contaminated with heavy metals must be treated before released into the environment. Since recycling of treated water is the focus in the coming era, economic and environment-friendly approaches should be developed. Physicochemical methods of heavy metal removal are found costly, create secondary pollution and most of them are non-recyclable. As compared to other removal strategies, algae are found to be efficient, cheaper, and recyclable. Algal bioremediation mechanisms such as biosorption and bioaccumulation were found effective in the removal of any kind of heavy metal from various industrial wastewater. Dried biomass of algae was found more effective than live algal cultures, and the former process was not affected by the physical conditions and nutrients of the media. The required density of algal biomass can be maintained with dry biomass of algae as compared to live cells for the highest removal. Metal removal by live algal culture was found slow; however, metal removal through biosorption and bioaccumulation is evident. The algal immobilization technique can increase the removal rate of heavy metals by protecting algal cells from fluctuating microenvironment. Batch and continuous systems of algal cultivation look promising for heavy metal remediation, and these systems need to be chosen based on the type of feed and metal concentration. Bioremoval of various heavy metals by algae was found comparable to commercial adsorbents. HCl is the widely used chemical agent for desorption and recovery of heavy metals from algal biomass. The other desorption agents such as EDTA and NaOH were also found effective and comparable to HCL, but the selection of the best desorbing agent should be based on the non-toxicity to algal cells and cost of the agent. The overall study indicates that the phycoremediation in batch or continuous cultivation system with recycling strategies of biomass would be ideal for economic and effective remediation of heavy metals.

ABBREVIATIONS

RO, reverse osmosis; ED, electrodialysis; ppm, parts per million; MgO, magnesia; NRAMP, natural resistance associated macrophage protein; FTR, Fe transporter; MT-2, Class-II metallothionein; BBM, Box–Behnken Methodology; RSM, response surface methodology; CCD, central composite design

CONFLICT OF INTEREST STATEMENT

The authors declare no conflict of interest.

REFERENCES

Abu-Eishah, S.I., (2008). Removal of Zn, Cd, and Pb ions from water by Sarooj clay. Appl. Clay Sci., 42 (1–2):201–205.

Agrawal, K., Bhatt, A., Bhardwaj, N., Kumar, B., Verma, P. (2020b). Algal Biomass: Potential Renewable Feedstock for Biofuels Production–Part I. In: Biofuel Production Technologies: Critical Analysis for Sustainability (pp. 203–237). Springer, Singapore.

Agrawal, K., Bhatt, A., Bhardwaj, N., Kumar, B., Verma, P. (2020c). Integrated Approach for the Treatment of Industrial Effluent by Physico-chemical and Microbiological Process for Sustainable Environment. In: Combined Application of Physico-Chemical & Microbiological Processes for Industrial Effluent Treatment Plant (pp. 119–143). Springer, Singapore.

Agrawal, K., Bhatt, A., Chaturvedi, V., Verma, P. (2020a). Bioremediation: An Effective Technology Toward a Sustainable Environment Via the Remediation of Emerging Environmental Pollutants. In: Emerging Technologies in Environmental Bioremediation (pp. 165–196). Elsevier.

Agrawal, K., Verma, P. (2020). Degradation and Detoxification of Waste Via Bioremediation: A Step Toward Sustainable Environment. In: Emerging Technologies in Environmental Bioremediation (pp. 67–83). Elsevier.

Agarwal, M., Chaudhry, K., Kumara, P., (2015). Heavy metal sources impacts & removal technologies. Int. J. Eng. Res. Tech., 3 (3):1–7.

Agarwal, R., Singh, K., (2017). Heavy metal removal from wastewater using various adsorbents: a review. J. Water Reuse Desal., 7(4):387–419.

Ahluwalia, S.S., Goyal, D., (2006). Microbial and plant derived biomass for removal of heavy metals from wastewater. Bioresour. Technol., 98 (12):2243–2257.

Ahmady-Asbchin, S., Andre's, Y., Ge´rente, C., Le Cloirec, P., (2008). Biosorption of Cu (II) from aqueous solution by *Fucus serratus*: surface characterization and sorption mechanisms. Bioresour. Technol., 99 (2008):6150–6155.

Algureiri, A.H., Abdulmajeed, Y.R., (2016). Removal of heavy metals from industrial wastewater by using RO membrane. IJCPE., 17 (4):125–136.

Al Ketife, A.M.D., Al Momani, F., Judd, S., (2020). A bioassimilation and bioaccumulation model for the removal of heavy metals from wastewater using algae: New strategy. Process Saf. Environ. Prot., 144:52–64.

Ajayan, K.V., Muthu, S., Unnikannan, P., Palliyath, S., (2015). Phycoremediation of tannery wastewater using microalgae *Scenedesmus* species. Int. J. Phytoremediation., 17 (10).

Ajayan, K.V., Selvaraju, M., Thirugnanamorrthy, K., (2011). Growth and heavy metals accumulation potential of microalgae grown in sewage wastewater petrochemical effluents. Pak. J. Biol. Sci., 14 (16):805–811.

Ajmal, M., Rao, R.A.K., Anwar, S., Ahmad, J., Ahmad, R., (2003). Adsorption studies on rice husk: removal and recovery of Cd (II) from wastewater. Bioresour. Technol., 86 (2003):147–149.

Akhtar, N., Iqbal, J., Iqbal, M. (2003). Microalgal-luffa sponge immobilized disc: A new efficient biosorbent for the removal of Ni (II) from aqueous solution. Lett. Appl. Microbiol., 37:149–153.

Aksu, Z. 1998. Biosorption of heavy metals by microalgae in batch and continuous systems. In: Wastewater Treatment with Algae (pp. 37– 52). Wong, Y.-S., and Tam, N. F. Y., Eds., Springer-Verlag and Landes Bioscience.

Cetinkaya Donmez, G., Aksu, Z., Ozturk, A., Kutsal, T., (1999). A comparative study on heavy metal biosorption of some algae. Process. Biochem., 34(9): 885–892.

Aksu, Z, Kutsal, T., (1991). A bioseparation process for removing lead (II) ions from wastewater by using *Chlorella vulgaris*. J. Chem. Tech. Biotech., 52:109–118.

Alam, M.D., Wan, C., Zhao, X., Chen, L., Chang, J., Bai, F., (2015). Enhanced removal of $Zn2^+$ or $Cd2^+$ by the flocculating *Chlorella vulgaris* JSC-7. J. Hazard. Mater., 289, 38–45.

Ali, I., (2014). Water treatment by adsorption columns: evaluation at ground level. Sep. Purif. Rev., 43 (3):175–205.

Alinnor, J., (2007). Adsorption of heavy metal ions from aqueous solution by fly ash. Fuel., 86 (2007):853–857.

Al-Rub, F., El-Naas, M., Benyahia, F., Ashour, I., (2004). Biosorption of nickel on blank alginate beads, free and immobilized algal cells. Proc. Biochem., 39 (11):1767–1773.

Amana, T., Kazi, A.A., Sabri, M.U., Bano, Q., (2008). Potato peels as solid waste for the removal of heavy metal copper (II) from waste water/industrial effluent. Colloids Surf. B: Biointerfaces., 63 (2008):16–121.

Anastopoulos, I., Kyzas, G., (2015). Progress in batch biosorption of heavy metals onto algae. J. Mol. Liq., 209: 77–86.

Ansari, F, Ravindran, B, Gupta, S, Nasr, M, Rawat, I, Bux, F., (2019). Techno-economic estimation of wastewater phycoremediation and environmental benefits using *Scenedesmus obliquus* microalgae. J. Environ. Manage., 240:293–302.

Arad, S., Ontman, O., (2010). Red microalgal cell-wall polysaccharides: biotechnological aspects. Curr. Opin. Biotechnol., 21 (3):358–364.

Arora, R., (2019). Adsorption of heavy metals–A review. Mater. Today Proc., 18 (7): 4745–4750.

Ballen-Segura, M., Hernández Rodríguez, L., Parra Ospina, D., Bolaños, A.V., Pérez, K., (2016). Using Scenedesmus sp. for the phycoremediation of tannery wastewater. TECCIENCIA., 12 (21):69–75.

Baumann, H.A., Morrison, L., Stengel, D.B., (2009). Metal accumulation and toxicity measured by PAM - chlorophyll fluorescence in seven species of marine macroalgae. Ecotox. Environ. Saf., 72:1063–1075.

Bayramoglu, G., Tuzun, I., Celik, G., Yilmaz, M., Arica, M.Y., (2006). Biosorption of mercury (II), cadmium (II) and lead (II) ions from aqueous system by microalgae *Chlamydomonas reinhardtii* immobilized in alginate beads. Int. J. Miner. Process,. 81:35–43.

Blumreisinger, M., Loos, D., (1983). Cell wall composition of chlorococcal algae. Phytochemistry., 22 (7):1603–1604.

Bozeman, J., Koopman, B., Bitton, G., (1989). Toxicity testing using immobilized algae. Aquat. Toxicol., 14: 345–352.

Brboot, M.M., Abi, B.A., Najah, M., Al-Shuwaik, N., (2011). Removal of heavy metals using chemicals precipitation. Eng. Tech. J., 29 (3).

Brinza, L., Dring, M.J., Gavrilescu, M., (2007). Marine micro and macro algal species as biosorbents for heavy metals. Environ. Eng. Manag. J., 6 (3):237–251.

Brouers, M., de Jong, H., Shi, D.J., and Hall, D.O., (1989). Immobilized cells: An appraisal of the methods and applications of cell immobilization techniques. In: Algae and Cyanobacterial Biotechnology (pp. 272–290). Cresswell, R. C., Rees, T. A. V., and Shah, N., Eds., Longman Scientific and Technical Publishers.

Barquilha, C.E.R., Cossich, E.S., Tavares, C.R.G., Silva, E.A., (2017). Biosorption of nickel(II) and copper(II) ions in batch and fixed-bed columns by free and immobilized marine algae *Sargassum* sp. J. Clean. Prod., 150: 58–64.

Canizares-Villanueva, R.O., Martinez-Jeronimo, F., Espinoza-Chavez, F., (2000). Acute toxicity to *Daphnia magna* of effluents containing Cd, Zn, and a mixture Cd–Zn, after metal removal by *Chlorella vulgaris*. Environ. Toxicol., 15, 160–164.

Chaturvedi, V., Bhange, K., Bhatt, R., Verma, P., (2013a). Biodetoxification of high amounts of malachite green by a multifunctional strain of *Pseudomonas mendocina* and its ability to metabolize dye adsorbed chicken feathers. J. Environ. Chem. Eng., 1(4), 1205–1213.

Chaturvedi, V., Chandravanshi, M., Rahangdale, M., Verma, P., (2013b). An integrated approach of using polystyrene foam as an attachment system for growth of mixed culture of cyanobacteria with concomitant treatment of copper mine waste water. J. Waste Manag., Article ID 282798:1–7.

Chekroun, K., Baghour, M., (2013). The role of algae in phytoremediation of heavy metals: A review. J. Mater. Environ. Sci., 4 (6):873–880.

Chella, S., Kollu, P., Komarala, E.V.P.R., Doshi, S., Saranya, M., Felix, S., Ramachandran, R., Saravanan, P., Koneru, V.L., Venugopal, V., Jeong, S.K., Grace, A.N., (2015) Solvothermal synthesis of MnFe2O4-graphene composite—investigation of its adsorption and antimicrobial properties. Appl. Surf. Sci., 327:27–36.

Chen, G., (2004). Electrochemical technologies in wastewater treatment. Sep. Purif. Technol., 38 (1):11–41.

Cheng, J., Yin, W., Chang, Z., Lundholm, N., Jiang, Z., (2017). Biosorption capacity and kinetics of cadmium (II) on live and dead *Chlorella vulgaris*. J. Appl. Phycol., 29:211–221.

Chmielewska, E., Medved, J., (2001). Bioaccumulation of Heavy Metals by Green Algae *Cladophora glomerata* in a Refinery Sewage Lagoon. Croatica Chemica Acta., 74 (1):135–145.

Chojnacki, A., Chojnacka, K., Hoffmanna, J., Górecki, H., (2004). The application of natural zeolites for mercury removal: from laboratory tests to industrial scale, Miner. Eng., 17 (7–8):933–937.

Crist, R., Oberholser, K., Shank, N., Nguyen, M., (1981). Nature of bonding between metallic ions and algal cell walls. Environ. Sci. Technol., 15 (10): 1212–1217.

da Costa, A.C.A., Leite, S.F.G., (1991). Metals biosorption by sodium alginate immobilized *Chlorella homosphaera* cells. Biotechnol. Lett., 13:559–562.

Da Kleinübing, S., Da Silva, E., Silva, M., Guibal, E., (2011). Equilibrium of Cu (II) and Ni (II) biosorption by marine alga *Sargassum filipendula* in a dynamic system: Competitiveness and selectivity. Bioresour. Technol., 102:4610–4617.

Dahman, Y., (2017). Nanopolymers**By Yaser Dahman, Kevin Deonanan, Timothy Dontsos, and Andrew Iammatteo. In: Nanotechnology and Functional Materials for Engineers (pp. 121–144).

Darnall, D.W., Greene, B., Henzi, M.T., Hosea, J.M., McPherson, R.A., Sneddon, J., Alexander, M.D., (1986). Selective recovery of gold and other metal ions from an algal biomass. Environ. Sci. Technol., 20:206–208.

Deng, L., Zhu, X., Su, Y., Su, H., Wang, X., (2008). Biosorption and desorption of Cd2? from wastewater by dehydrated shreds of *Cladophora fascicularis*. Chin. J. Oceanol. Limnol., 26:45–49.

Devars, S., Avilés, C., Cervantes, C., Moreno-Sanchez, R., (2000). Mercury uptake and removal by *Euglena gracilis*. Arch. Microbiol., 174:175–180.

Dixit, S., Singh, D., (2014). An evaluation of phycoremediation potential of cyanobacterium Nostoc muscorum: characterization of heavy metal removal efficiency. J. Appl. Phycol., 26:1331–1342.

Du, R., Liu, L., Wang, A., Wang, Y., (2013). Effects of temperature, algae biomass and ambient nutrient on the absorption of dissolved nitrogen and phosphate by Rhodophyte *Gracilaria asiatica*. Chin. J. Ocean. Limnol., 31:353–365.

Duan, C., Ma, T., Wang, J., Zhou, Y., (2020). Removal of heavy metals from aqueous solution using carbon-based adsorbents: A review. J. Water Process Eng., 37:101339.

El-Naggar, N. E.-A., Hamouda, R.A., Mousa, I.E., Abdel-Hamid, M.S., Rabei, N.H., (2018). Biosorption optimization, characterization, immobilization and application of *Gelidium amansii* biomass for complete Pb2+ removal from aqueous solutions. Sci. Rep., 8:13456.

El-Sikaily, A., El Nemr, A.K., Abdelwehab, O., (2007). Removal of toxic chromium from wastewater using green alga *Ulva lactuca* and its activated carbon. J. Hazard. Mater., 148 (1–2):216–228.

Elumalai, S., Saravanan, G.K., Ramganesh, S., Sakthivel, R., Prakasam, V., (2013). Phycoremediation of textile dye industrial effluent from Tirupur district, Tamil Nadu, India. IJSID., 3 (1):31–37.

Feng, N., Guo, X., Liang, S., Zhu, Y., Liu, J, (2011). Biosorption of heavy metals from aqueous solutions by chemically modified orange peel. J. Hazard. Mater., 185 (1):49–54.

Fiyadh, S.S., AlSaadi, M.A., Jaafar, W.Z., AlOmar, M.K., Fayaed, S.S., Mohd, N.S., Hin, L.S., El-Shafie, A., (2019). Review on heavy metal adsorption processes by carbon nanotubes. J. Clean. Prod., 230: 783–793.

Flouty, R., Estephane, G., (2012). Bioaccumulation and biosorption of copper and lead by a unicellular algae *Chlamydomonas reinhardtii* in single and binary metal systems: A comparative study. J. Environ. Manag., 111:106–114.

Ghafoor, A., Raur, A., Arif, M., Muzaffar, W., (1994). Chemical composition of effluents from different industries of the Faisalabad city. Pak. J. Agri. Sci., 31 (4).

Gin, K.Y.-H., Tang, Y.-Z., Aziz, M.A., (2002). Derivation and application of a new model for heavy metal biosorption by algae. Water Res., 36:1313–1323.

Goswami, R.K., Mehariya, S., Verma, P., Lavecchia, R., Zuorro, A., (2021). Microalgae-based biorefineries for sustainable resource recovery from wastewater. J. Water Process Eng., 40:101747.

Gottipati, R., Mishra, S., (2012). Application of response surface methodology for optimization of Cr (III) and Cr (VI) adsorption on commercial activated carbons. Res. J. Chem. Sci., 2 (2):40–48.

Granham, G.W., Codd G.A., Gadd G.M. (1992). Accumulation of cobalt, zinc and manganese by the esturine green microalga *Chlorella salina* immobilized in alginate microbeads. Environ. Sci. Tech., 26:1764–1770.

Guibal, E., Lorenzelli, R., Vincent, T., Cloirec, P., (1995). Application of silica gel to metal ion sorption: static and dynamic removal of uranyl ions. Environ. Tech., 16:101–114.

Gupta, S.K., Chabukdhara, M., Singh, J., Bux, F. (2015a). Evaluation and potential health hazard of selected metals in water, sediments, and fish from the Gomti River. Hum. Ecol. Risk Assess., 21:227–240.

Gupta, S.K., Shriwastav, A., Kumari, S., Ansari, F.A., Malik, A., Bux, F. (2015b). Phycoremediation of emerging contaminants. In: Algae: Role in Sustainable Energy Production and Pollution Remediation (pp. 129–146). Singh, B., Bauddh, K., Bux, F., Eds., Springer International Publishing AG Cham.

Gupta, V., Rastogi, A., (2008). Biosorption of lead from aqueous solutions by green algae *Spirogyra* species: Kinetics and equilibrium studies. J. Hazard. Mater., 152 407–414.

Gupta, V.K., Rastogi, A. (2008a). Equilibrium and kinetic modeling of cadmium (II) biosorption by nonliving algal biomass *Oedogonium* sp. from aqueous phase. J. Hazard Mater., 153:759–766.

Gupta, V.K., Rastogi, A., (2008b). Biosorption of lead from aqueous solutions by nonliving algal biomass *Oedogonium* sp. and *Nostoc* sp. - a comparative study. Coll. Surfaces B., 64:170–178.

Gupta, V.K., Rastogi, A. (2008c). Sorption and desorption studies of chromium (VI) from nonviable cyanobacterium *Nostoc muscorum* biomass. J. Hazard. Mater., 154:347–354.

Gupta, V.K., Rastogi, A. (2009). Biosorption of hexavalent chromium by raw and acid-treated green alga *Oedogonium hatei* from aqueous solutions. J. Hazard. Mater., 163: 396–402.

Gupta, V.K., Rastogi, A., Nayak, A., (2010). Biosorption of nickel onto treated alga (*Oedogonium hatei*): Application of isotherm and kinetic models. J. Colloid Interface Sci., 342:533–539.

Hadiyanto, H., Pradana, A.B., Buchori, L., Sri Budiyati, C., (2014). Biosorption of heavy metal Cu2+ and Cr2+ in textile wastewater by using immobilized algae. Res. J. Appl. Sci. Eng. Technol., 7(17): 3539–3543.

He, K., Chen, Y., Tang, Z., Hu, Y., (2016). Removal of heavy metal ions from aqueous solution by zeolite synthesized from fly ash. Environ. Sci. Poll. Res. Int., 23 (3):2778–2788.

He, M., Wang, Z., Tang, H., (1998). The chemical, toxicological and ecological studies in assessing the heavy metal pollution in Le an river, China. Water Res., 32:510–518.

Henriques, B., Teixeira, A., Figueira, P., Reis, A.T., Almeida, J., Vale, C., Pereira, E., (2019). Simultaneous removal of trace elements from contaminated waters by living Ulva lactuca. Sci. Total Environ., 652:880–888.

Herrero, R., Lodeiro, C.P., Rey-Castro, C., Sastre de Vicente M.E., (2006). Interactions of cadmium(II) and protons with dead biomass of marine algae *Fucus* sp. Mar. Chem., 99 (1–4):106–116.

Herrero, R., Lodeiro, P., García-Casal, L.J., Vilariño, T., Rey-Castro, C., David, C, Rodríguez, P., (2011). Full description of copper uptake by algal biomass combining an equilibrium NICA model with a kinetic intraparticle diffusion driving force approach. Biores. Technol., 102 (3):2990–2997.

Ibrahim, W.M., (2011). Biosorption of heavy metal ions from aqueous solution by red macroalgae. J. Hazard. Mater., 192:1827–1818.

Ince, M., Ince, O.K., (2019). Heavy Metal Removal Techniques Using Response Surface Methodology: Water/Wastewater Treatment. In: Biochemical Toxicology - Heavy Metals and Nanomaterials. IntechOpen.

Iqbal, M., Saeed, A., Edyvean, R., (2013). Bioremoval of antimony(III) from contaminated water using several plant wastes: Optimization of batch and dynamic flow conditions for sorption by green bean husk (Vigna radiata). Chem. Eng. J., 225:192–201.

Isam, M., Baloo, L., Rahman, S., Kutty, M. Yavari, S., 2019. Optimisation and modelling of Pb(II) and Cu(II) biosorption onto red algae (*Gracilaria changii*) by using response surface methodology. Water. 11:2325.

Ives, K.J., (1959). The significance of surface electric charge on algae in water purification. J. Biochem. Microbiol Tech. Eng., 1(1):37–47.

Iye, O.J. (2015). Bioremediation of heavy metal polluted water using immobilised freshwater green microalga, *Botryococcus* sp. A thesis submitted to University of Malaysia.

Jahan, M.A.A., Akhtar, N., Khan, N.M.S., Roy, C.K., Islam, R., Nurunnabi., (2014). Characterization of tannery wastewater and its treatment by aquatic macrophytes and algae. Bangladesh J. Sci. Ind. Res., 49 (4):233–242.

Jain, C., Malik, D., Yadav, A., (2016). Applicability of plant based biosorbents in the removal of heavy metals: a review. Environ. Process., 3:495–523.

Jalali, M., Aboulghazi, F., (2013). Sunflower stalk, an agricultural waste, as an adsorbent for the removal of lead and cadmium from aqueous solutions. J. Mater. Cycles Waste Manag., 15:548–555.

Jalali, R., Ghafourian, H., Asef, Y., Davarpanah, S., Sepehr, S., (2002). Removal and recovery of lead using nonliving biomass of marine algae. J. Hazard. Mater., 92 (3):253–262.

Jarup, L., (2003). Hazards of heavy metal contamination. Br. Med. Bull., 68:167–182.

Ji, L., Xie, S., Feng, J., Li, Y., (2012). Heavy metal uptake capacities by the common freshwater green alga *Cladophora fracta*. J. Appl. Phycol., 24, 979–983.

Jiang, N., Shang, R., Heijman, S., Rietveld, L., (2018). High-silica zeolites for adsorption of organic micro-pollutants in water treatment: A review. Water Res., 144:145–161.

Jin-fen, P., Rong-gen, L., Li, M., (2000). A review of heavy metal adsorption by marine algae. Chin. J. Ocean. Limnol., 18: 260–264.

Kabbashi, N., Atieh, M., Al-Mamun, A., Mirghami, M., Alam, M., Yahya, N., (2009). Kinetic adsorption of application of carbon nanotubes for Pb(II) removal from aqueous solution. J. Environ. Sci. (China)., 21 (4):539–544.

Kanchana, S., Jeyanthi, J., Kathiravan, R, Suganya, K., (2014). Biosorption of heavy metals using algae: a review. Int. J. Pharm. Med. Biol. Sci., 3 (2):1.

Karnib, M., Kabbani, A., Holail, H., Olama, Z., (2014). Heavy metals removal using activated carbon, silica and silica activated carbon composite. Energy Procedia., 50:113–120.

Kanamarlapudi, S.L.R.K., Chintalpudi, V.K., Muddada, S. (2018). Application of biosorption for removal of heavy metals from wastewater. IntechOpen.

Khan, T., Isa, M.H., Ul Mustafa, M.R., Yeek-Chia, H., Baloo, L., Abd Manan, T.S.B., Saeed, M.O., (2016), Cr(vi) adsorption from aqueous solution by an agricultural waste based carbon. RSC Adv., 6:56365.

Khan, N.A., Shaaban, M.G., Jamil, Z., (2003). Chromium removal from wastewater through adsorption process. In: Proc. UM Research Seminar 2003 organized by Institute of Research Management and Consultancy (IPPP). University of Malaya, Kuala Lumpur.

Kumar, B., Agrawal, K., Bhardwaj, N., Chaturvedi, V., Verma, P. (2018). Advances in Concurrent Bioelectricity Generation and Bioremediation Through Microbial Fuel Cells. In: Microbial Fuel Cell Technology for Bioelectricity (pp. 211–239). Springer, Cham.

Kumar, B., Agrawal, K., Bhardwaj, N., Chaturvedi, V., Verma, P. (2019). Tech-no-Economic Assessment of Microbe-Assisted Wastewater Treatment Strategies for Energy and Value-Added Product Recovery. In: Microbial Technology for the Welfare of Society (pp. 147–181). Springer.

Kumar, B., Agrawal, K., Verma, P., (2021). Current perspective and advances of microbe assisted electrochemical system as a sustainable approach for mitigating toxic dyes and heavy metals from wastewater. J. Hazard. Toxic Radioact. Waste., 25 (2):04020082.

Kumar, B., Verma, P. (2021). Techno-Economic Assessment of Biomass-Based Integrated Biorefinery for Energy and Value-Added Product. In: Biorefineries: A Step Towards Renewable and Clean Energy (pp. 581–616). Springer.

Kumar, J.I.N., Oommen, C., Kumar, R.N., (2009). Biosorption of heavy metals from aqueous solution by green marine macroalgae from Okha Port, Gulf of Kutch, India. American-Eurasian J. Agric. Environ. Sci., 6:317–323.

Kumar, S.K., Dahms, H.U., Won, E., Lee, J., Shin, K., (2015). Microalgae – A promising tool for heavy metal remediation. Ecotox. Environ. Saf., 113:329–352.

Kurniasih, Ariesyady, H.D., Sulaeman, A., Kardena, E., (2013). Biosorption of chromium (VI) using immobilized algal-bloom biomass: Kinetics and equilibrium studies. Int. J. Environ. Res., 2 (1):24–31.

Kyzas, G., Deliyanni, E., Matis, K., (2013). Graphene oxide and its application as adsorbent to wastewater treatment. J. Chem. Technol. Biotechnol., 89 (2):192–205.

Lakherwal, D., (2014). Adsorption of heavy metals: A review. Int. J. Environ. Res. Dev., 4 (1):41–48.

Lata, S., Singh, P.K., Samadder, S.R., (2015). Regeneration of adsorbents and recovery of heavy metals: A review. Int. J. Environ. Sci. Technol., 12 (4):1461–1478.

Latiffi, N.A.A., Mohamed, R.M.S.R., Apandi, N.M., Kassim A.H.M., (2015). Application of phycoremediation using microalgae *Scenedesmus* sp. as wastewater treatment in removal of heavy metals from food stall wastewater. Appl. Mech. Mater., 773:1168–1172.

Latinwo, G.K., aJimoda, L.A., Agarry S.E., bAdeniran J.A., (2015). Biosorption of some heavy metals from textile wastewater by Green Seaweed Biomass. Univers. J. Environ. Res. Technol., 5 (4):210–219.

Lee, M., Lim, J., Kam, S., (2002). Biosorption characteristics in the mixed heavy metal solution by biosorbents of Marine Brown Algae. Korean J. Chem. Eng., 19:277–284.

Lesmana, S.O., Febriana, N., Soetaredjo, F.E., Sunarso, J., Ismadji, S., (2009). Studies on potential applications of biomass for the separation of heavy metals from water and wastewater. Biochem Eng J., 44:19.

Leung. H.M., Leung, A.O.W., Wang, H.S., Ma, K.K., Liang, Y., Ho, K.C., Cheung, K.C., Tohidi, F., Yung, K.K.L., (2014). Assessment of heavy metals/metalloid (As, Pb, Cd, Ni, Zn, Cr, Cu, Mn) concentrations in edible fish species tissue in the Pearl River Delta (PRD), China. Mar. Pollut. Bull., 78:235–245.

Lewis, B.M., (1996). Removal of Heavy Metals from Water with Microalgal Bioresins. Water Treatment Technology Program Report No. 74.

Liehr, S.K., Chen, H.-J., Lin, S.-H., (1994). Metal removal by algal biofilms. Water Sci. Technol., 30:59–68.

Lopez, A., Lazora, N., Marques, A.M., (1997). The interphase technique: a simple method of cell immobilization in gel-beads. J. Microbial. Method., 30:231–234.

Mabeau, S., Kloareg, B., Joseleau, J., (1990). Fractionation and analysis of fucans from brown algae. Phytochemistry., 29 (8):2441–2445.

Mahan, C.A., Holcombe, J.A. (1992). Analytical chemistry., 64:1933.

Mallick, N, Rai, L.C., (1992). Removal and assessment of toxicity of *Anabena doliolum* and *Chlorella vulgaris* using free and immobilized cells. World J. Microbiol. Biotechnol., 8:110–114.

Mallick, N., Rai, L.C., (1993). Influence of culture density, pH, organic acids and divalent cations on the removal of nutrients and metals by immobilized *Anabaena doliolum* and *Chlorella vulgaris*. World J. Microbiol. Biotechnol., 9:196–201.

Marella, T., Saxena, A., Tiwari, A., (2020). Diatom mediated heavy metal remediation: A review. Biores. Technol., 305:123068.

Martins, B.L., Cruz, C.C.V., Luna, A.S., Henriques, C.A., (2006). Sorption and desorption of Pb2+ ions by dead *Sargassum* sp. biomass. Biochem. Eng. J., 27:310–314.

Mata, Y., Blázquez, M., Ballester, A., González, F., Muñoz, J., (2008). Characterization of the biosorption of cadmium, lead and copper with the brown alga *Fucus vesiculosus*. J. Hazard. Mater., 158 (2–3):316–323.

Matheickal, J., Yu, Q., (1999). Biosorption of lead(II) and copper(II) from aqueous solutions by pre-treated biomass of Australian marine algae. Biores. Technol., 69 (3):223–229.

Mehariya, S., Kumar, P., Marino, T., Casella, P., Lovine, A., Verma, P., Musuma, A.M. (2021a). Aquatic Weeds: A Potential Pollutant Removing Agent from Wastewater and Polluted Soil and Valuable Biofuel Feedstock, Bioremediation using weeds (pp. 59–77). Springer.

Mehariya, S., Goswami, R., Verma, P., Lavecchia, R., Zuorro, A., (2021b). Integrated approach for wastewater treatment and biofuel production in microalgae biorefineries. Energies., 14 (8):2282.

Mehta, S., Gaur, J., (2001). Removal of Ni and Cu from single and binary metal solutions by free and immobilized Chlorella vulgaris. Eur. J. Protistol., 37 (3):261–271.

Mehta, S., Gaur, J., (2005). Use of algae for removing heavy metal ions from wastewater: Progress and prospects. Crit. Rev. Biotechnol., 25 (3):113–52.

Michalak, I., Chojnacka, K., Witek-Krowiak, A., (2013). State of the art for the biosorption process—a review. Appl. Biochem. Biotechnol., 170:1389–1416.

Mirghaffari, N., Moeini, E., Farhadian, O., (2015). Biosorption of Cd and Pb ions from aqueous solutions by biomass of the green microalga, *Scenedesmus quadricauda*. J. Appl. Phycol,. 27:311–320.

Mofeed, J., 2017. Biosorption of heavy metals from aqueous industrial effluent by non-living biomass of two marine green algae *Ulva lactuca and Dunaliella salina* as biosorpents. Catrina., 16 (1):43–52.

Moradi, O., (2011). The removal of ions by functionalized carbon nanotube: Equilibrium, isotherms and thermodynamic studies. Chem. Biochem. Eng. Q., 25 (2):229–240.

Mubashar, M., Naveed, M., Mustafa, A., Ashraf, S., Shehzad Baig, K., Alamri, S., Siddiqui, M.H., Zabochnicka-Swiatek, M., Szota, M., Kalaji, H.M., (2020) Experimental investigation of *Chlorella vulgaris* and *Enterobacter* sp. MN17 for decolorization and removal of heavy metals from textile wastewater. Water., 12:3034.

Muñoz, R., Alvarez, M.T., Muñoz, A., Terrazas, E., Mattiasson, B.G.B., (2006). Sequential removal of heavy metals ions and organic pollutants using an algal-bacterial consortium. Chemosphere., 63 (6): 903–911.

Napan, K., (2014). Distribution of Heavy Metals from Flue Gas in Algal Bioreactor. Utah State University. All Graduate Theses and Dissertations. 4018.

Negm, N., Wahed, M., Hassan, A., Kana, M., (2018). Feasibility of metal adsorption using brown algae and fungi: Effect of biosorbents structure on adsorption isotherm and kinetics. J. Mol. Liq., 63.

Olguin, E.J. (2003). Phycoremediation: key issues for cost effective nutrient removal processes. Biotechnol. Adv., 22:81–91.

Olgun, A., Atar, N. (2012). Equilibrium, thermodynamic and kinetic studies for the adsorption of lead(II) and nickel(II) onto clay mixture containing boron impurity. J. Ind. Eng. Chem., 18 (5):1751–1757.

Omar, H., (2002). Bioremoval of zinc ions by *Scenedesmus obliquus* and *Scenedesmus quadricauda* and its effect on growth and metabolism. Int. Biodeterior Biodegrad., 50 (2):95–100.

Onyancha, D., Mavura, W., Ngila, J.C., Ongoma, P., (2008). Studies of chromium removal from tannery wastewaters by algae biosorbents, *Spirogyra condensata* and *Rhizoclonium hieroglyphicum*. J. Hazard. Mater., 158 (2–3):605–614.

Oskarsson, A., Widell, A., Olsson, I.M., Grawe, K.P., (2004). Cadmium in the food chain and health effects in sensitive population groups. Biometals., 17:531–534.

Ozer, D., Aksu, Z., Kutsal, T., Caglar, A., (1994). Adsorption isotherms of lead (II) and chromium (VI) on *Cladophora crispata*. Environ. Technol.,15:439–448.

Özgür, A., Uluozlü, D., Tüzen, M., (2011). Equilibrium, thermodynamic and kinetic investigations on biosorption of arsenic from aqueous solution by algae (*Maugeotia genuflexa*) biomass. Chem. Eng. J.,167 (1):155–161.

Palaniswamy, R., Veluchamy, C., (2017). Biosorption of heavy metals by *Spirulina platensis* from electroplating industrial effluent. Environ. Sci. Ind. J., 13(4):139.

Pandi, M., Shashirekha, V., Swamy, M., (2007). Bioabsorption of chromium from retan chrome liquor by cyanobacteria. Microbiol. Res., 164:420–428.

Pfenninger A., (1999). Manufacture and use of zeolites for adsorption processes. In: Structures and Structure Determination. Molecular Sieves (Science and Technology) (pp. 163–198). Baerlocher C., et al. Eds., Springer, Berlin, Heidelberg.

Phang, S.M., Chu, W.L., Rabiei, R., (2015). Phycoremediation. In: The Algae World. Cellular Origin, Life in Extreme Habitats and Astrobiology (Vol. 26). Sahoo D., Seckbach J., Eds., Springer, Dordrecht.

Philippis, R., Colica, G., Micheletti, E., (2011). Exopolysaccharide-producing cyanobacteria in heavy metal removal from water: molecular basis and practical applicability of the biosorption process. Appl. Microbiol. Biotechnol., 92:697–708.

Porwal, M., Rastogi, V., Kumar, A. (2017). An overview on carbon nanotubes. MOJ Bioequiv Availab., 3 (5):114–116.

Prasher, S., Beaugeard, M., Hawari, J., Bera, P., Patel, R., Kim, S., (2004). Biosorption of heavy metals by red algae (*Palmaria palmata*). Environ. Technol., 25:1097–1106.

Pugazhendhi, A., Shobana, S., Bakonyi, P., Nemestóthy, N., Xia, A., Banu, R., Kumar, G., (2019). A review on chemical mechanism of microalgae flocculation via polymers. Biotechnol. Rep., 21:e00302.

Quraishi, D., Abbas, I., (2019). Removing heavy metals by diatoms nitzschia palea and navicula incerta in their aqueous solutions. Plant Arch., 19 (Supplement 1):272–278.

Rabsch, U., Elbrachter, M. (1980). Cadmium and zinc uptake, growth, and primary production in *Coscinodiscus granii* cultures containing low levels of cells and dissolved organic carbon. Helgolainder Meeresunters., 33:79–88.

Rangabhashiyama, S., Balasubramanian, P., (2019). Characteristics, performances, equilibrium and kinetic modeling aspects of heavy metal removal using algae. Biores. Technol. Rep., 5:261–279.

Rao, P., Saisha, V., Bhavikatti, S.S., (2013). Removal of Chromium (VI) from synthetic waste water using immobilized algae. International Journal of Current Engineering and Technology.

Rangsayatorn, N., Upatham, E., Kruatrachue, M., Pokethitiyook, P., Lanza, G., (2002). Phytoremediation potential of *Spirulina* (*Arthrospira*) *platensis*: biosorption and toxicity studies of cadmium. Environ. Poll., 119:45–53.

Rao, L.N., Prabhakar G., (2011). Removal of heavy metals by biosorption–an overall review. JERS., II (IV):17–22.

Rawat, I., Kumar, R.R., Mutanda, T., Bux, F. 2011. Dual role of microalgae: Phycoremediation of domestic wastewater and biomass production for sustainable biofuels production. Appl. Energy., 88:3411–3424.

Richards, S., Dawson, J., Stutter, M., (2019). The potential use of natural vs commercial biosorbent material to remediate stream waters by removing heavy metal contaminants. J. Environ. Manag., 231:275–281.

Robinson, P.K., Wilkinson, S.C. (1994). Removal of aqueous mercury and phosphate by gel-entrapped *Chlorella* in packed-bed reactors. Enzyme Microb. Technol., 16:802–807.

Romera, E., González, F., Ballester, A., Blázquez, M., Muñoz, J., (2007). Comparative study of biosorption of heavy metals using different types of algae. Biores. Technol., 98 (17):3344–3353.

Sakulsak, N., (2012). Metallothionein: An overview on its metal homeostatic regulation in mammals. Int. J. Morphol., 30(3):1007–1012.

Salam, K., (2019). Towards sustainable development of microalgal biosorption for treating effluents containing heavy metals. Biofuel Res. J., 22, 948–961.

Salama, E., Roh, H., Dev, S. Khan, M., Abou-Shanab, R., Chang, S., Jeon, B., (2019). Algae as a green technology for heavy metals removal from various wastewater. World J. Microbiol. Biotechnol., 35:75.

Sbihi, K., Cherifi, O., El-gharmali, A., Oudra, B., Aziz, F., (2012). Accumulation and toxicological effects of cadmium, copper and zinc on the growth and photosynthesis of the freshwater diatom *Planothidium lanceolatum*: A laboratory study. J. Mater. Environ. Sci., 3 (3):497–506.

Schiewer, S., Volesky, B. (2000). Biosorption by marine algae. In: Bioremediation (pp. 139–169). Valdes, J.J. Eds., Springer, Dordrecht.

Selvi, K., Pattabhi, S., Kadirvelu, K., (2001). Removal of Cr (VI) from aqueous solution by adsorption onto activated carbon. Biores. Technol., 80 (2001):87–89.

Shafeeyan, M., Daud, W., Houshmand, A., Shamiri, A., (2010). A review on surface modification of activated carbon for carbon dioxide adsorption. J. Anal. Appl. Pyrolysis., 89 (2):143–151.

Sheng, P., Wee, K., Ting, Y., Chen, J., (2008). Biosorption of copper by immobilized marine algal biomass. Chem. Eng. J., 136 (2–3):156–163.

Sheoran, A.S., Bhandari, S., (2005). Treatment of mine water by a microbial mat: Bench-scale experiments. Mine Water Environ., 24:38–42.

Sibi, G., (2016). Biosorption of chromium from electroplating and galvanizing industrial effluents under extreme conditions using *Chlorella vulgaris*. Green Energy Environ., 1 (2):172–177.

Singh, J., Singh, D., Dixit, S., (2012). Cynobacteria: An agent of heavy metal removal. In: Bioremediation of Pollutants, Bioremediation of Pollutants (pp. 223–243). Maheshwari, D. K., and Dubey, R. C., Eds., I K International Publisher Co.

Siwi, W.P., Rinanti, A., Silalahi, M.D.S., Hadisoebroto, R., Fachrul, M.F., (2018). Effect of immobilized biosorbents on the heavy metals ($Cu2^{+}$) biosorption with variations of temperature and initial concentration of waste. Earth Environ. Sci., 106:012113.

Soeprobowati, T.R., Hariyati, R., (2017). The phycoremediation of textile wastewater discharge by *Chlorella pyrenoidosa* H. Chick, *Arthrospira platensis* Gomont, and *Chaetoceros calcitrans* (Paulson) H. Takano. AACL Bioflux., 10(3).640–651.

Solisio, C., Lodi, A., Torre, P., Converti, A., Del Borghi, M., (2006). Copper removal by dry and re-hydrated biomass of Spirulina platensis. Biores. Technol., 97 (14):1756–1760.

Song, H.L., Liang, L., Yang, K.Y., (2014). Removal of several metal ions from aqueous solution using powdered stem of Arundo donax L as a new biosorbent. Chem. Eng. Res. Des., 92:1915.

Su, H., Ye, Z., Hmidi, N., (2017). High-performance iron oxide–graphene oxide nanocomposite adsorbents for arsenic removal. Colloids Surf. A: Physicochem. Eng. Asp., 522:161–172.

Subashini, P.S., Rajiv, P., (2018). *Chlorella vulgaris* DPSF 01: A unique tool for removal of toxic chemicals from tannery wastewater. Afr. J. Biotechnol., 17(8):239–248.

Suemitsu, R., Venishi, R., Akashi, I., Nakano, M., (1986). The use of dyestuff-treated rice hulls for removal of heavy metals from wastewater. J. Appl. Polym. Sci., 31 (1986):75–83.

Synytsya, A., Čopíková, J., Kim, W.J., Park, Y.I., (2015). Cell wall polysaccharides of marine algae. (pp. 543–590). In: Springer Handbook of Marine Biotechnology. Kim, S. K. Ed,. Springer, Berlin, Heidelberg.

Terry, P., Stone, W., (2002). Biosorption of cadmium and copper contaminated water by *Scenedesmus abundans*. Chemosphere., 47 (3):249–255.

Ting, Y.P., Prince, I.G., Lawson, F., (1991). Uptake of cadmium and zinc by the alga *Chlorella vulgaris*: II. Multi-ion situation. Biotechnol. Bioeng., 37 (5):445–455.

Tonon, A.P., Oliveira, M.C., Soriano, E.M., Colepicolo, P. (2011). Absorption of metals and characterization of chemical elements present in three species of Gracilaria (Gracilariaceae) Greville: a genus of economical. Rev. Bras. Farmacogn., 21:355–360.
Tripathi, A., Ranjan, M., (2015). Heavy metal removal from wastewater using low-cost adsorbents. J. Bioremed. Biodeg., 6:315.
Tsezos, M. 1986. Adsorption by microbial biomass as a process for removal of ions from process or waste solutions. In: Immobilization of Ions by Bio-Sorption (pp. 201–218). Eccles, H., and Hunt, S., Eds., Ellis Harwood, Chichester, UK.
Tuzun, I., Bayramoglu, G., Yalcin, E., Basaran, G., Celik, G., Arica, M.Y., (2005). Equilibrium and kinetic studies on biosorption of Hg(II), Cd(II) and Pb(II) ions onto microalgae *Chlamydomonas reinhardtii*. J. Environ. Manag., 77:85–92.
Veglio, F., Beolchini, F., (1997). Removal of metals by biosorption: a review, Hydrometallurgy., 44 (3):301–316.
Vijayaraghavan, K., Jegan, J., Palanivenu, K., Velan, M., (2005). Biosorption of copper, cobalt and nickel by marine green alga *Ulva reticulate* in a packed column. Chemosphere., 60:419–426.
Volesky, B., Holan, Z.R., (1995). Biosorption of heavy metals. Biotechnol. Prog., 11:235–250.
Wan Maznah, W.O., Al-Fawwaz, A.T., Surif, M., (2012). Biosorption of copper and zinc by immobilized and free algal biomass, and the effects of metal biosorption on the growth and cellular structure of *Chlorella* sp and *Chlamydomonas* sp. isolated from rivers in Penang, Malaysia. J. Environ. Sci., 24 (8):1386–1393.
Wang, L., Wang, N., Zhu, L., Yu, H., Tang, H., (2008). Photocatalytic reduction of Cr (VI) over different TiO2 photocatalysts and the effects of dissolved organic species. J. Hazard. Mater., 152 (1):93–99.
Wang, N., Qiu, Y., Xiao, T., Wang, J., Chen, Y., Xu, X., Kang, Z., Fan, Li., Yu, H., (2019). Comparative studies on Pb(II) biosorption with three spongy microbe-based biosorbents: High performance, selectivity and application. J. Hazard. Mater., 373, 39–49.
Wang, Y., Shi, L., Gao, L., Wei, Q., Cui, L., Hu, L., Yan, L., Du, B., (2015). The removal of lead ions from aqueous solution by using magnetic hydroxypropyl chitosan/oxidized multiwalled carbon nanotubes Adv Compos Hybrid Mater composites. J. Colloid. Interface Sci., 451:7–14.
Wilan, T., Astuti Lieswito, N., Suwardi, A., Hadisoebroto, R., Fachrul, M.F., Rinanti, A., (2020). The biosorption of copper metal ion by tropical microalgae beads biosorbent. Int. J. Sci. Technol. Res., 9 (1).
Wilde, E.W., Benemann, J.R., (1993). Bioremoval of heavy metals by the use of microalgae, Biatech. Adv., 11:781–812.
Wilkinson, S.C., Goulding, K.H., Robinson, P.K., (1989). Mercury accumulation and volatilization in immobilized algal cell systems. Biotechnol. Lett. 11: 861–864.
Wilkinson, S.C., Goulding, K.H., Robinson, P.K., (1990). Mercury removal by immobilized algae in batch culture sytems. J. Appl. Phycol., 2:223–230.
Wong, M.H., Pak, D.C.H., (1992). Removal of copper and nickel by free and immobilized microalgae. Biomed. Environ. Sci., 5:99–108.
Yang, J., Cao, J., Xing, G., Yuan, H., (2015). Lipid production combined with biosorption and bioaccumulation of cadmium, copper, manganese and zinc by oleaginous microalgae *Chlorella minutissima* UTEX2341. Biores. Technol., 175:537–544.
Yang, X., Wu, X., Hao, H., He, Z., (2008). Mechanisms and assessment of water eutrophication. J. Zhejiang Univ. Sci. B., 9:197–209.
Ye, J., Xiao, H., Xiao, B., Xu, W., Gao, L., Lin, G., (2015). Bioremediation of heavy metal contaminated aqueous solution by using red algae *Porphyra leucosticte*, Water Sci Technol., 72 (9):1662–1666.

5 Microalgae-Mediated Elimination of Endocrine-Disrupting Chemicals

Chandra Prakash, Komal Agrawal, Pradeep Verma, Venkatesh Chaturvedi

CONTENTS

5.1 INTRODUCTION

Endocrine disrupters (EDs), or endocrine-disrupting chemicals (EDCs), have been shown to produce estrogenic effects in animals and humans by interfering with their endocrine functions with different modes of action (Schug et al. 2011). According to the US Environmental Protection Agency (EPA), an EDC is an exogenous compound that may interfere with the synthesis, secretion, transport, metabolism, receptor binding, or elimination of endogenous hormones, altering the endocrine and homeostatic systems (Kabir et al. 2015; Mnif et al. 2011). About 70 chemicals are currently suspected of being environmental EDs, including insecticides, herbicides, fungicides, antiseptics, detergents, plastic reinforcements, and hormone disrupters. The disposal of these can cause serious environmental problems due to the chemical toxicity of the active ingredients in the formulations and, sometimes, of their decomposition products (Jobling et al. 1998). Because of the wide use of EDs, their safety has received much attention. Most research has highlighted the identification and detection of EDs in aquatic environments or has evaluated their biological toxicity. EDs cause a wide variety of detrimental effects

DOI: 10.1201/9781003155713-5

on animals, which include genotoxicity by hormonal dysregulation, cytotoxicity, oxidative stress, histone modification, apoptosis, etc. They also cause DNA damage and chromosomal aberrations; they also affect the normal functioning of tissues and organs such as liver, kidney, and spleen (Kiyama and Wada-Kiyama, 2015). Water quality management is vital for the protection of aquatic and human health. It is mainly focused on organic pollutants, metal ions, and pathogenic microorganisms (Pal et al. 2014). However, in recent years, focus has shifted to emerging contaminants (ECs) such as EDs present in aquatic environments in alarming concentrations and their impact on aquatic ecosystems and human health. Most of the EDs enter the environment through anthropogenic activities such as discharge of raw or un-treated municipal water, landfill leachate, surface, and groundwater runoff, or industrial waste discharge (Ali et al. 2017; Tran and Gin, 2017; Tran et al. 2018). Municipal wastewater treatment plants(WWTPs) release a large amount of EDs into the environment (Tran and Gin, 2017). Among EDs, pharmaceuticals are frequently found in the aquatic environment; they consist of antibiotics, antimicrobial agents, neuroactive drugs, hormones, painkillers, insect repellents, etc. (Norvill et al. 2016). Pharmaceuticals are mainly reported at a very low concentration in different water bodies, but their presence in these systems can have alarming consequences on human and animal health. Pesticides and related agrochemicals are commonly used in agriculture, and thus, they are reported in huge amounts in agricultural fields and are frequently transported to nearby water bodies through runoff (Pal et al. 2014). Pesticides, due to their stable structure and low degradability, have been known to bioaccumulate in aquatic organisms, there by disordering the food chain and posing a hazard to human health (Knillmann et al. 2018; Machado and Soares, 2019).

Approximately 45,000 to 100,000 species of algae exist on earth (Chisti, 2018). Removal of EDs from the environment is very important; therefore, the use of microalgae is an appealing biotreatment for wastewater, due to their ability to remove heavy metals and organic compounds (Hom-Diaz et al. 2015). Microalgae have gained particular interest of researchers mainly because they can utilize wastewater organic matters as carbon and nutrient sources (Mehariya et al. 2021a,b). In addition, microalgae can treat wastewater, ultimately rendering algal cultivation cost-effective (Barros et al. 2015; Bhardwaj et al. 2020). Moreover, microalgae efficiently convert solar energy and carbon dioxide into algal biomass and incorporate nitrogen and phosphorus, which further helps in preventing eutrophication (Barros et al. 2015; Goswami et al. 2020). Recent studies have demonstrated the consumption and degradation of organic contaminants, pharmaceuticals, heavy metals, surfactants, pesticides, and plasticizers by microalgae. The use of microalgae has shown promise as a bioremediation technology (Shao et al. 2016). The utilization of microalgae to remove nutrients, pesticides, toxic elements, pharmaceutical chemicals, and oils from wastewater has been widely reported (Goswami et al. 2021a; Kumar et al. 2019; Ummalyma et al. 2018) (Figure 5.1). Moreover, microalgae have a wide range of applications, including in the pharmaceutical industry, food industry, animal feed, environmental detection, biotechnology, renewable energy, cosmetic industry, etc. (Agrawal et al. 2020; Bhardwaj et al. 2020; Goswami et al. 2021b, c;

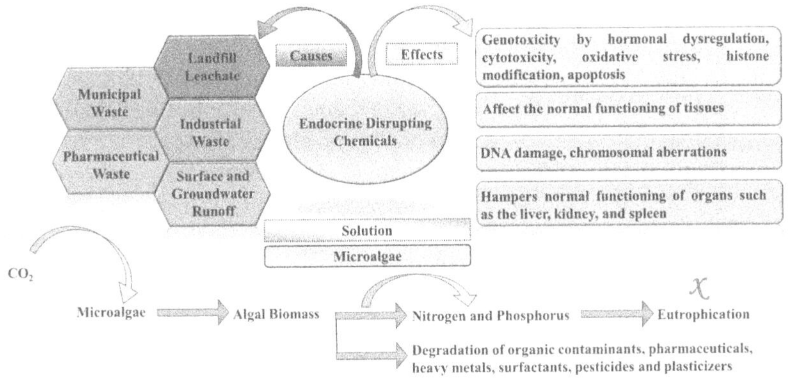

FIGURE 5.1 Various causes, effects, and solution for the removal of endocrine-disrupting chemicals (EDCs).

Mehariya et al. 2021c; Ravindran et al. 2016; Saini et al. 2021). In this chapter, we discuss algal-mediated elimination of common EDCs from wastewater.

5.2 REMOVAL OF VARIOUS EDCs USING MICROALGAE

Microalgae have been reported for the degradation and removal of various EDCs, discussed as follows (Figure 5.2):

5.2.1 Estrogens

17α-ethynylestradiol (EE2) and 17β-estradiol (E2) are two estrogens that cause various responses that may ultimately prompt impacts on reproduction, for instance, feminization of male fish and imposex of different creatures (Chikae et al. 2004). Estrogens are difficult to eliminate from wastewater by primary and secondary treatments. Photodegradation with the help of microalgae is a potential technique

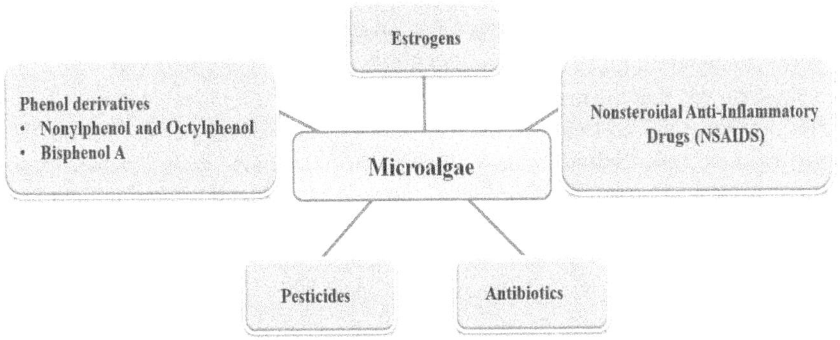

FIGURE 5.2 The potential of microalgae in the removal of various endocrine-disrupting chemicals (EDCs).

for the obliteration of aquatic environmental pollutants. Microalgae, due to their huge biomass, broad environmental habitat, and diversity, assume a significant part in the destiny of organic compounds present in oceanic biological systems (Dalrymple et al. 2007). Zepp and Schlotzhauer (1983) observed that microalgae could speed up the photodegradation of organic chemicals. Ge et al. (2009) studied the photodegradation of EE2 and E2 by three algal species, i.e., *Chlorella vulgaris*, *Anabaena cylindrica,* and *Microcystis aeruginosa kutz.* All three microalgae displayed the potential to degrade both the chemicals. Reactive oxygen species like hydroxyl radicals and singlet oxygen were believed to be the principal entities behind the degradation of endocrine disturbing chemicals. There are numerous new reports on algal-intervened degradation of estrogens. *Selenastrum capricornutum* and *Chlamydomonas reinhardtii* were tried for conceivable biodegradation of the chemicals E2 and EE2 when grown in the anaerobic digester. Following 7 days of treatment, the high pace of E2 was eliminated by *S. capricornutum.* For EE2, removal effectiveness was low. With *C. reinhardtii*, complete degradation of E2 and EE2 was noticed (Hom-Diaz et al. 2015).

Liu et al. (2018) studied the removal of E2 and diethylstilbestrol (DES) by microalgae *Raphidocelis subcapitata.* The results showed that *R. subcapitata* exhibited a quick capacity to eliminate E2 and DES in both single and mixed treatments by biodegradation. The removal was principally ascribed to the biodegradation or biotransformation by the microalgae cells as opposed to basic sorption and accumulation. The microalgae *R. subcapitata* showed a high capacity for the removal of E2 and DES. Wang et al. (2019) examined degradation of E2 and EE2 by four microalgae *Haematococcus pluvialis*, *Selenastrum capricornutum*, *Scenedesmus quadricauda*, and *Chlorella vulgaris*, respectively. It was discovered that *H. pluvialis*, *S. capricornutum,* and *S. quadricauda* could all viably eliminate all the three estrogens in engineered wastewater effluent. The estrogenic activity of E2, as dictated by yeast estrogenic screening, showed generous decreases in estrogenic activity after biotransformation by *H. pluvialis*, *S. capricornutum*, and *S. quadricauda.* Microalgae growth in wastewater is considered an affordable practice since microalgae can be utilized for toxin removal and also as energy assets (Ruksrithong and Phattarapattamawong, 2019). In their study, removal of estrone (E1) and E2 by *Chlorella vulgaris* and *Scenedesmus obliquus* was contemplated. *S. obliquus* showed a high rate of E1 and E2 degradation when compared with *Chlorella vulgaris.* In a study by Huang et al. (2019), E2 removal by *Chlorella* species was researched. The outcomes showed that the microalgae had a high rate of E2 removal at 0.5 mg/L concentration. Wu et al. 2021 studied the role of three microalgae species, *Selenastrum capricornutum*, *Scenedesmus quadricauda,* and *Chlorella vulgaris*, in a batch study. The role of Algal Extracellular Organic Matter (AEOM) on the elimination of two estrogens (E2 and EE2) was also tested (Wu et al. 2021). All three microalgae had the capacity to eliminate E2 and EE2, and *S. capricornutum* showed the most noteworthy E2 and EE2 removal efficiency of 91% and 83%, respectively, compared to the decrease in anticipated estrogenic removal of 86%. AEOM from three microalgae could initiate photodegradation of estrogens, and AEOM from *S. capricornutum* and *C. vulgaris* accomplished 100% of E2 and EE2 degradation under noticeable light illumination (Table 5.1).

TABLE 5.1
Removal of Estrogens by Microalgae

Microalgae	Estrogen	Concentration	Removal %	Reference
Raphidocelis subcapitata	E2 and DES	0.1, 0.5, and 1.5 mg/L	88.5%, 82.9%, 84.3% for E2 and 87.2%, 71.8%, 51.1%for DES	Liu et al. 2018
Chlorella vulgaris and *Scenedesmus obliquus*	E1 and E2	E1 and E2-5 µg/L	*S. obliquus*-91% and 99%, *C. vulgaris*-52% and 99%	Ruksrithong and Phattarapattamawong, 2019
Chlorella sp.	E2	0.5 mg/L	92%	Huang et al. 2019
Selenastrum capricornutum and *Chlamydomonas reinhardtii*	E2 and EE2		*S. capricornutum*-E2- 88% and 100%, EE2-60% and 95%, *C. reinhardtii*-100% for E2 and EE2	Hom-Diaz et al. 2015
Selenastrum capricornutum, *Scenedesmus quadricauda*, and *Chlorella vulgaris*	E2, EE2	3 mg/L	*Selenastrumcapricornutum*, E2 and EE2-91% and 83%	Wu et al. 2021

5.2.2 Phenol Derivatives

5.2.2.1 Nonylphenol and Octylphenol

Nonionic surfactants, such as alkylphenolethoxylates (APEs), are generally utilized in different domestic and industrial cleansers and pesticide formulations (Ying et al. 2002). APEs can be transformed into alkylphenols, for example, nonylphenol (NP) and octylphenol (OP) in WWTPs lastly enter the environment in the form of sludge. Recently, NP and OP were regularly identified in different natural systems such as streams, lakes, seas, groundwater, and soils (Soares et al. 2008). A high amount of NP and OP were identified in five German waterways (Bolz et al. 2001). Alkylphenols NP and OP are known to be endocrine disruptors, which have the capacity to imitate characteristic estrogens and upset the endocrine frameworks of higher organisms by cooperating with the estrogen receptors (Vazquez-Duhalt et al. 2005), subsequently, conceivably creating unfriendly impacts on the reproductive system. For instance, when exposed to NP or OP, male rainbow trout produced vitellogenin and exhibited hindrance of testicular development (Jobling et al. 1998). They can also affect plankton community when released into the environment. For instance, because of the presence of NP, the number of *cladocera* and *copepoda* were decreased, and rotifer number was increased, further impacting phytoplankton population through changes in zooplankton grazing (Hense et al. 2005). Thinking about the negative impacts of NP and OP on aquatic organisms, it is important to comprehend their metabolism in aquatic ecosystems. In a study by Zhou et al. (2013), the elimination of NP and OP by the freshwater microalgae *Scenedesmus obliquus* was studied. The outcomes exhibited that the microalgae eliminated a high amount of NP and OP. The removal was because of biodegradation. He et al. (2016) studied the removal of NP by four freshwater microalgae, including three green algae (*Scenedesmus quadricauda*, *Chlorella vulgaris*, and *Ankistrodesmus acicularis*) and one cyanobacterium (*Chroococcus minutus*) individually. All four algal species showed a quick and high capacity to eliminate NP (counting bioaccumulation and biodegradation). Among these species, *A. acicularis* had the most elevated NP expulsion rate followed by *C. vulgaris*, *S. quadricauda,* and *C. minutus*. *C. vulgaris* had the most elevated NP biodegradation rate, followedby *A. acicularis*, *S. quadricauda*, and *C. minutus*. Wang et al. (2019) considered biodegradation of NP by four types of marine microalgae: *Phaeocystis globosa*, *Nannochloropsis oculata*, *Dunaliella salina,* and *Platymonas sub cordiformis* individually. The outcomes showed a sharp decrease in NP in the medium containing the four micro-algal species during the initial 24 hours of growth. After 120 hours of exposure to NP, the four species could biodegrade a higher amount of NP in the medium, with efficiencies going from 43.43% to 90.94%. López-Pacheco et al. (2019) examined the removal of NP by a consortium of two microorganisms, *Arthrospira maxima* and *Chlorella vulgaris*. The consortia diminished up to 96% of 4-NP in water in the initial 48 hours of culture.

5.2.2 Bisphenol A

In industry, bisphenol A(BPA) is utilized in the creation of epoxy resin and polycarbonate (PC) plastics. It is generally utilized in different kinds of food and drink packaging, baby bottles, and dental sealants (Staples et al. 1998). In many studies, it was shown that BPA acts like E2 and is considered as an estrogen-like EDC or a xenoestrogen. BPA caused DNA damage in eukaryotes, caused aneuploidy, and underlying chromosomal abnormalities (Naik and Vijayalaxmi, 2009). BPA was shown to elicit altered gene and protein expression (Fernandez et al. 2012). BPA may cause variations in DNA methylation, alterations in histones, and changes in the expression of non-coding RNAs. These progressions can up-control or down-manage distinctive gene expressions, which may thus bring about lasting health impacts like neural and immune disorders, infertility, and early onset of complex diseases (Ferreira et al. 2015). BPA is an endocrine disruptor that can lead to carcinogenesis (Crain et al. 2007). BPA may expand the danger of cancer by influencing different cell processes like DNA methylation and chromatin remodeling during development. BPA exposure caused epigenetic adjustments in the rat prostate that lead to malignant growth (Doherty et al. 2010). BPA has been commonly utilized notwithstanding that it is hazardous. Subsequently, there was an increase in BPA utilization worldwide. It has been shown that human exposure to BPA in the environment can be partitioned into three principal ways: intake, inhalation, and skin contact (Xiao et al. 2011). Furthermore, studies have shown that BPA can affect aquatic organisms even in conditions where the concentration of BPA is under 1 mg/L. The public concern encompassing BPA for the most part centers on children in light of the fact that BPA could be gradually released during typical utilization of infant bottles (Fasano et al. 2012). There are a number of reports pertaining to microalgae-based removal of BPA. Li et al. (2009) examined toxicity and elimination of BPA by a diatom *Stephanodiscus hantzschii*. The results showed that the diatom was impervious to high convergences of BPA and had the ability to degrade the pollutant. Solé and Matamoros (2016) studied the impact of free and immobilized microalgae on the elimination of six EDCs by blending them in 2.5 L reactors with treated wastewater. Both the free and immobilized microalgae reactors had the capacity to eliminate up to 80% of the EDCs within 10 days of incubation. Free microalgae were found to increase the kinetic removal rate for BPA by 25%, and immobilizing the microalgae in alginate beads also improved the removal rate for bisphenol. Biogenic manganese oxides (BioMnOx) have stood out as dynamic oxidants, adsorbents, and catalyst. In an examination by Wang et al. (2017), green algae-created BioMnOx was utilized for removal of BPA. BioMnOx was produced by *Desmodesmus* sp. WR1. It showed high efficiency in eliminating BPA. Ben Ouada et al. (2018a) studied the degradation of BPA by an alkaliphilic chlorophyta, *Picocystis* sp. This microalga had the ability to endure high concentrations of BPA in the development medium and could degrade a high amount of BPA (Table 5.2).

5.2.3 NSAIDS

Rapid advancement of industries, enormous exploitation of natural resources, urbanization, and increased consumer demands have prompted a decrease in the quality of the environment and an increment in environmental pollution (Tyumina et al. 2020). The most potent among the arising pollutants are pharmaceuticals or drugs having stable structures and increased biological activity due to the presence of functional

TABLE 5.2
Removal of Phenol Derivatives by Microalgae

Microalgae	Chemicals	Concentration	Removal %	Reference
Scenedesmus obliquus	NP and OP	4 mg/L	>89%,>58%	Zhou et al. 2013
P. globosa, *N. oculata*, *D. salina* and *P. subcordiformis*	NP	0.5–2.5 mg/L	66.37%, 74.82%, 69.86%, and 82.38%	Wang et al. 2019
Scenedesmus quadricauda, *Chlorella vulgaris*, *Ankistrodesmus acicularis*, *Chroococcus minutus*	NP	0.5–2.5 mg/L	68.8%, 65.6%, 63.1%, 34.9%	He et al. 2016
Arthrospira maxima and *Chlorella vulgaris*	NP		96%	López-Pacheco et al. 2019
Stephanodiscus hantzschii	BPA	5,7,9 mg/L	48%,28%,26%	Li et al. 2009
Desmodesmus sp. WR1	BPA		78%	Wang et al. 2017
Picocystis sp.	BPA	25, 75 mg/L	72%, 42%	b
Picocystis sp.	BPA	10 mg/L	91.36%	Ben Ali et al., 2020
Picocystis sp. and *Graesiella* sp.	BPA	25 mg/L	72%, 52.6%	Ben Ouada et al. 2018a

groups (Moreira et al. 2018). Until now, more than 600 pharmaceuticals have been identified in the water bodies of 71 nations (aus der Beek et al. 2016). They are present in low concentrations (from several ng/L to many µg/L), they are mainly anti-infection agents, non-steroidal anti-inflammatory drugs (NSAIDs), chemicals, antispasmodics, lipid-lowering drugs, anticancer medications, antiepileptics and antidepressants, and so forth (Palli et al. 2017). NSAIDs are the most often and universally ecologically identified medications of different synthetic structure with calming, antipyretic, and pain-relieving impacts (Liu et al. 2017). NSAID atoms have both high reactivity and stability. This decides their protection from biodegradation, ecotoxicity, steadiness, and subsequent danger to the environment (Pi et al. 2017). The most notable illustration of ecotoxic NSAID impact is the enormous decrease in population of three vulture species in the Indian subcontinent brought about by inebriation and renal failure of birds because of their feeding on bodies of diclofenac(DCF)-treated cattle (Swan et al. 2006). A number of reports have confirmed the removal of NSAIDs by microalgae. Ismail et al. (2016) examined that *Chlorella* sp. was the most impervious to ketoprofen. A characterized bacterial consortium (K2) removed 5 mM ketoprofen as a sole carbon source both in dark or illumination. Ketoprofen didn't go through photo-degradation. In dark conditions, biodegradation was quicker with a lag period of 10 hours, 41% chemical oxygen demand (COD) removal, and 82% decrease in toxicity. The consortium degraded up to 16 mM ketoprofen. By integrating *Chlorella* sp. to the K2 consortium, ketoprofen was degraded within 7 days under a diurnal pattern of 12 h light/12 h dark. Ding et al. (2017) contemplated the removal of naproxen by *Cymbella* sp. and *Scenedesmus quadricauda* individually. Removal of naproxen was shown by both algal species. *Cymbella* sp. showed a more satisfactory removal of naproxen with higher removal proficiency. A sum of 12 metabolites was recognized by LC-MS/MS analysis, and the degradation pathways of naproxen in two microalgae were proposed. Hydroxylation, decarboxylation, demethylation, tyrosine combination, and glucuronidation added to naproxen change in algal cells. Escapa et al. (2018) considered the proficiency of three distinctive microalgae strains, *Chlorella sorokiniana*, *Chlorella vulgaris*, and *Scenedesmus obliquus*, in the bioremediation of DCF-polluted water. For this reason, microalgae were cultured in bubbling column photobioreactors (PBRs) under batch mode until the end of the exponential growth phase. For the three strains, the amount of DCF in the PBR medium diminished along with microalgae growth, which highlighted biodegradation as the primary removal mechanism. Among the three strains, *S. obliquus* was the most able to reduce DCF concentration (99% expulsion from concentration of 25,000 μg/L). In any case, a large amount of removal doesn't ensure a proficient treatment due to the high concentration of by-products. Consequently, for a complete assessment of the microalgae removal proficiency, the last effluents from the PBRs were tried for their impacts on the early-stage advancement of zebrafish. Once more, the *S. obliquus* treatment was the most proficient in decreasing toxicity, with the relating effluents affecting the incipient organism's mortality. Ben Ouada et al. (2019) examined algal removal of DCF by two microalgal strains *Picocystis* sp. and *Graesiella* sp., respectively. They were grown under various DCF concentrations and their development, photosynthetic activity, and DCF removal proficiency were observed. Results showed that DCF had slight inhibitory consequences for the microalgal growth, which didn't surpass 21% for *Picocystis* and 36% for *Graesiella* after 5 days of incubation. The two species showed different patterns of DCF removal. In the presence of *Picocystis* sp., the amount of eliminated DCF was up to 73%, 43%, and 25% of 25, 50, and 100 mg/L, respectively, while just 52%, 28%, and 24% were removed by *Graesiella* at the same DCF concentration. DCF removal was done essentially by biodegradation. Silva et al. (2020) considered the valorization of the green microalga *Scenedesmus obliquus* biomass, which was utilized for the biosorption of two non-steroidal mitigating drugs, specifically salicylic acid and ibuprofen, from water. Results of adsorption studies were found to fit the pseudo-second-order kinetics and the Langmuir isotherm model, respectively. The Langmuir maximum capacity for adsorption for salicylic acid (63 mg/g) was found to be higher than for ibuprofen (12 mg/g), which was compared with activated carbon as reference (250 and 147 mg/g, respectively). For the two drugs, the adsorption onto microalgae biomass was spontaneous, favorable, and exothermic. It was found that the removal of salicylic acid and ibuprofen by *Scenedesmus obliquus* biomass happened by physical interactions. Encarnação et al. (2020) contemplated the performance of free and immobilized cells of microalgae *Nannochloropsis* sp. in eliminating four drugs, i.e., paracetamol, ibuprofen, olanzapine, and simvastatin. The outcomes showed that free microalgae cells stay alive for an extended time than the immobilized ones, recommending the restraint of cell multiplication by the polymeric matrix polyvinyl alcohol. Both free and immobilized cells react distinctively to each drug. The expulsion of paracetamol and ibuprofen by *Nannochloropsis* sp., after 24 hours of culture, was higher in immobilized cells. Free cells eliminated an essentially higher concentration of olanzapine than immobilized ones, recommending a higher affinity to

TABLE 5.3
Removal of NSAIDS by Microalgae

Microalgae	Chemical	Concentration	Removal %	Reference
Chlorella sp.	Ketoprofen	5 mM	82	Ismail et al. 2016
Scenedesmus obliquus	DCF	25,000 μg/L	99	Escapa et al. 2018
Picocystis sp.	DCF	25 mg/L	73	Ben Ouada et al. 2019
Graesiella sp.	DCF	25 mg/L	52	Ben Ouada et al. 2019
Navicula sp.	Atenolol, carbamazepine, ibuprofen and naproxen	–	>90	Ding et al. 2020

this particle than to paracetamol and ibuprofen. The outcomes show the adequacy of *Nannochloropsis* sp. free cells for eliminating olanzapine and *Nannochloropsis* sp. immobilized cells for eliminating paracetamol and ibuprofen. Ding et al. (2020) studied the removal of mixed drugs by freshwater diatom *Navicula* sp. Results showed that *Navicula* sp. could proficiently eliminate atenolol, carbamazepine, ibuprofen, and naproxen with efficiencies of >90% after 21 days of incubation. When compared with the expulsion efficiencies of every drug in the individual drug medicines, the removal of sulfamethoxazole (SMX), bezafibrate, and naproxen was improved in the mixed treatment, though the removal efficiencies of carbamazepine and atenolol diminished. Also, the presence of hydrophobic drugs (i.e., ibuprofen and naproxen) speed up the removal of carbamazepine and SMX and restrained the removal of atenolol in the mixture of six drugs (Table 5.3.).

5.2.4 Antibiotics

Products and wastes from the pharmaceutical industries have played an important role in the deterioration of the environment. Pharmaceutical compounds, when present in the environment, can have very deleterious consequences (Xiong et al. 2017). A variety of pharmaceutical compounds have been detected in high concentrations across various water bodies around the world (Martinez, 2009). In India and abroad, after the use of pharmaceuticals, companies discharge their effluents in open places and into streams directly or after efficient treatment. Pharmaceutical compounds reaching the environment are primarily attributed to discharge of effluents, either treated or untreated, and inappropriate dumping of expired or unused drugs, etc. (Kraemer et al. 2019). The global use of antibiotics increased steadily over the past decades, both due to an augmentation of antibiotic use in human medicine and other sectors of commercial activity. Antibiotic consumption in livestock reached 63,151 tons in 2010 and is predicted to increase by another 67% by 2030 (Kabir et al. 2015). Antibiotic use is also rising in aquaculture, the fastest-growing food sector worldwide due to intensive farming. For this reason, antibiotics of pharmaceutical origin are now found in large quantities in human-made environments such as sewage and WWTPs (Larsson, 2014). One of the most noted consequences of antibiotic misuse and antibiotic pollution is the increased frequency of bacteria harboring antibiotic resistance genes (ARGs) in different environments (Danner et al. 2019). Therefore, the removal of antibiotics from different environments is of utmost importance. deGodos et al. (2012) examined the components of tetracycline removal in High Rate Algal Ponds (HRAP) frameworks. For this reason, two HRAPs were taken care of with manufactured wastewater for 46 days before tetracycline was added at 2 mg/L to the influent of one of the reactors (Te-HRAP). From the 62nd day, tetracycline removal was around 69±1% in the Te-HRAP, and this elimination was due to photo-degradation and biosorption. Tetracycline addition in the HRAP was followed by the deflocculation of the Te-HRAP biomass, but it had no effect on COD and biomass productivity. The outcomes of the study exhibited that shallow geometry of HRAPs is advantageous to support the photo-degradation of antibiotics during wastewater biological treatment. Guo et al. (2016) examined the possibility of utilizing lipid-accumulating microalgae to eliminate cephalosporin antibiotic, 7-amino cephalosporanic acid (7-ACA) from wastewater with the extra advantage of biofuels creation. Three microalgal strains (*Chlorella* sp. Cha-01, *Chlamydomonas* sp. Tai-03, and *Mychonastes* sp. YL-02) were grown under 7-ACA stress, and their biomass efficiency, lipid creation, and N-NO_3 - utilization were observed. It was tracked down that 7-ACA had slight hindrance on the microalgal development in the ratio of 12.0% (Cha-01), 9.6% (YL-02), and 11.7% (Tai-03). Nonetheless, lipid accumulation in the three microalgae was not impacted by the presence of 7-ACA. The examination of the 7-ACA removal during microalgal development shows that 7-ACA was predominantly eliminated by adsorption just as hydrolysis and photolysis responses. Xiong et al. (2017) observed that biodegradation of levofloxacin(LEV) was higher by fresh water microalgae *Chlorella vulgaris*, under moderately saline conditions Among the six wild species (*Chlamydomonas mexicana*, *Chlamydomonas pitschmannii*, *Chlorella vulgaris*, *Ourococcus multisporus*, *Micractinium resseri*, *Tribonema aequale*), *C. vulgaris* showed the most elevated removal rate (12%) of LEV at 1 mg/L. The acclimated *C. vulgaris*, which was incubated in 200 mg/L of LEV for 11 days, showed improved removal of 1 mg LEV L^{-1} by 16% following 11 days of growth. The addition of 1% (w/v) sodium chloride into the microalga media essentially improved LEV removal by >80% in the *C. vulgaris* culture. The bioaccumulation of LEV at day 11 in *C. vulgaris* cells without NaCl was 34 mg/g, which was raised to 101 mg/g LEV at 1% NaCl. The bioconcentration factor for LEV was 34 and 1004 out of 0 and 1% NaCl, respectively. The mass equilibrium investigation of LEV showed that over 90% of LEV was biodegraded by *C. vulgaris* at day 11 with the addition of 1% NaCl. Xie et al. (2020) contemplated that ciprofloxacin (CIP) and sulfadiazine (SDZ) removal was accomplished at 100% and 54.53%, respectively, with starch efficiency of >1000 mg/L/d by *Chlamydomonas* sp. Tai-03. The end courses showed that CIP removal was mostly accomplished by biodegradation (65.05%) while SDZ was basically taken out by photolysis (35.60%). Kiki et al. (2020) studied the removal capability of four microalgae viz. *Haematococcus pluvialis, Selenastrum capricornutum, Scenedesmus quadricauda,* and *Chlorella vulgaris* against ten antibiotics (sulfamerazine, SMX, sulfamonomethoxine, trimethoprim, clarithromycin, azithromycin, roxithromycin (ROX), lomefloxacin, LEV, and flumequine) in a progression of synthetic wastewater batch culture tests, kept at 20 µg,

50 μg and 100 μg/L initial concentration levels and over a time of 40 days. Biodegradation was the main mechanism of antibiotic removal, with minor role of bio adsorption, bioaccumulation, and abiotic factors. Antibiotic elimination followed the pseudo-first-order kinetics, with the fastest antibiotic removal rate accomplished by *H. pluvialis*. Monod energy was effectively applied to clarify the connection between algal growth and expulsion of antibiotics and nutrients in batch culture. *S. capricornutum* and *C. vulgaris* showed more affinity for macrolides and fluoroquinolones than for sulfonamides, while *H. pluvialis* and *S. quadricauda* showed a moderately higher inclination for sulfonamides than for the other anti-microbial groups. Yang et al. (2020) examined that *Scenedesmus obliquus* was cultured in the presence of varying concentrations of ofloxacin (OFL) in BG11 medium. In this culture system, *S. obliquus* could adequately eliminate OFL with a concentration of 10 mg/L; the removal efficiency was limited under higher dosages (20–320 mg/L). Then, the lipid content essentially increased by 21.10–49.63%, which was brought about via carbon being changed over from carbohydrate to lipid. The best lipid productivity (7.53 mg/L/d) happened at an OFL concentration of 10 mg/L, which was roughly 1.5 times more than the control. Additionally, *S. obliquus* cultured with OFL had the potential to improve the biodiesel quality because of an increment in saturated fatty acids and a reduction in unsaturated fatty acids. This investigation shows that the microalgae-antibiotic framework is a promising answer to accomplish antibiotic removal and biofuel production. Da Silva Rodrigues et al. (2020) examined that SMX has been usually found in WWTP effluents. SMX and different antibiotics can be considered as environmental contaminants of emerging concern. Because of their toxicity and their potential for the advancement of bacterial resistance, their presence in the aquatic system poses a threat to human wellbeing. This investigation assessed the bioremediation of SMX in WWTP effluents utilizing a tertiary therapy made by a microalgae-bacteria consortium under low power LED illumination, and furthermore the evaluation of sulfonamide resistance gene (sul1). The expulsion of SMX from WWTP effluents was 54.34±2.35%, in which the microalgae-bacteria consortium improved the performance of SMX removal. The fundamental interaction of SMX removal can be ascribed to the biodegradation by bacteria because of the increase in oxygen delivered by microalgae photosynthesis. Hence, the microalgae-bacteria consortium utilized in this study was shown to be a promising option for bioremediation of SMX, with the potential for expulsion of other foreign substances from wastewater effluent. Hena et al. (2020) examined the removal of metronidazole (MDZ) from liquid media by *C. vulgaris*. Two distinctive starting sizes of inoculum (0.05 and 0.5 g L−1) were tried for a wide concentration range of MDZ (1–50 μM). The impact of MDZ concentration on biomass production was evaluated for 20 days. Exopolymeric substances (EPS) were measured and connected with the removal of antibiotics from culture media. In particular, MDZ invigorated the creation of EPS in *C. vulgaris*, which assumed a significant part in the adsorption of this antibiotic. Additionally, MDZ altogether affected the zeta potential of *C. vulgaris* in the test culture. This abatement in surface negative charge caused auto-flocculation at a stationary phase. Toxicity experiments showed that MDZ was toxic to *C. vulgaris* at the stationary phase. Results from this investigation would propel our insight into the treatment of MDZ-contaminated waters with *C. vulgaris*. Li et al. (2020) considered

that ROX has received concern because of its enormous use, ubiquitous nature, and high ecotoxicology hazard. This study researched the chronic effects of ROX on the development, chlorophyll content, antioxidant enzymes, and malonaldehyde (MDA) content of *Chlorella pyrenoidosa* and also the removal efficiency of ROX during microalgae development. After 96 hours of exposure, 1.0–2.0 mg/L of ROX altogether restrained the synthesis of chlorophyll and increased the activities of Superoxide dismutase (SOD) and Catalase (CAT ($p < 0.05$). The MDA content expanded with the ROX concentration, increasing from 0.5–1.0 mg/L, and afterward diminished to 105.76% of the control exposure to 2.0 mg/L. During the 21 days of exposure, low amount of ROX (0.1 and 0.25 mg/L) showed no critical impact on the development and chlorophyll content of microalgae during the initial 14 days; however, it fundamentally hindered the development of microalgae and the amount of chlorophyll at 21 days. ROX (1.0 mg/L) fundamentally hindered the development of microalgae during 3–21 days and the amount of chlorophyll at 7–21 days. During the first 14 days, abiotic removal of ROX played a more important role, contributing about 12.21–21.37% of ROX removal. After 14 days, the biodegradation of ROX by *C. pyrenoidosa* gradually became a more important removal mechanism, contributing about 45.99–53.30% of ROX removal at 21 days. Bio adsorption and bioaccumulation both played minor roles in the removal of ROX during algae cultivation (Table 5.4).

5.2.5 Pesticides

Pesticides include herbicides, fungicides, insecticides, nematicides, rodenticide, and anti-rodent drugs, etc. (Tsaboula et al. 2018). In modern agriculture, pesticides occupy an indispensable position and are widely used in industries such as forestry and animal husbandry to enhance crop production and increase economic benefits (Foo and Hameed, 2010). However, the indiscriminate utilization of pesticides could lead to a sharp increase in and accumulation of pesticides in the environment, resulting

TABLE 5.4
Removal of Antibiotics by Microalgae

Microalgae	Chemical	Concentration	Removal %	Reference
Chlorella sp. Cha-01, *Chlamydomonas* sp. Tai-03 and *Mychonastes* sp. YL-02	Cephalosporin antibiotic, 7-ACA	–	–	Guo et al. 2016
Chlorella vulgaris	LEV	200 mgL^{-1}	90%	Xiong et al. 2017
Chlamydomonas sp. Tai-03	CIP and SDZ	>1000 mgL^{-1}	100% 54.53%	Xie et al. 2020
S. obliquus	OFL	10 mg/L		Yang et al. 2020
Microalgae-bacteriaconsortium	SMX		54.34±2.35%	da Silva Rodrigues et al. 2020
C. vulgaris	MDZ			Hena et al. 2020
C. pyrenoidosa	ROX	1.0 mg/L	12.21–21.37%	Li et al. 2020

in serious pesticide contamination (Song et al. 2019). Due to the popularization of mechanized agricultural production, pesticides are mostly applied via spraying by machines and even drones, which could inevitably contaminate the atmosphere, soil, and water (Huizhen et al. 2014). Agricultural runoff also contributes to the contamination of adjacent water bodies. The entry of pesticides into the environment is associated with multiple environmental and health hazards like the toxicity of the pesticide itself will inhibit the growth of plants and animals. Pesticides can enter the food chain and gradually bio-accumulate in higher trophic level organisms or even the human body via the bio-magnification process (Kabra et al. 2014). After exposure to pesticides, a series of symptoms such as itchy skin, inflammation of the nose and throat, headache, rash and blisters on the skin, nausea, vomiting, diarrhea, dizziness, blindness, blurred vision, and very few deaths are possible (usually within 24 hours), but in most cases, the acute effects of pesticides are not serious and do not require medical attention every time (Boudh and Singh, 2019).

Matamoros and Rodríguez (2016) examined the impact of microalgae on pesticide removal and also the impact of continuous vs batch feeding. The following pesticides were considered: mecoprop, atrazine, simazine, diazinone, alachlor, chlorfenvinphos, lindane, malathion, pentachlorobenzene, chlorpyrifos, endosulfan, and clofibric. In this study, 2L batch reactors and 5L continuous reactors were supplemented with 10 μgL(−1) of each pesticide. Also, three different hydraulic retention times (HRTs) viz. 2, 4, and 8 days were assessed in the continuous feeding reactors. Batch-feeding experiments showed that the microalgae removed the pesticides, namely lindane, alachlor, and chlorpyrifos, by 50%. Continuous feeding reactors exhibited higher removal power for pesticides, namely pentachlorobenzene, chlorpyrifos, and lindane. It was observed that longer HRTs increased the removal of pesticides, a low HRT of 2 days eliminated malathion, pentachlorobenzene, chlorpyrifos, and endosulfan by up to 70%. Hultberg and Bodin (2018) studied the use of microalga *Chlorella vulgaris*, the fungus *Aspergillus niger*, and biopellets made out of these microorganisms in the elimination of pesticides from polluted water. A combination of 38 pesticides was tried, and the concentration of 17 of these was discovered to be diminished altogether in the biopellet treatment, as compared to control. After incubation, the concentration of all pesticides in the algal treatment didn't reduce essentially from that in the control. Notwithstanding, in the fungal and biopellet treatment, the removal was essentially lower (59.6±2.0 μg/L and 56.1±2.8 μg/L, respectively) than in the control (66.6±1.0 μg/L). In this manner, algal and fungal treatment through biopellet can likewise give an extension to eliminating organic pollutants from wastewater. Hu et al. (2021) reported the removal efficiency and bioaccumulation of atrazine in the microalgae after an 8-day exposure to a solution containing 40 μg/L and 80 μg/L of atrazine and also its degradation products. The microalgae were also exposed to pure atrazine solution of similar concentration. During incubation, atrazine was photocatalytically degraded and 31.4% of atrazine was removed after 60 minutes, and three degraded metabolites, namely, desisopropyl-atrazine (DIA), desethyl-atrazine (DEA), and desethyl-desisopropyl-atrazine (DEIA), were recognized. After 8 days of incubation, 83.0% and 64.3% of atrazine were eliminated from solutions containing 40 μg/L and 80 μg/L of atrazine, respectively. In correlation with the control, i.e., pure atrazine solution with equivalent concentration, *Chlorella* sp.

in the degraded atrazine solution showed lower removal efficiency and development rate. It was due to the inhibitory effect of degraded metabolites on photosynthetic parameters. There was a potential inhibitory effect of degraded products on the growth of the microalgae. Atrazine and the metabolites hindered algal photosynthesis by means of low light absorption and electron transport, and decreased use of light energy through energy dispersal. These outcomes exhibited that microalgae *Chlorella* sp. had an empowering atrazine-removing potential and the metabolites of atrazine may restrain algal growth and removal ability. Wan et al. (2020) examined that the toxicity of trichlorfon (TCF) to freshwater green algae *Chlamydomonas reinhardtii* and also its biodegradation and metabolic fate. The growth of *C. reinhardtii* diminished with increasing TCF concentration, with growth inhibition of 51.3% at 200 mg/L TCF compared with the control. In terms of pigment content, chlorophyll content, and antioxidant enzymes, *C. reinhardtii* exhibited resistance and acclimatization towards TCF. A 100% biodegradation rate was accomplished at a concentration of 100 mg/L TCF. Ten metabolites were distinguished by GC-MS, and the degradation pathways of TCF by the microalgae were proposed. This study demonstrated that *C. reinhardtii* can be utilized to eliminate TCF from characteristic natural water environments and to treat TCF-infected wastewater. Avila et al. (2021) chose three hydrophobic pesticides, chlorpyrifos, cypermethrin, and oxadiazon, to study the potential for their removal from the aquatic system by a microalgae consortium. An abiotic and a killed control containing thermally inactivated algal biomass were utilized to explain their removal pathways, and pesticide content was evaluated in liquid and biomass stages for 7 days. At incubation, complete removal (biodegradation in addition to photodegradation) contributed to the removal of 55% of oxadiazon, 35% of chlorpyrifos, and 14% of cypermethrin. Moreover, over 60% of chlorpyrifos and cypermethrin were removed by sorption by microalgae biomass. By and large, the three pesticides showed a high rate of elimination from the aqueous stage. O,O-diethyl thiophosphate was identified as a degraded metabolite of chlorpyrifos by microalgae elimination. Phycoremediation was combined with anaerobic degradation of the microalgae biomass containing the pesticides by sorption. Anaerobic degradation was not restrained by the pesticides as checked by methane formation yields. The removal effectiveness of the pesticides in the digestate was as per the following: chlorpyrifos>cypermethrin>oxadiazon. These results show the capability of minimal effort algal-based frameworks for the treatment of wastewater or effluents from agrochemical enterprises. The coordination of wastewater treatment with biogas creation through anaerobic digestion is a biorefinery approach that encourages the economic feasibility of the process. Weis et al. (2020) contemplated that the pesticide bifenthrin, having moderate toxicity (class II), is generally utilized as a bug spray in tobacco growth. Microalgae were recovered from a characteristic water source in the city of Santa Cruz do Sul, RS, Brazil, which is an artificial reservoir utilized for public water supply. Biodegradation, biosorption, the impact of pH, level of inoculum, and photoperiod were assessed in batch experiments for 20 days. The results indicated that microalgae isolated from the water of the lake was able to remove roughly 99% of bifenthrin through biodegradation and biosorption processes. Photodegradation was also distinguished, and the best condition for phycoremediation was 20%inoculum with a photoperiod of 18:6 hours (Table 5.5.).

TABLE 5.5
Removal of Pesticides by Microalgae

Microalgae	Chemical	Concentration	Removal %	Reference
Chlorella sp.	Atrazine	40 μg/L and 80 μg/L	83.0% and 64.3%	Hu et al. 2021
Chlamydomonas reinhardtii	TCF	100 mg L^{-1}	100%	Wan et al. 2020
Inactivated dead microalgae biomass	Chlorpyrifos, cypermethrin, and oxadiazon	–	55% of oxadiazon, 35% of chlorpyrifos, and 14% of cypermethrin	Avila et al. 2021
Microalgae	Bifenthrin	–	99%	Weis et al. 2020
Microalgae	Terbutryn, diuron, and imidacloprid	–	100%	García-Galán et al. 2020

5.3 CONCLUSION

There is a worldwide increase in the release of EDCs in the environment. These chemicals disrupt the endocrine system of the body and create havoc in the lives of animals who consume them. Therefore, the removal of these chemicals from the environment is of utmost importance. Microalgae-based treatments have gained impetus due to the fact that they can utilize organic matter and nutrients from wastewater and can also remove a number of organic pollutants from wastewater by a number of methods, which include adsorption, accumulation, and degradation/assimilation. Microalgae can remove a diverse array of pollutants such as hormones, pesticides, antibiotics, and NSAIDs from wastewater. They have shown immense potential for the removal of these chemicals with a concomitant increase in algal biomass, which can be used in the production of value-added products such as biofuels. The only drawback of this process is the very slow rate of removal of organic pollutants as compared to other microorganisms such as bacteria and fungi, which is attributed to their slow growth rate. It is anticipated that with the use of genetic engineering-based techniques, algal growth can be enhanced so that the full potential of microalgae in the removal of organic contaminants can be harnessed.

ABBREVIATIONS

7-ACA, 7-amino cephalosporanic acid; AEOM, algal extracellular natural matter; BPA, bisphenol A; CIP, ciprofloxacin; DCF, diclofenac; DEA, desethyl-atrazine; DEIA, desethyl-desisopropyl-atrazine; DES, diethylstilbestrol; DIA, desisopropyl-atrazine; E1, estrone; E2, 17β-estradiol; ED, endocrine disrupters; EDC, endocrine-disrupting chemicals; EE2, 17α-ethynylestradiol; EPA, Environmental Protection Agency; HRT, hydraulic retention time; LEV, levofloxacin; MDA, malonaldehyde; NP, nonylphenol; NSAIDS, nonsteroidal anti-inflammatory drugs; OFL, ofloxacin; OP, octylphenol; PBRs, photobioreactors; ROX, roxithromycin; SDZ, sulfadiazine; SMX, sulfamethoxazole; TCF, trichlorfon; WWTP, wastewater treatment plants

CONFLICT OF INTEREST STATEMENT

The authors declare no conflict of interest.

REFERENCES

Agrawal K, Bhatt A, Bhardwaj N, Kumar B, Verma P. Algal biomass: potential renewable feedstock for biofuels production–part I. In Biofuel production technologies: critical analysis for sustainability. 2020, pp. 203–237. Springer, Singapore.

Ali M., Wang W., Chaudhry N., Geng Y. Hospital waste management in developing countries: A mini review. Waste Manag. Res. 2017;35:581–592. doi: 10.1177/0734242X17691344.

Amaral Mendes JJ. The endocrine disrupters: a major medical challenge. Food Chem Toxicol. 2002;40:781–788. DOI: 10.1016/s0278-6915(02)00018-2

aus der Beek T, Weber FA, Bergmann A, Hickmann S, Ebert I, Hein A, Küster A. Pharmaceuticals in the environment–global occurrences and perspectives. Environ Toxicol Chem. 2016;35:823–835. DOI: 10.1002/etc.3339

Avila R, Peris A, Eljarrat E, Vicent T, Blánquez P. Biodegradation of hydrophobic pesticides by microalgae: Transformation products and impact on algae biochemical methane potential. Sci Total Environ. 2021;754:142114. DOI: 10.1016/j.scitotenv.2020.142114

Barros A I, Gonçalves A L, Simões M, Pires J C M. Harvesting techniques applied to microalgae: A review. Renew Sustain Energy Rev. 2015;41(C):1489–1500. DOI: 10.1016/j.rser.2014.09.037.

Ben Ali R, Ben Ouada S, Leboulanger C, Ammar J, Sayadi S, Ben Ouada H. Bisphenol A removal by the Chlorophyta Picocystis sp.: optimization and kinetic study. Int J Phytoremediation. 2020;1–11. DOI: 10.1080/15226514.2020.1859985.

Ben Ouada S, Ben Ali R, Cimetiere N, Leboulanger C, Ben Ouada H, Sayadi S. Biodegradation of diclofenac by two green microalgae: Picocystis sp. and Graesiella sp. Ecotoxicol Environ Saf. 2019;186:109769.

Ben Ouada S, Ben Ali R, Leboulanger C, et al (a). Effect and removal of bisphenol A by two extremophilic microalgal strains (Chlorophyta). J Appl Phycol. 2018;30:1765–1776. DOI: 10.1007/s10811-017-1386-x.

Ben Ouada S, Ben Ali R, Leboulanger C, Ben Ouada H, Sayadi S(b). Effect of Bisphenol A on the extremophilic microalgal strain *Picocystis* sp. (Chlorophyta) and its high BPA removal ability. Ecotoxicol Environ Saf. 2018;158:1–8. DOI: 10.1016/j.ecoenv.2018.04.008.

Bhardwaj N, Agrawal K, Verma P. Algal biofuels: an economic and effective alternative of fossil fuels. In Microbial strategies for techno-economic biofuel production. 2020, pp. 207–227. Springer, Singapore.

Bolz U, Hagenmaier H, Körner W. Phenolic xenoestrogens in surface water, sediments, and sewage sludge from Baden-Württemberg, south-west Germany. Environ Pollut. 2001;115(2):291–301. doi: 10.1016/s0269-7491(01)00100-2.

Boudh S, Singh J S. Pesticide contamination: environmental problems and remediation strategies. In Emerging and eco-friendly approaches for waste management. 2019, pp. 245–269. Springer.

Chikae M, Ikeda R, Hasan Q, Morita Y, Tamiya E. Effects of tamoxifen, 17-ethynylestradiol, flutamide, and methyl testosterone on plasma vitellogenin levels of male and female Japanese medaka (Oryzias latipes). Environ Toxicol Pharmacol. 2004;17:29–33. DOI: 10.1016/j.etap.2004.02.002

Chisti Y. Society and microalgae: Understanding the past and present. In Microalgae in health and disease prevention. I.A. Levine, J. Fleurence(Eds.). 2018, pp. 11–21. AcademicPress.

Crain D A, Eriksen M, Iguchi T, Jobling S, Laufer H, Leblanc G A, Guillette L J Jr. An ecological assessment of bisphenol-A: Evidence from comparative biology. Reprod Toxicol. 2007;24:225–239. DOI: 10.1016/j.reprotox.2007.05.008

da Silva Rodrigues DA, da Cunha CCRF, Freitas MG, de Barros ALC, E Castro PBN, Pereira AR, de Queiroz Silva S, da Fonseca Santiago A, de Cássia Franco Afonso RJ. Biodegradation of sulfamethoxazole by microalgae-bacteria consortium in wastewater treatment plant effluents. Sci Total Environ. 2020;749:141441. DOI: 10.1016/j.scitotenv.2020.141441

Dalrymple OK,Yeh DH, Trotz MA. Review: removing pharmaceuticals and endocrine-disrupting compounds from wastewater by photocatalysis. J Chem Technol Biotechnol. 2007;82:121–134.

Danner MC, Robertson A, Behrends V, Reiss J. Antibiotic pollution in surface fresh waters: Occurrence and effects. Sci Total Environ. 2019;664:793–804. DOI: 10.1016/j.scitotenv.2019.01.406.

deGodos I, Muñoz R, Guieysse B. Tetracycline removal during wastewater treatment in high-rate algal ponds. J Hazard Mater. 2012;229–230:446–449. DOI: 10.1016/j.jhazmat.2012.05.106

Ding T, Lin K, Yang B, Yang M, Li J, Li W, Gan J. Biodegradation of naproxen by freshwater algae *Cymbella* sp. and *Scenedesmus quadricauda* and the comparative toxicity. Bioresour Technol. 2017;238:164–173. DOI: 10.1016/j.biortech.2017.04.018

Ding T, Wang S, Yang B, Li J. Biological removal of pharmaceuticals by *Navicula* sp. and biotransformation of bezafibrate. Chemosphere. 2020;240:124949. DOI: 10.1016/j.chemosphere.2019.124949

Doherty LF, Bromer JG, Zhou Y, Aldad TS, Taylor HS. In utero exposure to diethylstilbestrol (DES) or bisphenol-A (BPA) increases EZH2 expression in the mammary gland: An epigenetic mechanism linking endocrine disruptors to breast cancer. Horm Cancer. 2010;1:146–155. DOI: 10.1007/s12672-010-0015-9

Encarnação T, Palito C, Pais AACC, Valente AJM, Burrows HD. Removal of pharmaceuticals from water by free and immobilised microalgae. Molecules. 2020;25:3639.

Escapa C, Torres T, Neuparth T, Coimbra RN, García AI, Santos MM, Otero M. Zebrafish embryo bioassays for a comprehensive evaluation of microalgae efficiency in the removal of diclofenac from water. Sci Total Environ. 2018;640–641:1024–1033. DOI: 10.1016/j.scitotenv.2018.05.353

Fasano E, Bono-Blay F, Cirillo T, Montuori P, Lacorte S. Migration of phthalates, alkylphenols, bisphenol A and di(2-ethylhexyl) adipate from food packaging. Food Control. 2012;27:132–138. DOI: 10.1016/j.foodcont.2012.03.005

Fernandez SV, Huang Y, Snider KE, Zhou Y, Pogash TJ, Russo J. Expression and DNA methylation changes in human breast epithelial cells after bisphenol A exposure. Int J Oncol. 2012;41:369–377. DOI: 10.3892/ijo.2012.1444.

Ferreira LL, Couto R, Oliveira PJ. Bisphenol A as epigenetic modulator: Setting the stage for carcinogenesis? Eur J Clin Invest. 2015;45(1):32–36. DOI: 10.1111/eci.12362.

Foo, KY, Hameed, BH. Detoxification of pesticide waste via activated carbon adsorption process. J Hazard Mater. 2010;175(1–3):1–11. DOI: 10.1016/j.jhazmat.2009.10.014.

García-Galán MJ, Monllor-Alcaraz LS, Postigo C, Uggetti E, López de Alda M, Díez-Montero R, García J. Microalgae-based bioremediation of water contaminated by pesticides in peri-urban agricultural areas. Environ Pollut. 2020;265(B):114579.

Ge L, Deng H, Wu F, Deng N. Microalgae-promoted photodegradation of two endocrine disrupters in aqueous solutions. J Chem Technol Biotechnol 2009;84:331–336. DOI: 10.1002/jctb.2043

Goswami RK, Agrawal K, Mehariya S, Molino A, Musmarra D, Verma P. Microalgae-based biorefinery for utilization of carbon dioxide for production of valuable bioproducts. In Chemo-Biological Systems for CO_2 Utilization. 2020, pp. 199–224. Taylor & Francis.

Goswami RK, Mehariya S, Verma P, Lavecchia R, Zuorro A. Microalgae-based biorefineries for sustainable resource recovery from wastewater. JWater Process Eng. 2021a;40:101747. DOI: 10.1016/j.jwpe.2020.101747

Goswami RK, Mehariya S, Karthikeyan OP, Verma P. Advanced microalgae-based renewable biohydrogen production systems: A review. Bioresour Technol. 2021b;320(A):124301. DOI: 10.1016/j.biortech.2020.124301

Goswami RK, Agrawal K, Verma P. An overview of microalgal carotenoids: advances in the production and its impact on sustainable development. In Bioenergy research: evaluating strategies for commercialization and sustainability. 2021c, pp.105–128. Taylor & Francis.

Guo WQ, Zheng HS, Li S, Du JS, Feng XC, Yin RL, Wu QL, Ren NQ, Chang JS. Removal of cephalosporin antibiotics 7-ACA from wastewater during the cultivation of lipid-accumulating microalgae. Bioresour Technol. 2016;221:284–290. DOI: 10.1016/j.biortech.2016.09.036

He N, Sun X, Zhong Y, Sun K, Liu W, Duan S. Removal and biodegradation of nonylphenol by four freshwater microalgae. Int J Environ Res Public Health. 2016;13(12):1239. DOI: 10.3390/ijerph13121239.

Hena S, Gutierrez L, Croué JP. Removal of metronidazole from aqueous media by C. vulgaris. J Hazard Mater. 2020;384:121400. DOI: 10.1016/j.jhazmat.2019.121400

Hense BA, Welzl G, Severin GF, Schramm KW. Nonylphenol induced changes in trophic web structure of plankton analysed by multivariate statistical approaches. Aquat Toxicol. 2005 Jun 15;73(2):190–209. doi: 10.1016/j.aquatox.2005.03.010.

Hom-Diaz A, Llorca M, Rodríguez-Mozaz S, Vicent T, Barceló D, Blánquez P. Microalgae cultivation on wastewater digestate: β-estradiol and 17α-ethynylestradiol degradation and transformation products identification. J Environ Manage. 2015;155:106–113. DOI: 10.1016/j.jenvman.2015.03.003.

Hu N, Xu Y, Sun C, Zhu L, Sun S, Zhao Y, Hu C. Removal of atrazine in catalytic degradation solutions by microalgae *Chlorella* sp. and evaluation of toxicity of degradation products via algal growth and photosynthetic activity. Ecotoxicol Environ Saf. 2021;207:111546. DOI: 10.1016/j.ecoenv.2020.111546.

Huang B, Tang J, He H, Gu L, Pan X. Ecotoxicological effects and removal of 17β-estradiol in chlorella algae. Ecotoxicol Environ Saf. 2019;174:377–383. DOI: 10.1016/j.ecoenv.2019.01.129.

Huizhen L, Zeng EY, Jing, Y. Mitigating pesticide pollution in China requires law enforcement, farmer training, and technological innovation. Environ Toxicol Chem. 2014;33(5):963–971. DOI: 10.1002/etc.2549

Hultberg M, Bodin H. Effects of fungal-assisted algal harvesting through biopellet formation on pesticides in water. Biodegradation. 2018;29(6):557–565. DOI: 10.1007/s10532-018-9852-y

Ismail MM, Essam TM, Ragab YM, Mourad FE. Biodegradation of ketoprofen using a microalgal-bacterial consortium. BiotechnolLett. 2016;38(9):1493–1502. DOI: 10.1007/s10529-016-2145-9

Jobling S, Nolan M, Tyler CR, Brighty GC, Sumpter JP. Widespread sexual disruption in wild fish. Environ SciTechnol. 1998;32:2498–2506.

Kabir ER, Rahman MS, Rahman I. A review on endocrine disruptors and their possible impacts on human health. Environ Toxicol Pharmacol. 2015;40:241–58. DOI: 10.1016/j.etap.2015.06.009.

Kabra AN, Ji MK, Choi J, Kim JR, Govindwar SP, Jeon BH. Toxicity of atrazine and its bioaccumulation and biodegradation in a green microalga *Chlamydomonas mexicana*. Environ Sci Pollut Res Int. 2014;21(21):12270–12278. DOI: 10.1007/s11356-014-3157-4

Kiki C, Rashid A, Wang Y, Li Y, Zeng Q, Yu CP, Sun Q. Dissipation of antibiotics by microalgae: Kinetics, identification of transformation products and pathways. J Hazard Mater. 2020;387:121985. DOI: 10.1016/j.jhazmat.2019.121985

Kiyama R, Wada-Kiyama Y. Estrogenic endocrine disruptors: Molecular mechanisms of action. Environ Int. 2015;83:11–40. DOI: 10.1016/j.envint.2015.05.012.

Knillmann S, Orlinskiy P, Kaske O, Foit K, Liess M. Indication of pesticide effects and recolonization in streams. Sci Total Environ. 2018 Jul 15;630:1619–1627. doi: 10.1016/j.scitotenv.2018.02.056.

Kraemer SA, Ramachandran A, Perron GG. Antibiotic pollution in the environment: From microbial ecology to public policy. Microorganisms. 2019;7(6):180. DOI: 10.3390/microorganisms7060180.

Kumar B, Agrawal K, Bhardwaj N, Chaturvedi V, Verma P. Techno-economic assessment of microbe-assisted wastewater treatment strategies for energy and value-added product recovery. In Microbial Technology for the Welfare of Society. 2019, pp.147–181. Springer.

Larsson DG. Antibiotics in the environment. Upsala J Med Sci. 2014;119(2):108–112. DOI: 10.3109/03009734.2014.896438.

Li R, Chen GZ, Tam NF, Luan TG, Shin PK, Cheung SG, Liu Y. Toxicity of bisphenol A and its bioaccumulation and removal by a marine microalga *Stephanodiscus hantzschii*. Ecotoxicol Environ Saf. 2009;72(2):321–328. DOI: 10.1016/j.ecoenv.2008.05.012.

Li J, Min Z, Li W, Xu L, Han J, Li P. Interactive effects of roxithromycin and freshwater microalgae, *Chlorella pyrenoidosa*: Toxicity and removal mechanism. Ecotoxicol Environ Saf. 2020;191:110156. DOI: 10.1016/j.ecoenv.2019.110156

Liu W, Chen Q, He N, Sun K, Sun D, Wu X, Duan S. Removal and biodegradation of 17β-estradiol and diethylstilbestrol by the freshwater microalgae *Raphidocelis subcapitata*. Int J Environ Res Public Health. 2018;15(3):452. DOI: 10.3390/ijerph15030452.

Liu Y, Wang L, Pan B, Wang C, Bao S, Nie X. Toxic effects of diclofenac on life history parameters and the expression of detoxification-related genes in Daphnia magna. Aquat Toxicol. 2017;183:104–113. DOI: 10.1016/j.aquatox.2016.12.020

López-Pacheco IY, Salinas-Salazar C, Silva-Núñez A, Rodas-Zuluaga LI, Donoso-Quezada J, Ayala-Mar S, Barceló D, Iqbal HMN, Parra-Saldívar R. Removal and biotransformation of 4-nonylphenol by *Arthrospira maxima* and *Chlorella vulgaris* consortium. Environ Res. 2019;179(B):108848. DOI: 10.1016/j.envres.

Machado, M.D., Soares, E.V. Sensitivity of freshwater and marine green algae to three compounds of emerging concern. J Appl Phycol 31, 399–408 (2019).

Martinez JL. Environmental pollution by antibiotics and by antibiotic resistance determinants. Environ Pollut. 2009;157(11):2893–2902. DOI: 10.1016/j.envpol.2009.05.051.

Matamoros V, Rodríguez Y. Batch vs continuous-feeding operational mode for the removal of pesticides from agricultural run-off by microalgae systems: A laboratory scale study. J Hazard Mater. 2016;309:126–132. DOI: 10.1016/j.jhazmat.2016.01.080.

Mehariya S, Kumar P, Marino T, CasellaP, Lovine A, Verma P,Musmarra D,Molino A. Aquatic weeds: A potential pollutant removing agent from wastewater and polluted soil and valuable biofuel feedstock. In:Bioremediation using weeds. 2021a, pp. 59–77. Springer.

Mehariya S, Goswami R, Verma P, Lavecchia R, Zuorro A. Integrated approach for wastewater treatment and biofuel production in microalgae biorefineries Energies. 2021b;14(8):2282. DOI: 10.3390/en14082282

Mehariya S, Goswami RK, Karthikeysan OP, Verma P. Microalgae for high-value products: A way towards green nutraceutical and pharmaceutical compounds. Chemosphere. 2021c;280:130553. DOI: 10.1016/j.chemosphere.2021.130553

Mnif W, Hassine AI, Bouaziz A, Bartegi A, Thomas O, Roig B. Effect of endocrine disruptor pesticides: a review. Int J Environ Res Public Health. 2011;8:2265–2303. DOI: 10.3390/ijerph8062265.

Moreira IS, Bessa VS, Murgolo S, Piccirillo C, Mascolo G, Castro PML. Biodegradation of diclofenac by the bacterial strain Labrys portucalensis F11. Ecotoxicol Environ Saf. 2018;152:104–113. DOI: 10.1016/j.ecoenv.2018.01.040

Naik P, Vijayalaxmi KK. Cytogenetic evaluation for genotoxicity of bisphenol-A in bone marrow cells of Swiss albino mice. Mutat Res. 2009;676:106–112. DOI: 10.1016/j.mrgentox.2009.04.010

Norvill ZN, Shilton A, Guieysse B. Emerging contaminant degradation and removal in algal wastewater treatment ponds: Identifying the research gaps. J Hazard Mater. 2016 Aug 5;313:291–309. doi: 10.1016/j.jhazmat.2016.03.085.

Pal A, He Y, Jekel M, Reinhard M, Gin KY (2014) Emerging contaminants of public health significance as water quality indicator compounds in the urban water cycle. Environ Int 71:46–62.

Palli L, Castellet-Rovira F, Pérez-Trujillo M, Caniani D, Sarrà-Adroguer M, Gori R. Preliminary evaluation of *Pleurotus ostreatus* for the removal of selected pharmaceuticals from hospital wastewater. Biotechnol. Prog. 2017;33:1529–1537.

Pi N, Ng JZ, Kelly BC. Bioaccumulation of pharmaceutically active compounds and endocrine disrupting chemicals in aquatic macrophytes: results of hydroponic experiments with *Echinodorus horemanii* and *Eichhornia crassipes*. Sci Total Environ. 2017;601–602:812–820. DOI: 10.1016/j.scitotenv.2017.05.137

Ravindran B, Gupta SK, Cho WM, Kim JK, Sang RL, Jeong KH, Dong JL, Choi HC. Microalgae potential and multiple roles—Current progress and future prospects—An overview. Sustainability. 2016;8(12):1215.

Ruksrithong C, Phattarapattamawong S. Removals of estrone and 17β-estradiol by microalgae cultivation: kinetics and removal mechanisms. Environ Technol. 2019;40(2):163–170. DOI: 10.1080/09593330.2017.1384068.

Saini KC, Yadav DS, Mehariya S, Rathore P, Kumar B, Marino T, Leone GP, Verma P, Musmarra D, Molino A. Overview of extraction of astaxanthin from *Haematococcus pluvialis* using CO2 supercritical fluid extraction technology vis-a-vis quality demands. In Global Perspectives on Astaxanthin. 2021, pp. 341–354. Academic Press.

Schug TT, Janesick A, Blumberg B, Heindel JJ. Endocrine disrupting chemicals and disease susceptibility. J Steroid Biochem Mol Biol. 2011;127:204–215. DOI: 10.1016/j.jsbmb.2011.08.007

Shao J, He Y, Li F, Zhang H, Chen A, Luo S, Gu JD. Growth inhibition and possible mechanism of oleamide against the toxin-producing cyanobacterium *Microcystis aeruginosa* NIES-843. Ecotoxicology. 2016;25(1):225–233. DOI: 10.1007/s10646-015-1582-x.

Silva A, Coimbra RN, Escapa C, Figueiredo SA, Freitas OM, Otero M. Green microalgae *Scenedesmus obliquus* utilization for the adsorptive removal of nonsteroidal anti-inflammatory drugs (NSAIDs) from water samples. Int J Environ Res Public Health. 2020;17(10):3707. DOI: 10.3390/ijerph17103707

Solé A, Matamoros V. Removal of endocrine disrupting compounds from wastewater by microalgae co-immobilized in alginate beads. Chemosphere. 2016;164:516–523. DOI: 10.1016/j.chemosphere.2016.08.047.

Song B, Xu P, Chen M, Tang W, Zeng G, Gong J, Zhang P, Ye S. Using nanomaterials to facilitate the phytoremediation of contaminated soil. Crit Rev Environ Sci Tech. 2019;49(9):791–824.

Soares, A, Guieysse, B, Jefferson, B, Cartmell, E, Lester, JN. Nonylphenol in the environment: a critical review on occurrence, fate, toxicity and treatment in wastewaters Environ. Int., 34 (2008), pp. 1033–1049.

Staples CA, Dorn PB, Klecka GM, O'Block ST, Harris LR. A review of the environmental fate, effects, and exposures of bisphenol A. Chemosphere. 1998;36:2149–2173. DOI: 10.1016/s0045-6535(97)10133-3

Swan GE, Cuthbert R, Quevedo M, Green RE, Pain DJ, Bartels P, Cunningham AA, Duncan N, Meharg AA, Oaks JL, Parry-Jones J, Shultz S, Taggart MA, Verdoorn G, Wolter, K. Toxicity of diclofenac to Gyps vultures. Biol Lett. 2006;2:279–282.

Tran NH, Gin KY. Occurrence and removal of pharmaceuticals, hormones, personal care products, and endocrine disrupters in a full-scale water reclamation plant. Sci Total Environ. 2017 Dec 1;599-600:1503–1516. doi: 10.1016/j.scitotenv.2017.05.097.

Tran NH, Reinhard M, Gin KY. Occurrence and fate of emerging contaminants in municipal wastewater treatment plants from different geographical regions-a review. Water Res. 2018 Apr 15;133:182–207. doi: 10.1016/j.watres.2017.12.029.

Tsaboula A, Papadakis E-N, Vryzas Z, Kotopoulou A, Kintzikoglou K, Papadopoulou-Mourkidou E. Assessment and management of pesticide pollution at a river basin level part I: Aquatic ecotoxicological quality indices. Sci Total Environ. 2019;653:1597–1611. DOI: https://doi.org/10.1016/j.scitotenv.2018.08.240.

Tyumina EA, Bazhutin GA, Cartagena Gómez, A dP, Ivshina, IB. Nonsteroidal anti-inflammatory drugs as emerging contaminants. Microbiology. 2020;89:148–163.

Ummalyma SB, Pandey A, Sukumaran RK, Sahoo D. Bioremediation by microalgae: Current and emerging trends for effluents treatments for value addition of waste streams. In Biosynthetic Technology and Environmental Challenges. 2018, pp. 355–375. Springer, Singapore.

Vazquez-Duhalt R, Marquez Rocha F, Ponce E, Licea AF, Viana MT. Nonylphenol, an integrated vision of a pollutant. Scientific review Appl. Ecol. Environ. Res., 4 (2005), pp. 1–25.

Wan L, Wu Y, Ding H, Zhang W. Toxicity, biodegradation, and metabolic fate of organophosphorus pesticide trichlorfon on the freshwater algae *Chlamydomonas reinhardtii*. J Agric Food Chem. 2020;68(6):1645–1653. DOI: 10.1021/acs.jafc.9b05765

Wang Y, Sun Q, Li Y, Wang H, Wu K, Yu CP. Biotransformation of estrone, 17β-estradiol and 17α-ethynylestradiol by four species of microalgae. Ecotoxicol Environ Saf. 2019;180:723–732. DOI: 10.1016/j.ecoenv.2019.05.061.

Wang R, Wang S, Tai Y, Tao R, Dai Y, Guo J, Yang Y, Duan S. Biogenic manganese oxides generated by green algae Desmodesmus sp. WR1 to improve bisphenol A removal. J Hazard Mater. 2017;339:310–319. DOI: 10.1016/j.jhazmat.2017.06.026.

Wang L, Xiao H, He N, Sun D, Duan S. Biosorption and biodegradation of the environmental hormone nonylphenol by four marine microalgae. Sci Rep. 2019;9(1):5277. DOI: 10.1038/s41598-019-41808-8.

Weis L, de Cassiade Souza Schneider R, Hoeltz M, Rieger A, Tostes S, Lobo EA. Potential for bifenthrin removal using microalgae from a natural source. Water Sci Technol. 2020;82(6):1131–1141.

Wu PH, Yeh HY, Chou PH, Hsiao WW, Yu CP. Algal extracellular organic matter mediated photocatalytic degradation of estrogens. Ecotoxicol Environ Saf. 2021;209:111818. DOI: 10.1016/j.ecoenv.2020.111818

Xiao S, Diao H, Smith MA, Song X, YeX. Preimplantation exposure to bisphenol A (BPA) affects embryo transport, preimplantation embryo development, and uterine receptivity in mice. Reprod Toxicol. 2011;32:434–441. DOI: 10.1016/j.reprotox.2011.08.010

Xie P, Chen C, Zhang C, Su G, Ren N, Ho SH. Revealing the role of adsorption in ciprofloxacin and sulfadiazine elimination routes in microalgae. Water Res. 2020;172:115475. DOI:10.1016/j.watres.2020.115475

Xiong JQ, Kurade MB, Jeon BH. Biodegradation of levofloxacin by an acclimated freshwater microalga, *Chlorella vulgaris*. Chem Eng J. 2017;313:1251–1257. DOI: 10.1016/j.cej.2016.11.017

Yang L, Ren L, Tan X, Chu H, Chen J, Zhang Y, Zhou X. Removal of ofloxacin with biofuel production by oleaginous microalgae *Scenedesmus obliquus*. Bioresour Technol. 2020;315:123738. DOI: 10.1016/j.biortech.2020.123738

Ying GG, Williams B, Kookana R. Environmental fate of alkylphenols and alkylphenol ethoxylates – a review Environ. Int., 28 (2002), pp. 215–226.

Zepp RG, Schlotzhauer PF. Influence of algae on photolysis rates of chemicals in water. Environ Sci Technol. 1983;17:462–468. DOI: 10.1021/es00114a005

Zhou GJ, Peng FQ, Yang B, Ying GG. Cellular responses and bioremoval of nonylphenol and octylphenol in the freshwater green microalga *Scenedesmus obliquus*. Ecotoxicol Environ Saf. 2013;87:10–16. DOI: 10.1016/j.ecoenv.2012.10.002.

6 The Application of Microalgae for Bioremediation of Pharmaceuticals from Wastewater: Recent Trend and Possibilities

Prithu Baruah and Neha Chaurasia

CONTENTS

DOI: 10.1201/9781003155713-6

6.1 INTRODUCTION

About less than 1% of the total freshwater of the earth (2.5%) is accessible by human beings (Coimbra et al. 2021). Among these, groundwater and surface water form an essential source for human use such as domestic purposes, agriculture, and industrial activities (Shah and Shah 2020). Thus, the quality and quantity of water are key requisites for human survival (Coimbra et al. 2021). However, contamination of water by various micropollutants such as steroid hormones, agrochemicals, pharmaceuticals, and personal care products has become a major threat to water quality (Coimbra et al. 2021). In recent times, pharmaceuticals have been recognized as a contaminant of emerging environmental concern due to their ecotoxic effects and adverse health impacts (Agrawal et al. 2020a, Agrawal and Verma 2021; Kumar et al. 2019; Xiong et al. 2018). Pharmaceuticals are compounds that are biologically active, manufactured for the treatment and prevention of various diseases in both humans and animals. Earlier, plants were the primary source of medicine. However, due to the high requirement of drugs and medicines during the First World War, research and manufacture of synthetic analogs have been given preference as they require less time for production. Since then, production and consumption of pharmaceuticals have accelerated (Chhaya and Prajapati 2020). Every year, a huge amount of pharmaceuticals are utilized in human and veterinary medicine for curing various ailments such as infections, fever, prevention of pregnancy, physical and mental stress, and also for accelerating agriculture (Sofowora et al. 2013). Pharmaceutical medicine plays a pivotal role in improving longevity and the quality of human life (Zaied et al. 2020). Some of the important pharmaceuticals which are consumed worldwide include antibiotics, antidepressants, antipyretic, analgesic, hormones, and chemotherapy products (Chhaya and Prajapati 2020). According to studies, the defined daily intake of pharmaceuticals such as antidepressants, cholesterol-lowering, antihypertensive, antidiabetic drugs has increased by two folds in Organisation for Economic Co-operation and Development(OECD) member countries (OECD Indicators 2017). After the consumption of these compounds, humans and animals excrete the metabolites and parent compounds in feces and urine, which then enter the sewage streams (Al Aukidy et al. 2014; Chabukdhara et al. 2019; Daughton 2001).

After consumption, pharmaceuticals go through several processes of metabolism such as cleavage, glucuronidation, and hydroxylation (Beausse 2004). Since humans and animals are unable to metabolize and assimilate many of the pharmaceuticals, they excrete the same in unchanged form or with minor modifications in feces and urine (Dębska et al. 2004). Based on the chemical nature of the compound, 5–90% of the consumed pharmaceuticals are released as parent compounds or metabolites in excreted urine and feces (Mojiri et al., 2020; Vumazonke et al. 2020). Due to improper disposal, through toilets these substances enter municipal wastewater treatment plants (WWTPs), which are equipped to remove only regulated parameters but not such pollutants (Pereira et al. 2020, Santos et al. 2010). Due to their recalcitrant nature, pharmaceuticals released into the municipal wastewater may enter other aquatic habitats (Mojiri et al. 2019; Zhou and Broodbank 2014). It is important to note that major sources of wastewater contaminated with pharmaceuticals are the industries involved in pharmaceutical manufacturing (Zaied et al.

2020). In the pharmaceutical industry, water is an important raw material used for several purposes such as washing of equipment, solid cake, or extraction. During the time of processing or manufacturing, this water comes in contact with a product, raw materials, by-products, waste products, or intermediates, and thereby it gets contaminated (Gadipelly et al. 2014). These pharmaceutical compounds finally enter the environment after incomplete removal by the WWTPs (Chabukdhara et al. 2019). In recent times, a large amount of pharmaceutical contaminants have been observed by researchers in various wastewaters (surface and ground waters) as well as in sources of drinking water (Yang et al. 2014). This is possible due to the development of analytical techniques such as gas chromatography-mass spectrometry (GC-MS) and liquid chromatography (LC), which are equipped to detect even very low concentrations of pharmaceuticals in wastewater (Hena et al. 2020b). Although pharmaceuticals enter the environment in low concentration, they can harm human health, ecosystem, and drinking water quality, making them a concern for researchers worldwide (Sirés and Brillas 2012; Yuan et al. 2009, Zaied et al. 2020). Present WWTPs, based on activated-sludge processes, are not efficient enough to remove pharmaceuticals completely (Kim et al. 2007). Although there are several advanced treatment methods for pharmaceutical remediation (Section 6.3), their limitations and low efficiency encouraged researchers to look for a superior alternative (Xiong et al. 2018). In recent times, microalgae have received tremendous scientific interest as a potential candidate for pharmaceutical bioremediation. The major advantages of microalgae-based remediation include environment-friendliness, dependence on sunlight as an energy source, low operational cost, and capacity of carbon fixation. Besides, microalgae can simultaneously produce numerous valuable products such as cosmetics, nutraceuticals, biomass for biofuel production, low-cost food for aquaculture, etc. (Goswami et al.2021a; Kumar et al. 2010; Mehariya et al. 2021a; Raja et al. 2008; Wijffels et al. 2013). Mixotrophic microalgae have high flexibility to survive in adverse environments because they can shift their metabolism between autotrophic and heterotrophic mode based on the presence of nutrients and carbon sources in the environment. This capability of mixotrophic microalgae makes them superior to other bioremediation agents like bacteria and fungi which demand nutrients and carbon for growth and biodegradation (Goswami et al. 2021b). The characteristic feature of mixotrophic microalgae to adapt to harsh environments makes them an ideal fit for wastewater treatment (Kumar et al. 2010; Li et al. 2016; Raja et al. 2008; Subash Chandra bose et al. 2013; Wijffels et al. 2013). This chapter aims to provide a comprehensive overview of the recent trends in microalgae-based bioremediation of pharmaceuticals and their possible prospects.

6.2 PHARMACEUTICALS IN THE ENVIRONMENT

6.2.1 Source and Entry of Pharmaceuticals into the Environment

Pharmaceuticals can enter the aquatic environment in numerous ways such as discharges from hospitals, domestic and household wastewater, sewage treatment plants, improper disposal from manufacturing plants, and WWTPs (Leung et al. 2012; Liu and Wong 2013). Figure 6.1 shows the routes of pharmaceutical entry

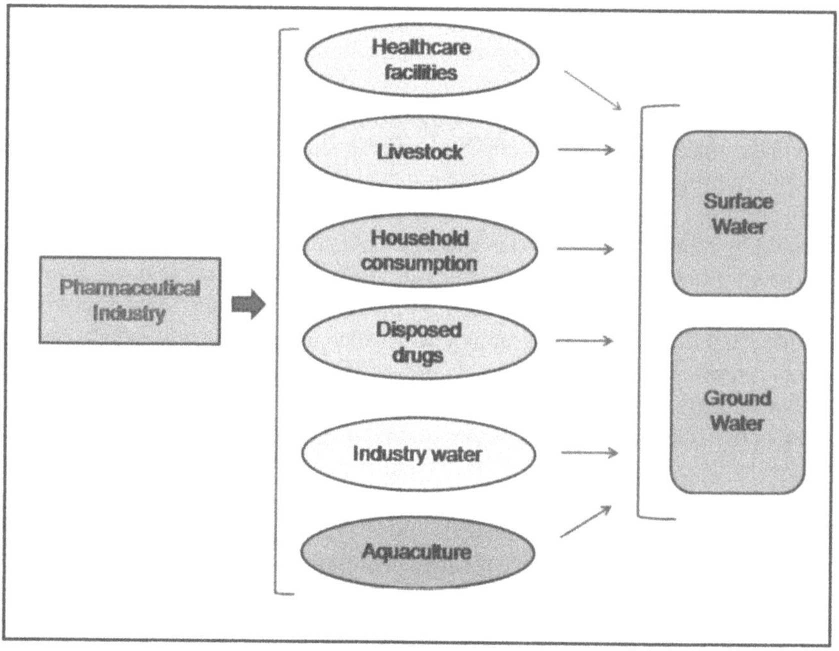

FIGURE 6.1 The routes of entry of pharmaceuticals into the environment.

into the environment. Although pharmaceuticals are extensively used in healthcare facilities such as hospitals, they contribute quite less to the overall pharmaceutical load in municipal WWTPs (Herrmann et al. 2015; Mendoza et al. 2015; Seifrtová et al. 2008). According to studies, the contribution of hospitals in the total load of most pharmaceutical compounds in municipal WWTPs is under 10% and in some cases even less than 3% (Kümmerer 2010). One of the important sources of pharmaceuticals in the environment is domestic sewage. Medicines and drugs consumed by animals and humans are discharged into the surrounding environment directly or indirectly. Some of the compounds (e.g., methotrexate) which are dissolved or non-metabolized are excreted into the environment in urine and feces, which ultimately get released into the sewage treatment systems (Kim et al. 2011; Kimura et al. 2007; Montforts 1999). After a pharmaceutical is consumed, it gets metabolized in the body and gets eliminated in body wastes as a mixture of metabolites and its parent compound. Two mechanisms namely Phase I and Phase II metabolites come into play in the elimination process. During the first phase, hepatic metabolism is used, which enhances water solubility and the polarity of the metabolites through hydrolysis, biochemical oxidations, and reduction process. In the second phase, a conjugation step comprising glucuronidation and sulphation produces Phase II metabolites. This process involves the transfer of polar groups to parent compounds or metabolites, which increases the water solubility and hydrophilicity of the conjugated metabolites for easy elimination through feces and/or urine (Mompelat et al. 2009, Leclercq et al. 2009, Evgenidou et al. 2015). The sludge from sewage treatment plants containing pharmaceutical drugs is often used as a fertilizer on croplands (Topp et al. 2008; Xia et al. 2005). This can result in direct contamination of the soil. Apart from this, run-off with heavy rainfall can contribute to possible contamination of surrounding groundwater and surface water (Kemper 2008, Kay et al. 2005; Topp et al. 2008). Furthermore, pharmaceuticals used in aquaculture along with their degradation products and metabolites are released directly into surface waters, which results in its contamination (Lalumera et al. 2004; Le and Munekage 2004). In addition to this, expired and unused pharmaceuticals without any proper treatment are directly disposed of in the environment (Chhaya and Prajapati 2020). According to studies, pharmaceutical manufacturing facilities are more responsible for high pharmaceutical concentration in the environment compared to those resulting from drug utilization (Kessler 2010). The wastewater from pharmaceutical industries discharged to the WWTPs is inefficiently treated by the conventional treatment technologies, resulting in their unwanted accumulation in water bodies (Vieno et al. 2007). Besides, the natural hydrologic cycle also results in pharmaceutical entry into the environment (Mompelat et al. 2009; Petrović et al. 2003).

6.2.2 Environmental and Health Risks of Pharmaceuticals

In recent times, frequent detection of pharmaceutical compounds in the environment has become a worldwide concern. The development of analytical techniques such as liquid chromatography-mass spectrometry (LC-MS) and GC-MS has enabled researchers to detect even very low concentrations of pharmaceuticals in the environment. Table 6.1 shows the different analytical methods used for the detection of pharmaceuticals. Despite the concentration of most of the pharmaceuticals in the environment ranges from ng/l to mg/l levels, studies show that they can have a hazardous impact on both target and non-target organisms. For instance, exposure to pharmaceuticals can result in a change in microbial communities, a decrease in soil microbial activity, restrained microbial growth, and alteration in bacterial denitrification rate (Xiong et al. 2018). Discharge of effluent, re-utilization of treated sewage, and agricultural applications of sludge can result in entry and accumulation of pharmaceutical residues in the environment, which can have a deleterious effect on animal and human health (Rajapaksha et al. 2014; Vithanage et al. 2014). The action of pharmaceuticals can be through a combination of various mechanisms such as altering cellular metabolism, changing the expression of genes related to the cellular regulator, and modulating the concentration levels of intracellular ions. As a result of this, different tissues and organs can have different effects on the exposure to these chemicals (Carpenter et al. 2002). The use of reclaimed water and sewage sludge as manure can result in the uptake of pharmaceuticals by plants. Pharmaceuticals such as sulfonamides and trimethoprim having high hydrophilicity can undergo accumulation in plant roots due to their less permeability in the cell membranes (Tanoue et al. 2012). Certain pharmaceutical compounds can bioaccumulate in the body of fishes and other aquatic organisms, causing unexpected adverse effects on them (Yang et al. 2017b). For instance, liver enlargement can occur in fishes as a result of chronic exposure to compounds such as estrogenic contaminants (Gunnarsson et al. 2009). In addition to this, single and mixed residues of pharmaceuticals can have a

TABLE 6.1
Analytical Methods for Detection of Pharmaceuticals

Pharmaceutical Compound	Type/Class	Method	Reference
4-aminophenol	Intermediate in paracetamol synthesis.	Electrochemical	Shiroma et al. (2012)
Acetaminophen/ paracetamol	Analgesics and antipyretics	Raman spectroscopy	Craig et al. (2015)
Amino-penicillin acid	Antibiotics	Raman spectroscopy	Restaino et al. (2017)
Amlodipine	Calcium channel blocker	Electrochemical	Mofidi et al. (2017b)
Carbamazepine	Anticonvulsant	High-performance liquid chromatography (HPLC)	Kahkha et al. (2018)
Carvacrol	Antimicrobial	Gas chromatography	Ghiasvand et al. (2018)
Clomipramine	Antidepressants	Gas chromatography	Nojavan et al. (2018)
Clonidine	Antihypertensives	HPLC	Baharfar et al. (2017)
Colchicine	Anti-gout agents	HPLC	Bahrani et al. (2017)
Cyproheptadine	Antihistamines	Gas chromatography	Shamsayei et al. (2017)
Diclofenac	Nonsteroidal anti-inflammatory drugs	Electrochemical	Mofidi et al. (2017a)
Diethylstilbestrol	nonsteroidal estrogen	Gas chromatography	Huang and Lee (2015
Ephedrine	Alpha/Beta-Adrenergic Agonists	HPLC	Baharfar et al. (2017
17β-Estradiol	Estrane Steroid	Gas chromatography	González et al. (2017)
Estradiol valerate	Synthetic hormone	Electrochemical	Mofidi et al. (2018)
Estriol	Steroid	Gas chromatography	Huang and Lee (2015)
Estrone	Steroid	Gas chromatography	Gonzales et al. (2017)
Fenbufen	Non-steroidal anti-inflammatory drug	HPLC	Lyu et al. (2015)
Flucytosine	Antifungal	Raman spectroscopy	Berger et al. (2017)
Haloperidol	Antipsychotic	Capillary electrophoresis	Ryšavá et al. (2019)
Ibuprofen	Non-steroidal anti-inflammatory drug	HPLC	Lyu et al. (2015)
Imatinib mesylate	Kinase inhibitor	Capillary electrophoresis	Forough et al. (2017)
Imatinib	Kinase inhibitor	Electrochemical	Hatamluyi and Es' haghi (2007)
Ketoprofen	Nonsteroidal anti-inflammatory	HPLC	Lyu et al. (2015)
Ketotifen	Antihistamines	Gas chromatography	Shamsayei et al. (2017)
Loperamide	antidiarrheal agent	Capillary electrophoresis	Rysava et al. (2019)

Pharmaceutical Compound	Type/Class	Method	Reference
Methadone	synthetic opioid agonist	Capillary electrophoresis	Fakhari et al. (2017)
Methamphetamine	Stimulant	Capillary electrophoresis	Fakhari et al. (2017)
Nortriptyline	Antidepressants	Capillary electrophoresis	Šlampová and Kubáň (2017)
Papaverine	Antispasmodics	Capillary electrophoresis	Šlampová and Kubáň (2017)
Pramipexole	Dopamine agonists	HPLC	Fashi et al. (2017)
Propylthiouracil	Antithyroid agents	Electrochemical	Tahmasebi et al. (2018)
Pseudoephedrine	Nasal decongestants	Capillary electrophoresis	Aladaghlo et al. (2017)
Riluzole	Benzothiazoles	HPLC	Rahimi and Nojavan (2019)
Thymol	Antibacterial	Gas chromatography	Ghiasvand et al. (2018)
Tramadol	Opioid analgesic	Capillary electrophoresis	Aladaghlo et al. (2017)
Trimipramine	Tricyclic antidepressant	Gas chromatography	Tabani et al. (2019)
Vanillylmandelic acid	Clinical biomarker	Electrochemical	Hrdlicka et al. (2019)
Verapamil	Calcium-channel blockers	HPLC	Rahimi and Nojavan (2019)
Zolpidem	Sedative-hypnotics	HPLC	Yaripour et al. (2018)

negative effect on reproduction and histopathological alterations in zebrafish (Galus et al. 2013a, 2013b; Overturf et al. 2015). Pharmaceuticals in general have a longer half-life in the environment and are biodegradation persistent (Xiong et al. 2018). Steroid pharmaceuticals such as 17α-ethynylestradiol, estrone, and 17β-estradiol can trigger feminization of fish by endocrine disruption (Kramer et al. 1998). Bacteria (e.g. *Acinetobacter baumannii*) can develop genetic resistance due to exposure to pharmaceuticals such as antibiotics (Dijkshoorn et al. 2007). Transfer of resistant genes to man through bio-amplification in the food chain can result in compromised treatment of ailments caused by these bacteria, thereby affecting public health (Kelly et al. 2007).

6.3 MODERN METHODS OF PHARMACEUTICAL REMEDIATION

Several advanced methods have been developed to remove pharmaceuticals from wastewater. Advanced oxidation processes (AOPs) are effective methods for the remediation of pharmaceuticals. AOPs such as the Fenton-based process, ozonation, photocatalysis, dioxide chlorine, and strong oxidant use are effective in the degradation of contaminants (Agrawal and Verma 2020; Gou et al. 2014). AOPs can result in the generation of reactive oxygen species (ROS), which is helpful in the effective mineralization of contaminants in the water, carbon dioxide, and acids or inorganic ions (Dalrymple et al. 2007; Kanakaraju et al. 2018). Although AOPs are an effective remediation method for contaminants, they are not cost-effective, making their application on a large scale difficult. Also, byproducts generated during incomplete mineralization of the contaminant have high toxicity (Gou et al. 2014). Adsorption has been observed as another promising method for pollutant removal due to its eco-friendliness. In the last few years, researchers have used various materials as adsorbents such as activated carbon, bottom ash, ion exchange resin, powdered activated carbon, biochar, carbon nanotubes, and chitosan and organic resin for eliminating various pollutants. The major drawback of these absorbents is the high cost and low adsorption capacities (Quesada et al. 2019, Zhu et al., 2017). Quesada et al. (2019) discuss the use of low-cost adsorbent as an alternative method for the removal of pharmaceuticals. Bioremediation using bacteria has been studied by many researchers across the globe. The major problem of using bacteria to remove pharmaceuticals such as antibiotics is that they develop antibiotic resistance, which is dangerous for humanity. Furthermore, low kinetics of catabolism by fungi hamper their application for bioremediation on a large scale. Another disadvantage of bacteria and fungi-based mineralization of pollutants is that the carbon that is fixed is ultimately discharged into the atmospheric carbon dioxide pool (Afreen et al. 2017; Subash Chandra bose et al. 2013; Xiong et al. 2018).

6.4 REMOVAL OF PHARMACEUTICALS BY MICROALGAE

6.4.1 Ecological Role of Microalgae

Microalgae are a diverse group of single-celled eukaryotic organisms found in various environments such as lakes, ponds, rivers, soil as well as in extreme environmental conditions such as hypersaline conditions and ice (Blánquez et al. 2020). Algae are referred to as filamentous when the individual cells are joined in the end-to-end fashion (Martínez et al. 2006). Algae may occur singly or in colonial form. In the aquatic ecosystem, microalgae are the primary producers of food by the process of photosynthesis. Other heterotrophic organisms depend on algae for food directly or indirectly, making them the base of the food chain. Microalgae can survive a wide range of adverse conditions such as low nutrient supply, extreme temperature, and pH, making them superior to other organisms such as fungi (Subash Chandra bose et al. 2013). Microalgae's ability of acclimatization to variations in nutrients, light, temperature, and salinity enhances their tolerance as well as bioremediation capacity (Goswami et al. 2021b). Genetic alterations resulting from spontaneous mutation or adaptation at the physiological level are responsible for such adaptation mechanism (Chaturvedi et al. 2021; Cho et al. 2016; Osundeko et al. 2014). Mixotrophic microalgae have high flexibility to survive in adverse environments because they can shift their metabolism between autotrophic and heterotrophic mode based on the presence of nutrients and carbon sources in the environment. This capability of mixotrophic microalgae makes them superior to other bioremediation agents like bacteria and fungi which demand nutrients and carbon for growth and biodegradation. The characteristic feature of mixotrophic microalgae to adapt to unfavorable environments makes them an ideal fit for wastewater treatment (Chaturvedi et al. 2013; Kumar et al.

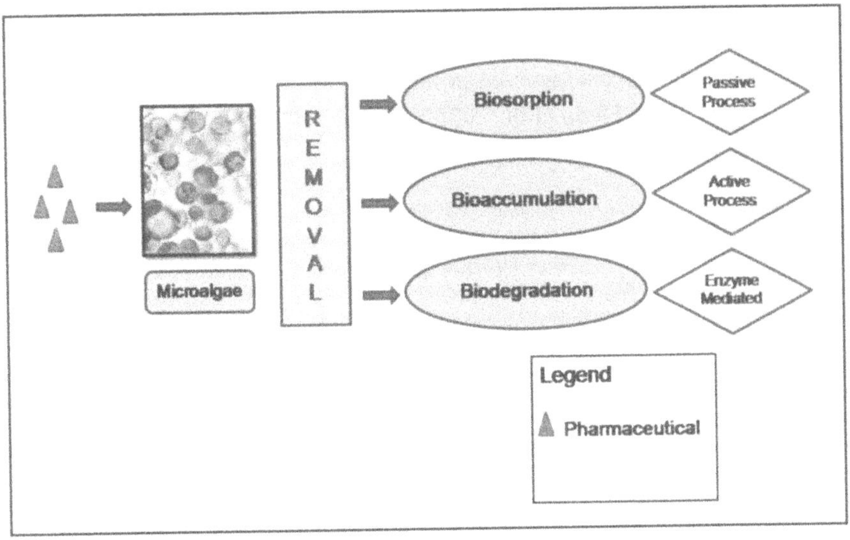

FIGURE 6.2 The basic steps involved in the removal of pharmaceuticals by microalgae.

2010; Li et al. 2016; Raja et al. 2008; Subash Chandra bose et al. 2013; Wijffels et al. 2013). Microalgae are interesting organisms from the environmental point of view since they contribute to carbon dioxide mitigation by its sequestration for supporting its growth (Silva et al. 2019).

6.4.2 Mechanism of Pharmaceuticals Removal by Microalgae

The mechanism of pharmaceutical removal by microalgae includes (i) biosorption, (ii) bioaccumulation, and (iii) biodegradation. Figure 6.2 shows the basic steps involved in pharmaceutical removal by microalgae. Depending on the physicochemical characterization of the pharmaceutical, microalgae may adopt a particular mechanism or a combination of mechanisms for contaminant removal from water (Hena et al. 2020b). Table 6.2 shows different microalgal species involved in the removal of pharmaceuticals.

6.4.2.1 Biosorption

Biosorption by nature is a passive mechanism that involves the distribution of soluble pollutants between the liquid phase and extracellular polymeric substances (EPSs) or the cell wall of microalgae. Such distribution occurs mainly due to the ionic or hydrophobic interaction between the contaminant and microalgal solid surface (Hena et al. 2020b). There are several studies in which adsorption of pharmaceuticals by microalgae has been reported. About 10% of the available norgestrel and progesterone was found to be absorbed by *Scenedesmus obliquus* and *Chlorella pyrenoidosa* dead cell biomass (Peng et al. 2014). *C. vulgaris* was found to remove 50–71% tetracycline in high-rate algal pond (HRAP) within 43 hours by adsorption (De Godos et al. 2012). They observed a variation in pH ranging from 7.5 to 8 of the culture which allowed

TABLE 6.2
Microalgal Species Involved in Removal of Pharmaceuticals (Adapted from Hena et al. 2020b; Leng et al. 2020 with permission)

Pharmaceutical Compound	Microalgae Species	Reference
7-ACA	*Chlorella* sp., *Chlamydomona* ssp., *Mychonastes* sp.	Guo et al. (2016)
7-ACA	*C. pyrenoidosa*	Yu et al. (2017)
Cefradine	*C. pyrenoidosa*	Li et al. (2015)
Ceftazidime	*C. pyrenoidosa*	Guo and Chen (2015)
Ciprofloxacin	*C. mexicana*	Xiong et al. (2017c)
Ciprofloxacin	*Nannochloris sp.*	Bai and Acharya (2017)
β-estradiol	*Selenastrum Capricornutum, Chlamydomonas reinhardtii*	Hom-Diaz et al. (2015)
17α-ethinylestradiol	*S. Capricornutum, C. reinhardtii*	Hom-Diaz et al. (2015)
17 α-estradiol	*S. dimorphus*	Zhang et al. (2014)
17β-estradiol	*S. dimorphus*	Zhang et al. (2014)
Estrone	*S. dimorphus*	Zhang et al. (2014)
Estriol	*S. dimorphus*	Zhang et al. (2014)
17α-ethinylestradiol	*Desmodesmus subspicatus*	Maes et al. (2014)
Florfenicol	*Chlorella sp.*	Song et al. (2019)
Levofloxacin	*C. vulgaris*	Xiong et al. (2017b)
Metronidazole	*C. vulgaris*	Hena et al. (2020a)
Norgestrel	*S. obliquus, C. pyrenoidosa*	Peng et al. (2014)
Norfloxacin	*C. vulgaris*	Zhang et al. (2012)
Progesterone	*S. obliquus, C. pyrenoidosa*	Peng et al. (2014)
Sulfamethazine	*Scenedesmus obliquus*	Xiong et al. (2019)
Sulfamethoxazole	*Scenedesmus obliquus*	Xiong et al. (2019)
Sulfamethoxazole	*Nannochloris sp.*	Bai and Acharya (2016)
Sulfamethoxazole	*Nannochloris sp.*	Bai and Acharya(2017)
Sulfamethoxazole	*C. pyrenoidosa*	Sun et al. (2017)
Tilmicosin	*Chlorella PY-ZU1*	Cheng et al. (2017)
Trimethoprim	*Nannochloris sp.*	Bai and Acharya (2017)
Trimethoprim	*Nannochloris sp.*	Bai and Acharya (2016)

tetracycline to remain in zwitterionic form, encouraging strong adsorption on the microalgal cell surface in HRAP. Guo et al. (2016) demonstrated the removal of the pharmaceutical 7-amino-cephalosporanic acid (7-ACA) from wastewater by three freshwater green microalgae (*Chlorella* sp., *Mychonastes* sp. and *Chlamydomonas* sp.). These three species followed the following order of adsorption capacity: *Mychonastes* sp. (2.95 mg/g) <*Chlamydomonas* sp. (3.09 mg/g) <*Chlorella* sp. (4.74 mg/g). A recent study (Daneshvar et al. 2018) demonstrated the high adsorption capability of tetracycline removal (295 mg/g) by *S. quadricauda*. *Chlorella* sp. biomass showed different removal efficiency of cefalexin from wastewater before and after lipid extraction. Removal of cefalexin was higher (82.7%) before lipid extraction compared to after lipid extraction (71.2%). Ding et al. (2017) reported adsorption of ibuprofen by freshwater diatom *Navicula* sp. The cell wall of algae has bonded EPS with predominant negative charge due to the presence of functional groups such as hydroxyl, carboxyl, and phosphoryl. Electrostatic interactions cause pharmaceutical pollutants with positively charged functional groups to get attracted towards the microalgal cells, favoring suitable biosorption of the pollutant (Hena et al. 2020b, Xiong et al. 2018).

6.4.2.2 Bioaccumulation

Bioaccumulation is an active process in which pollutants are accumulated against the concentration gradient using metabolic energy. According to Adams and Rowland (2003), bioaccumulation involves pollutant accumulation from the external environment into the cytoplasm of the cell. It is a slow process compared to adsorption in which microalgae take up contaminants such as pharmaceuticals along with nutrients for supporting their growth. Bioaccumulation is an intracellular process in contrast to adsorption which is an extracellular process for the elimination of contaminants from water (Bai and Acharya 2016; Davis et al. 2003). According to studies, bioaccumulation has been found to play a significant role in the removal of pharmaceutical contaminants such as sulfamethoxazole, trimethoprim, and doxycycline (Bai and Acharya 2017; Prata et al. 2018). *Cladophora* sp., a freshwater microalga, showed the ability to accumulate triclocarban, triclosan, and methyl-triclosan at a range of 50–400 ng/g biomass of the green alga (Coogan et al. 2007). In another study, 23% of the radiolabeled 17α-ethinylestradiol gets accumulated in the cells of the freshwater green alga *Desmodesmus subspicatus* in 24 hours (Maes et al. 2014). It has been reported that microalgal species like *C. mexicana* and *S. obliquus* can bioaccumulate the pharmaceutical carbamazepine (Xiong et al. 2016). On accumulation in the microalgal cells, some pharmaceuticals can evoke increased generation of ROS, which in normal concentration are involved in the control of cellular functions but can cause severe cell injury and ultimately cell death if present in excess (Kumar et al. 2016; Leng et al. 2020; Xiong et al. 2018).

6.4.2.3 Biodegradation

Biodegradation is a process of intracellular or extracellular enzymatic breakdown of pharmaceuticals with some broken-down products further utilized by the algae (Gao and Chi 2015; Ke et al. 2010; Naghdi et al. 2018; Zhou et al. 2013, Hena et al. 2020b). During biodegradation, complex compounds are converted into a simpler form. In one study, *S. obliquus* and *C. pyrenoidosa* were found to biodegrade 95% of the steroid hormone progesterone within 5 days (Peng et al. 2014). In another study, ceftazidime and 7-ACA, basic parent structure of ceftazidime, were intracellularly degraded by *C. pyrenoidosa*. The process involves the initial adsorption of ceftazidime followed by its transmission into the algal cell wall and finally, it gets degraded by microalgal enzymes (Yu et al. 2017). Biodegradation is mediated by several enzymatic reactions such as decarboxylation, hydrogenation, hydroxylation, dehydrogenation, dehydroxylation, carboxylation, hydrolysis, oxidation, reduction, and cleavage of ring, glycosylation, and demethylation (Hena et al. 2020b). This enzyme-mediated process consists of three phases (Ding et al. 2017; Maes et al. 2014; Matamoros et al. 2016; Peng et al. 2014; Xiong et al. 2016, 2017a). In the initial phase, cytochrome P450 (phase I enzyme) adds hydroxyl group to the xenobiotic compound, making it more hydrophilic. This step involves reactions such as oxidation, reduction, or hydrolysis. In the second phase, the enzymes glucosyltransferases or glutathione-S-transferases catalyze the formation of a conjugate bond between glutathione and the xenobiotic containing electrophilic groups (COOH, CONH2, epoxide ring) (Nakajima et al. 2007). In the third phase, the pharmaceutical compound gets bio-transformed into detoxified and simpler forms obtained by the action of numerous enzymes such as carboxylase, decarboxylase, dehydratase, mono(di)oxygenase, hydrolases, laccases, transferase, pyro-phosphatase dehydrogenase, glutamyl-tRNA reductase (dos Santos Ferreira et al. 2007; Foflonker et al. 2016; Mus et al. 2007; Otto and Schlosser 2014, Petroutsos et al. 2008; Wang et al. 2004). Ding et al. (2017) concluded that acylation, hydroxylation, glucuronidation, and demethylation contributed to the biotransformation of the pharmaceutical ibuprofen in cells of *Navicula* sp., a freshwater diatom. Similarly, levofloxacin, an antibiotic, was found to be removed by *S. obliquus* (Xiong et al. 2017d) and *C. vulgaris* (Xiong et al. 2017a) by biodegradation mechanism.

6.4.3 Factors Affecting Pharmaceutical Removal by Microalgae

The removal efficiency of pharmaceuticals from wastewater depends on both biotic and abiotic factors. Biotic factors include the type of algal species used for wastewater treatment and abiotic factors include light, temperature, and hydraulic retention time (HRT). A brief explanation about each factor is given below:

6.4.3.1 Microalgal Species

The removal of pharmaceutical contaminants from wastewater greatly depends on the algal species used for removal. Some species of algae can easily remove a particular pollutant while a different species can remove a different contaminant. *Chlorella* has been demonstrated to effectively remove pharmaceuticals such as antibiotics. Several researchers have noted that *C. pyrenoidosa* can remediate pharmaceuticals such as 7-ACA, amoxicillin, cefradine, and ceftazidime (Guo and Chen 2015; Li et al. 2015; Yang et al. 2017a; Yu et al. 2017). Xiong et al. (2017a) investigated the removal of levofloxacin by various microalgal species such as *C. vulgaris*, *Ourococcus multisporus*, *Tribonema aequale*, *Micractinium resseri*, *C. pitschmannii*, and *C. Mexicana* and found that *C. vulgaris* showed maximum removal efficiency. In another study, it was demonstrated that *C. vulgaris* showed better removal of enrofloxacin than *C. mexicana*, *S. obliquus*, *Micractinium resseri*, and *Ourococcus multisporus* (Xiong et al., 2017b). Escapa et al. (2016) noted that *C. sorokiniana* showed 29% removal while *C. vulgaris* showed 21% removal of diclofenac in 9 days. However, there are some species which show better removal of pharmaceuticals than *Chlorella sps*. For example, *S. Obliquus* can remove cefradine more efficiently than *C. pyrenoidosa*. *S. Obliquus* showed 60% removal while *C. pyrenoidosa* showed less than 30% removal of cefradine (Yang et al. 2017a).

6.4.3.2 Light

Microalgae show photo-heterotrophic and photo-autotrophic properties, and hence light is essential for their survival. The culture system for algae meant for wastewater treatment is often made to use maximum light. Light influences the growth of microalgae and thus consequently impacts pollutant removal from wastewater. It has been demonstrated that biomass production of microalgae is increased by an increase in the intensity of light (Hena et al. 2018; Luo et al. 2017). In addition, photodegradation of pharmaceuticals such as antibiotics can occur as a result of illumination which can be helpful in improvement in removal efficiency of the contaminant (Du et al. 2015). Exposure to strong light can evoke ROS generation, which can accelerate the removal of pharmaceutical such as tetracycline (Norvill et al. 2017).

6.4.3.3 Temperature

Temperature is another important factor for wastewater treatment using microalgae. Biodegradation is an enzyme-mediated process, and enzyme activity is regulated by temperature. Therefore, optimal temperature is essential for efficient bioremediation of pollutants such as pharmaceuticals. According to reports, the optimal temperature range is 25–30°C for microalgae, and growth gets negatively affected above 30°C (Bamba et al. 2015; Chinnasamy et al. 2009). In one study, it has been noted that with an increase in temperature from 15°C to 25°C, the rate of degradation of norfloxacin increased gradually. This shows that enhancement of norfloxacin degradation rate for the photocatalytic reaction of algae resulted from an increase in temperature. However, an increase in temperature above 45°C resulted in a reduction in norfloxacin degradation rate (Zhang et al. 2012). High temperature such as above 30°C can cause injury to microalgal cells, resulting in the release of organic acids of low molecular weight into the culture medium, which in turn improve photolysis degradation of pharmaceuticals indirectly (Zhang et al. 2012). Matamoros et al. (2015) noted that biodegradation of carbamazepine by microalgae was enhanced by 30% in warmer conditions compared to cold conditions. Gao et al. (2011) compared the ability of four *Chlorella* species namely *Chlorella* sp. (2f5aia), *Chlorella* sp. (1uoai), *C. miniata* (WW1), and *C. vulgaris* to remove nonylphenol. It was found that *C. vulgaris* showed maximum removal of nonylphenol with biodegradation efficiency significantly accelerated by temperature and light intensity.

6.4.3.4 HRT

HRT is an important parameter for the removal of organic contaminants having biodegradable properties from wastewater in the proper way (Luo et al. 2014). It has been shown in previous studies that the increase in HRT removal of organic contaminants increases due to enhancement in sorption and biodegradation (Gros et al. 2010; Matamoros and Bayona 2013). Chu et al. (2015) noted that *C. pyrenoidosa* showed the best pollutant removal performance at a longer HRT of 10 days in the summer season. Matamoros et al. (2015) demonstrated the effect of seasonality and HRT on organic micropollutant removal from wastewater in HRAPs. They concluded that the efficiency of microcontaminant removal from wastewater increased during warm conditions, while the effects of HRT on removal were observed only during the cold condition.

6.5 OTHER APPLICATION OF MICROALGAE

6.5.1 Production of Biofuel

The microalgae biomass generated after wastewater treatment can be used as a feedstock for biofuel production (Agrawal et al. 2020b; Mehariya et al. 2021c). Biodiesel can be produced by the conversion of lipid present in algae by methods used for converting vegetable oil into biodiesel (Ganesan et al. 2020). *Chlorella sp.* (19% lipids, 56% sugars), *Spirulina platensis* (8% lipids, 60% sugars), and *C. reinhardtii* (21% lipids, 48% sugars) are a few algae that are potentially subject of research for biofuel production (Demirbas and Demirbas 2011; Xu et al. 2018). Biohydrogen, bioethanol, biogas, and biodiesel are some of the biofuels which can be produced from the residual biomass of algae (Bhardwaj et al. 2020; Goswami et al. 2021c; Sivaramakrishnan and Incharoensakdi 2018; Zhang et al. 2016; 2018).

6.5.2 Biomitigation of Carbon Dioxide

Microalgae can fix carbon dioxide through photosynthesis with higher efficiency than land plants (Khan et al. 2009; Li et al. 2008; Rosenberg et al. 2011). Microalgae show tolerance towards high carbon dioxide concentration in feeding airstreams and are useful in biofuel production (Goswami et al. 2020; Chang and Yang 2003). *S. obliquus* was found to tolerate carbon dioxide concentration as high as 12% (v/v) with 67% removal efficiency (Li et al. 2011). Microalgal strains such as *Scenedesmus* sp., *Chlorella* sp., and *Botryococcus braunii* show a high capacity of biomitigation of carbon dioxide with 200–1300 mg/L/day consumption rate ((Rosenberg et al. 2011; Sydney et al. 2010; Yoo et al. 2010; Zhao et al. 2011). Considering these facts, it is clear that microalgae have a huge potential for multipurpose applications such as biofuel production, wastewater treatment, and mitigation of greenhouse gases (Kumar et al. 2010; Rizwan et al. 2018).

6.6 CONCLUSION AND PROSPECTS

There is an exponential rise in the use of pharmaceuticals across the globe, and consequently, the residues of these chemicals are frequently detected in the environment. The development of analytical methods such as HPLC and GC-MS that measure very low concentrations of pollutants contributed to the easy detection in the environment. Since pharmaceuticals have an adverse impact on animal and human health, it has become a major concern of researchers as well as a hot topic of scientific study. Considering the importance of the removal of these compounds from the environment, many treatment methodologies have been developed (e.g., AOP). However, most of these methods have several limitations such as the requirement of huge operational costs, lack of eco-friendliness, etc. Recently, microalgae have received attention from researchers across the globe as a promising candidate for pharmaceutical removal from wastewater. This chapter provided a comprehensive overview of the present status of microalgal utilization for wastewater treatment containing pharmaceuticals. Although considerable work has been done by numerous researchers across the globe on microalgae-based wastewater treatment, further exploration of new candidates is still needed to enhance the remediation efficiencies. Integration of microalgae-based methods with advanced treatment technologies such as AOPs, microbial fuel cells, and constructed wetlands can help improve the efficiency of wastewater treatment systems. In-depth studies based on molecular methods such as metatranscriptomics and metagenomics can help explore candidate genes that can be further utilized for strain improvement using genetic engineering. In addition to this, future research should focus more on the in-depth analysis of biodegradation pathways and the molecular basis of biodegradation.

ABBREVIATIONS

7-ACA, 7-amino-cephalosporanic acid; AOPs, advanced oxidation processes; EPSs, extracellular polymeric substances; GC-MS, gas chromatography-mass spectrometry; HRT, hydraulic retention time; HPLC, high-performance liquid chromatography; HRAP, high-rate algal pond; LC, liquid chromatography; LC-MS, liquid chromatography-mass spectrometry; ROS, reactive oxygen species; WWTPs, wastewater treatment plants

CONFLICT OF INTEREST

On behalf of all authors, the corresponding author states that there is no conflict of interest.

REFERENCES

Adams, W., & Rowland, C. (2003). Aquatic toxicology test methods. *Handbook of Ecotoxicology*, 2, 19–38.

Afreen, S., Shamsi, T. N., Baig, M. A., Ahmad, N., Fatima, S., Qureshi, M. I., ... & Fatma, T. (2017). A novel multicopper oxidase (laccase) from cyanobacteria: purification, characterization with potential in the decolorization of anthraquinonic dye. *PLOS ONE*, *12*(4), e0175144.

Agrawal, K., Bhatt, A., Bhardwaj, N., Kumar, B., Verma, P. (2020a). Integrated Approach for the Treatment of Industrial Effluent by Physico-chemical and Microbiological Process for Sustainable Environment. In *Combined Application of Physico-Chemical & Microbiological Processes for Industrial Effluent Treatment Plant* (pp. 119–143). Springer, Singapore.

Agrawal, K., Bhatt, A., Bhardwaj, N., Kumar, B., Verma, P. (2020b). Algal Biomass: Potential Renewable Feedstock for Biofuels Production–Part I. In *Biofuel Production Technologies: Critical Analysis for Sustainability* (pp. 203–237). Springer, Singapore doi: https://doi.org/10.1007/978-981-13-8637-4_8

Agrawal, K., & Verma, P. (2020). Advanced Oxidative Processes: An Overview of Their Role in Treating Various Wastewaters. In *Advanced Oxidation Processes for Effluent Treatment Plants* (pp. 87–102). Springer

Agrawal, K., & Verma, P. (2021) Metagenomics: A Possible Solution for Uncovering the "Mystery Box" of Microbial Communities Involved in the Treatment of Wastewater. In *Wastewater Treatment Cutting Edge Molecular Tools, Techniques and Applied Aspects* (pp. 41–53). Elsevier

Al Aukidy, M., Verlicchi, P., & Voulvoulis, N. (2014). A framework for the assessment of the environmental risk posed by pharmaceuticals originating from hospital effluents. *Science of the Total Environment*, *493*, 54–64.

Aladaghlo, Z., Fakhari, A. R., & Hasheminasab, K. S. (2017). Carrier assisted electromembrane extraction based on nonionic lipophilic surfactants for the determination of basic drugs in urine samples. *Analytical Methods*, *9*(38), 5659–5667.

Baharfar, M., Yamini, Y., Seidi, S., & Karami, M. (2017). Quantitative analysis of clonidine and ephedrine by a microfluidic system: On-chip electromembrane extraction followed by high performance liquid chromatography. *Journal of Chromatography B*, *1068*, 313–321.

Bahrani, S., Ghaedi, M., Dashtian, K., Ostovan, A., Mansoorkhani, M. J. K., & Salehi, A. (2017). MOF-5 (Zn)-Fe2O4 nanocomposite based magnetic solid-phase microextraction followed by HPLC-UV for efficient enrichment of colchicine in root of colchicium extracts and plasma samples. *Journal of Chromatography B*, *1067*, 45–52.

Bai, X., & Acharya, K. (2016). Removal of trimethoprim, sulfamethoxazole, and triclosan by the green alga *Nannochloris* sp. *Journal of Hazardous Materials*, *315*, 70–75.

Bai, X., & Acharya, K. (2017). Algae-mediated removal of selected pharmaceutical and personal care products (PPCPs) from Lake Mead water. *Science of the Total Environment*, *581*, 734–740.

Bamba, B. S. B., Lozano, P., Adjé, F., Ouattara, A., Vian, M. A., Tranchant, C., & Lozano, Y. (2015). Effects of temperature and other operational parameters on *Chlorella vulgaris* mass cultivation in a simple and low-cost column photobioreactor. *Applied Biochemistry and Biotechnology*, *177*(2), 389–406.

Beausse, J. (2004). Selected drugs in solid matrices: a review of environmental determination, occurrence and properties of principal substances. *Trends in Analytical Chemistry*, *23*(10–11), 753–761.

Berger, A. G., Restaino, S. M., & White, I. M. (2017). Vertical-flow paper SERS system for therapeutic drug monitoring of flucytosine in serum. *Analytica Chimica Acta*, *949*, 59–66.

Bhardwaj, N., Agrawal, K., Verma, P. (2020). Algal Biofuels: An Economic and Effective Alternative of Fossil Fuels. In *Microbial Strategies for Techno-economic Biofuel Production* (pp.59–83). Springer doi: https://doi.org/10.1007/978-981-15-7190-9_7

Blánquez P., Hom-Díaz A., Vicent T., Guieysse B. (2020) Microalgae-Based Processes for the Removal of Pharmaceuticals in Wastewater. In: Rodriguez-Mozaz S., Blánquez Cano P., Sarrà Adroguer M. (eds) *Removal and Degradation of Pharmaceutically Active Compounds in Wastewater Treatment. The Handbook of Environmental Chemistry*,108. Springer, Cham.

Carpenter, D. O., Arcaro, K., & Spink, D. C. (2002). Understanding the human health effects of chemical mixtures. *Environmental Health Perspectives*, *110*(suppl 1), 25–42.

Chabukdhara, M., Gogoi, M., & Gupta, S. K. (2019). Potential and Feasibility of the Microalgal System in Removal of Pharmaceutical Compounds from Wastewater. In *Application of Microalgae in Wastewater Treatment* (pp. 177–206). Springer, Cham.

Chang, E. H., & Yang, S. S. (2003). Some characteristics of microalgae isolated in Taiwan for biofixation of carbon dioxide. *Botanical Bulletin of Academia Sinica*, *44*, 43–52.

Chaturvedi, V., Chandravanshi, M., Rahangdale, M., & Verma, P. (2013). An integrated approach of using polystyrene foam as an attachment system for growth of mixed culture of cyanobacteria with concomitant treatment of copper mine waste water. *Journal of Waste Management*, Article ID 282798. https://doi.org/10.1155/2013/282798

Chaturvedi, V., Goswami, R. K., Verma, P. (2021) Genetic Engineering for Enhancement of Biofuel Production in Microalgae. In *Biorefineries: A Step Towards Renewable and Clean Energy* (pp.539–559). Springer doi: https://doi.org/10.1007/978-981-15-9593-6_2

Cheng, J., Ye, Q., Yang, Z., Yang, W., Zhou, J., & Cen, K. (2017). Microstructure and antioxidative capacity of the microalgae mutant *Chlorella* PY-ZU1 during tilmicosin removal from wastewater under 15% CO2. *Journal of Hazardous Materials*, *324*, 414–419.

Chhaya, Raychoudhury T., Prajapati S.K. (2020) Bioremediation of Pharmaceuticals in Water and Wastewater. In: Shah M. (eds) *Microbial Bioremediation & Biodegradation*. Springer, Singapore.

Chinnasamy, S., Ramakrishnan, B., Bhatnagar, A., & Das, K. C. (2009). Biomass production potential of a wastewater alga *Chlorella vulgaris* ARC 1 under elevated levels of CO2 and temperature. *International Journal of Molecular Sciences*, *10*(2), 518–532.

Cho, K., Lee, C. H., Ko, K., Lee, Y. J., Kim, K. N., Kim, M. K., ... & Oda, T. (2016). Use of phenol-induced oxidative stress acclimation to stimulate cell growth and biodiesel production by the oceanic microalga *Dunaliella salina*. *Algal Research*, *17*, 61–66.

Chu, H. Q., Tan, X. B., Zhang, Y. L., Yang, L. B., Zhao, F. C., & Guo, J. (2015). Continuous cultivation of *Chlorella pyrenoidosa* using anaerobic digested starch processing wastewater in the outdoors. *Bioresource Technology*, *185*, 40–48.

Coimbra, R. N, Escapa, C., & Otero, M. (2021). Removal of Pharmaceuticals from Water: Conventional and Alternative Treatments. *Water*, 13,.

Coogan, M. A., Edziyie, R. E., La Point, T. W., & Venables, B. J. (2007). Algal bioaccumulation of triclocarban, triclosan, and methyl-triclosan in a North Texas wastewater treatment plant receiving stream. *Chemosphere*, *67*(10), 1911–1918.

Craig, D., Mazilu, M., & Dholakia, K. (2015). Quantitative detection of pharmaceuticals using a combination of paper microfluidics and wavelength modulated Raman spectroscopy. *PLOS ONE*, *10*(5), e0123334.

Dalrymple, O. K., Yeh, D. H., & Trotz, M. A. (2007). Removing pharmaceuticals and endocrine- disrupting compounds from wastewater by photocatalysis. *Journal of Chemical Technology & Biotechnology*, *82*(2), 121–134.

Daneshvar, E., Zarrinmehr, M. J., Hashtjin, A. M., Farhadian, O., & Bhatnagar, A. (2018). Versatile applications of freshwater and marine water microalgae in dairy wastewater treatment, lipid extraction and tetracycline biosorption. *Bioresource Technology*, *268*, 523–530.

Daughton CG. (2001). Pharmaceuticals and personal care products in the environment: overarching issues and overview. In: Daughton CG, Jones-Lepp TL, editors. Pharmaceuticals and personal care products in the environment: scientific and regulatory issues. ACS Symposium Series 791, Washington, DC: American Chemical Society.

Davis, T. A., Volesky, B., & Mucci, A. (2003). A review of the biochemistry of heavy metal biosorption by brown algae. *Water Research*, *37*(18), 4311–4330.

de Godos, I., Muñoz, R., & Guieysse, B. (2012). Tetracycline removal during wastewater treatment in high-rate algal ponds. *Journal of Hazardous Materials*, *229*, 446–449.

Dębska, J., Kot-Wasik, A., & Namieśnik, J. (2004). Fate and analysis of pharmaceutical residues in the aquatic environment. *Critical Reviews in Analytical Chemistry*, *34*(1), 51–67.

Demirbas, A., & Demirbas, M. F. (2011). Importance of algae oil as a source of biodiesel. *Energy Conversion and Management*, *52*(1), 163–170.

Dijkshoorn, L., Nemec, A., & Seifert, H. (2007). An increasing threat in hospitals: multidrug-resistant *Acinetobacter baumannii*. *Nature Reviews Microbiology*, *5*(12), 939–951.

Ding, T., Yang, M., Zhang, J., Yang, B., Lin, K., Li, J., & Gan, J. (2017). Toxicity, degradation and metabolic fate of ibuprofen on freshwater diatom *Navicula* sp. *Journal of Hazardous Materials*, *330*, 127–134.

dos Santos Ferreira, V., Rocchetta, I., Conforti, V., Bench, S., Feldman, R., & Levin, M. J. (2007). Gene expression patterns in *Euglena gracilis*: insights into the cellular response to environmental stress. *Gene*, *389*(2), 136–145.

Du, Y., Feng, Y., Guo, R., & Chen, J. (2015). Enhancement by the artificial controlled culture for the algal treatment of antibiotic ceftazidime: a three-step response performance and high-removal efficiency. *RSC Advances*, *5*(89), 72755–72763.

Escapa, C., Coimbra, R. N., Paniagua, S., García, A. I., & Otero, M. (2016). Comparative assessment of diclofenac removal from water by different microalgae strains. *Algal Research*, *18*, 127–134.

Evgenidou, E. N., Konstantinou, I. K., & Lambropoulou, D. A. (2015). Occurrence and removal of transformation products of PPCPs and illicit drugs in wastewaters: a review. *Science of the Total Environment*, *505*, 905–926.

Fakhari, A. R., Asadi, S., Kosalar, H. M., Sahragard, A., Hashemzadeh, A., & Amini, M. M. (2017). Metal–organic framework enhanced electromembrane extraction–a conceptual study using basic drugs as model substances. *Analytical Methods*, *9*(38), 5646–5652.

Fashi, A., Khanban, F., Yaftian, M. R., & Zamani, A. (2017). The cooperative effect of reduced graphene oxide and Triton X-114 on the electromembrane microextraction efficiency of Pramipexole as a model analyte in urine samples. *Talanta*, *162*, 210–217.

Foflonker, F., Ananyev, G., Qiu, H., Morrison, A., Palenik, B., Dismukes, G. C., & Bhattacharya, D. (2016). The unexpected extremophile: tolerance to fluctuating salinity in the green alga Picochlorum. *Algal Research*, *16*, 465–472.

Forough, M., Farhadi, K., Eyshi, A., Molaei, R., Khalili, H., Kouzegaran, V. J., & Matin, A. A. (2017). Rapid ionic liquid-supported nano-hybrid composite reinforced hollow-fiber electromembrane extraction followed by field-amplified sample injection-capillary electrophoresis: an effective approach for extraction and quantification of Imatinib mesylate in human plasma. *Journal of Chromatography A*, *1516*, 21–34.

Gadipelly, C., Pérez-González, A., Yadav, G. D., Ortiz, I., Ibáñez, R., Rathod, V. K., & Marathe, K. V. (2014). Pharmaceutical industry wastewater: review of the technologies for water treatment and reuse. *Industrial & Engineering Chemistry Research*, *53*(29), 11571–11592.

Galus, M., Jeyaranjaan, J., Smith, E., Li, H., Metcalfe, C., & Wilson, J. Y. (2013a). Chronic effects of exposure to a pharmaceutical mixture and municipal wastewater in zebrafish. *Aquatic Toxicology*, *132*, 212–222.

Galus, M., Kirischian, N., Higgins, S., Purdy, J., Chow, J., Rangaranjan, S., … & Wilson, J. Y. (2013b). Chronic, low concentration exposure to pharmaceuticals impacts multiple organ systems in zebrafish. *Aquatic Toxicology*, *132*, 200–211.

Ganesan, R., Manigandan, S., Samuel, M. S., Shanmuganathan, R., Brindhadevi, K., Chi, N. T. L., Duc, P.A., & Pugazhendhi, A. (2020). A review on prospective production of biofuel from microalgae. *Biotechnology Reports*, *27*, e00509.

Gao, J., & Chi, J. (2015). Biodegradation of phthalate acid esters by different marine microalgal species. *Marine Pollution Bulletin*, *99*(1–2), 70–75.

Gao, Q. T., Wong, Y. S., & Tam, N. F. Y. (2011). Removal and biodegradation of nonylphenol by different *Chlorella* species. *Marine Pollution Bulletin*, *63*(5–12), 445–451.

Ghiasvand, A. R., Ghaedrahmati, L., Heidari, N., Haddad, P. R., & Farhadi, S. (2018). Synthesis and characterization of MIL-101 (Cr) intercalated by polyaniline composite, doped with silica nanoparticles and its evaluation as an efficient solid-phase extraction sorbent. *Journal of Separation Science*, *41*(20), 3910–3917.

González, A., Avivar, J., Maya, F., Cabello, C. P., Palomino, G. T., & Cerdà, V. (2017). In-syringe dispersive μ-SPE of estrogens using magnetic carbon microparticles obtained from zeolitic imidazolate frameworks. *Analytical and Bioanalytical Chemistry*, *409*(1), 225–234.

Goswami, R. K., Agrawal, K., Mehariya, S., Molino, A., Musmarra, D., & Verma, P. (2020). Microalgae-Based Biorefinery for Utilization of Carbon Dioxide for Production of Valuable Bioproducts. In *Chemo-Biological Systems for CO_2 Utilization* (pp. 199–224). Taylor & Francis doi: https://doi.org/10.1201/9780429317187

Goswami, R.K., Agrawal, K., & Verma, P. (2021a). An Overview of Microalgal Carotenoids: Advances in the Production and Its Impact on Sustainable Development. In *Bioenergy Research: Evaluating Strategies for Commercialization and Sustainability* (pp.105–128). John Wiley & Sons

Goswami, R. K., Mehariya, S., Karthikeyan, O. P. K., & Verma, P. (2021c). Advanced microalgae-based renewable biohydrogen production systems: a review. *Bioresource Technology, 320*(A), 124301.

Goswami, R. K., Mehariya, S., Verma, P, Lavecchia, R., & Zuorro, A. (2021b). Microalgae-based biorefineries for sustainable resource recovery from wastewater. *Journal of Water Process Engineering, 40*: 101747.

Gou, N., Yuan, S., Lan, J., Gao, C., Alshawabkeh, A. N., & Gu, A. Z. (2014). A quantitative toxicogenomics assay reveals the evolution and nature of toxicity during the transformation of environmental pollutants. *Environmental Science & Technology, 48*(15), 8855–8863.

Gros, M., Petrović, M., Ginebreda, A., & Barceló, D. (2010). Removal of pharmaceuticals during wastewater treatment and environmental risk assessment using hazard indexes. *Environment International, 36*(1), 15–26.

Gunnarsson, L., Adolfsson-Erici, M., Björlenius, B., Rutgersson, C., Förlin, L., & Larsson, D. G. J. (2009). Comparison of six different sewage treatment processes—reduction of estrogenic substances and effects on gene expression in exposed male fish. *Science of the Total Environment, 407*(19), 5235–5242.

Guo, R., & Chen, J. (2015). Application of alga-activated sludge combined system (AASCS) as a novel treatment to remove cephalosporins. *Chemical Engineering Journal, 260*, 550–556.

Guo, W. Q., Zheng, H. S., Li, S., Du, J. S., Feng, X. C., Yin, R. L., … & Chang, J. S. (2016). Removal of cephalosporin antibiotics 7-ACA from wastewater during the cultivation of lipid-accumulating microalgae. *Bioresource Technology, 221*, 284–290.

Hatamluyi, B., & Es'haghi, Z. (2017). A layer-by-layer sensing architecture based on dendrimer and ionic liquid supported reduced graphene oxide for simultaneous hollow-fiber solid phase microextraction and electrochemical determination of anti-cancer drug imatinib in biological samples. *Journal of Electroanalytical Chemistry, 801*, 439–449.

Hena, S., Gutierrez, L., & Croué, J. P. (2020a). Removal of metronidazole from aqueous media by *C. vulgaris*. *Journal of Hazardous Materials, 384*, 121400.

Hena, S., Gutierrez, L., & Croué, J. P. (2020b). Removal of pharmaceutical and personal care products (PPCPs) from wastewater using microalgae: A review. *Journal of Hazardous Materials, 403*, 124041.

Hena, S., Znad, H., Heong, K. T., & Judd, S. (2018). Dairy farm wastewater treatment and lipid accumulation by *Arthrospira platensis*. *Water Research, 128*, 267–277.

Herrmann, M., Olsson, O., Fiehn, R., Herrel, M., & Kümmerer, K. (2015). The significance of different health institutions and their respective contributions of active pharmaceutical ingredients to wastewater. *Environment International, 85*, 61–76.

Hom-Diaz, A., Llorca, M., Rodríguez-Mozaz, S., Vicent, T., Barceló, D., & Blánquez, P. (2015). Microalgae cultivation on wastewater digestate: β-estradiol and 17α-ethynylestradiol degradation and transformation products identification. *Journal of Environmental Management, 155*, 106–113.

Hrdlička, V., Navrátil, T., & Barek, J. (2019). Application of hollow fibre based microextraction for voltammetric determination of vanillylmandelic acid in human urine. *Journal of Electroanalytical Chemistry, 835*, 130–136.

Huang, Z., & Lee, H. K. (2015). Performance of metal-organic framework MIL-101 after surfactant modification in the extraction of endocrine disrupting chemicals from environmental water samples. *Talanta, 143*, 366–373.

Kahkha, M. R. R., Oveisi, A. R., Kaykhaii, M., & Kahkha, B. R. (2018). Determination of carbamazepine in urine and water samples using amino-functionalized metal–organic framework as sorbent. *Chemistry Central Journal*, *12*(1), 1–12.

Kanakaraju, D., Glass, B. D., & Oelgemöller, M. (2018). Advanced oxidation process-mediated removal of pharmaceuticals from water: a review. *Journal of Environmental Management*, *219*, 189–207.

Kay, P., Blackwell, P. A., & Boxall, A. B. (2005). Transport of veterinary antibiotics in overland flow following the application of slurry to arable land. *Chemosphere*, *59*(7), 951–959.

Ke, L., Luo, L., Wang, P., Luan, T., & Tam, N. F. Y. (2010). Effects of metals on biosorption and biodegradation of mixed polycyclic aromatic hydrocarbons by a freshwater green alga *Selenastrum capricornutum*. *Bioresource Technology*, *101*(18), 6950–6961.

Kelly, B. C., Ikonomou, M. G., Blair, J. D., Morin, A. E., & Gobas, F. A. (2007). Food web–specific biomagnification of persistent organic pollutants. *Science*, *317*(5835), 236–239.

Kemper, N. (2008). Veterinary antibiotics in the aquatic and terrestrial environment. *Ecological Indicators*, *8*(1), 1–13.

Kessler R. (2010). Pharmaceutical factories as a source of drugs in water. *Environmental health perspectives*, *118*(9), a383.

Khan, S. A., Hussain, M. Z., Prasad, S., & Banerjee, U. C. (2009). Prospects of biodiesel production from microalgae in India. *Renewable and Sustainable Energy Reviews*, *13*(9), 2361–2372.

Kim, S. D., Cho, J., Kim, I. S., Vanderford, B. J., & Snyder, S. A. (2007). Occurrence and removal of pharmaceuticals and endocrine disruptors in South Korean surface, drinking, and waste waters. *Water Research*, *41*(5), 1013–1021.

Kim, K. R., Owens, G., Kwon, S. I., So, K. H., Lee, D. B., & Ok, Y. S. (2011). Occurrence and environmental fate of veterinary antibiotics in the terrestrial environment. *Water, Air, & Soil Pollution*, *214*(1), 163–174.

Kimura, K., Hara, H., & Watanabe, Y. (2007). Elimination of selected acidic pharmaceuticals from municipal wastewater by an activated sludge system and membrane bioreactors. *Environmental Science & Technology*, *41*(10), 3708–3714.

Kramer, V. J., Miles-Richardson, S., Pierens, S. L., & Giesy, J. P. (1998). Reproductive impairment and induction of alkaline-labile phosphate, a biomarker of estrogen exposure, in fathead minnows (*Pimephales promelas*) exposed to waterborne 17β-estradiol. *Aquatic Toxicology*, *40*(4), 335–360.

Kumar, B., Agrawal, K., Bhardwaj, N., Chaturvedi, V., Verma, P. (2019). Techno-Economic Assessment of Microbe-Assisted Wastewater Treatment Strategies for Energy and Value-Added Product Recovery. In *Microbial Technology for the Welfare of Society* (pp.147–181). Springer doi: https://doi.org/10.1007/978-981-13-8844-6_7

Kumar, A., Ergas, S., Yuan, X., Sahu, A., Zhang, Q., Dewulf, J., … & Van Langenhove, H. (2010). Enhanced CO2 fixation and biofuel production via microalgae: recent developments and future directions. *Trends in Biotechnology*, *28*(7), 371–380.

Kumar, M. S., Kabra, A. N., Min, B., El-Dalatony, M. M., Xiong, J., Thajuddin, N., … & Jeon, B. H. (2016). Insecticides induced biochemical changes in freshwater microalga *Chlamydomonas mexicana*. *Environmental Science and Pollution Research*, *23*(2), 1091–1099.

Kümmerer, K. (2010). Pharmaceuticals in the environment. *Annual Review of Environment and Resources*, *35*, 57–75.

Lalumera, G. M., Calamari, D., Galli, P., Castiglioni, S., Crosa, G., & Fanelli, R. (2004). Preliminary investigation on the environmental occurrence and effects of antibiotics used in aquaculture in Italy. *Chemosphere*, *54*(5), 661–668.

Le, T. X., & Munekage, Y. (2004). Residues of selected antibiotics in water and mud from shrimp ponds in mangrove areas in Viet Nam. *Marine Pollution Bulletin*, *49*(11–12), 922–929.

Leclercq, M., Mathieu, O., Gomez, E., Casellas, C., Fenet, H., & Hillaire-Buys, D. (2009). Presence and fate of carbamazepine, oxcarbazepine, and seven of their metabolites at wastewater treatment plants. *Archives of Environmental Contamination and Toxicology, 56*(3), 408–415.

Leng, L., Wei, L., Xiong, Q., Xu, S., Li, W., Lv, S., ... & Zhou, W. (2020). Use of microalgae based technology for the removal of antibiotics from wastewater: A review. *Chemosphere, 238*, 124680.

Leung, H. W., Minh, T. B., Murphy, M. B., Lam, J. C., So, M. K., Martin, M., ... & Richardson, B. J. (2012). Distribution, fate and risk assessment of antibiotics in sewage treatment plants in Hong Kong, South China. *Environment International, 42*, 1–9.

Li, C., Xiao, S., & Ju, L. K. (2016). Cultivation of phagotrophic algae with waste activated sludge as a fast approach to reclaim waste organics. *Water Research, 91*, 195–202.

Li, F. F., Yang, Z. H., Zeng, R., Yang, G., Chang, X., Yan, J. B., & Hou, Y. L. (2011). Microalgae capture of CO_2 from actual flue gas discharged from a combustion chamber. *Industrial & Engineering Chemistry Research, 50*(10), 6496–6502.

Li, H., Pan, Y., Wang, Z., Chen, S., Guo, R., & Chen, J. (2015). An algal process treatment combined with the Fenton reaction for high concentrations of amoxicillin and cefradine. *RSC Advances, 5*(122), 100775–100782.

Li, Y., Horsman, M., Wu, N., Lan, C. Q., & Dubois-Calero, N. (2008). Biofuels from microalgae. *Biotechnology Progress, 24*(4), 815–820.

Liu, J. L., & Wong, M. H. (2013). Pharmaceuticals and personal care products (PPCPs): a review on environmental contamination in China. *Environment International, 59*, 208–224.

Luo, Y., Guo, W., Ngo, H. H., Nghiem, L. D., Hai, F. I., Zhang, J., ... & Wang, X. C. (2014). A review on the occurrence of micropollutants in the aquatic environment and their fate and removal during wastewater treatment. *Science of the Total Environment, 473*, 619–641.

Luo, Y., Le-Clech, P., & Henderson, R. K. (2017). Simultaneous microalgae cultivation and wastewater treatment in submerged membrane photobioreactors: a review. *Algal Research, 24*, 425–437.

Lyu, D. Y., Yang, C. X., & Yan, X. P. (2015). Fabrication of aluminum terephthalate metal-organic framework incorporated polymer monolith for the microextraction of non-steroidal anti-inflammatory drugs in water and urine samples. *Journal of Chromatography A, 1393*, 1–7.

Maes, H. M., Maletz, S. X., Ratte, H. T., Hollender, J., & Schaeffer, A. (2014). Uptake, elimination, and biotransformation of 17α-ethinylestradiol by the freshwater alga *Desmodesmus subspicatus*. *Environmental Science & Technology, 48*(20), 12354–12361.

Martínez, M., Bernal, P., Almela, C., Vélez, D., García-Agustín, P., Serrano, R., & Navarro-Aviñó, J. (2006). An engineered plant that accumulates higher levels of heavy metals than *Thlaspi caerulescens*, with yields of 100 times more biomass in mine soils. *Chemosphere, 64*(3), 478–485.

Matamoros, V., & Bayona, J. M. (2013). Removal of Pharmaceutical Compounds from Wastewater and Surface Water by Natural Treatments. In *Comprehensive Analytical Chemistry* (Vol. 62, pp. 409–433). Elsevier.

Matamoros, V., Gutiérrez, R., Ferrer, I., García, J., & Bayona, J. M. (2015). Capability of microalgae-based wastewater treatment systems to remove emerging organic contaminants: a pilot-scale study. *Journal of Hazardous Materials, 288*, 34–42.

Matamoros, V., Uggetti, E., García, J., & Bayona, J. M. (2016). Assessment of the mechanisms involved in the removal of emerging contaminants by microalgae from wastewater: a laboratory scale study. *Journal of Hazardous Materials, 301*, 197–205.

Mehariya, S., Goswami, R. K., Karthikeysan, O. P., & Verma, P. (2021a). Microalgae for high-value products: away towards green nutraceutical and pharmaceutical compounds. *Chemosphere, 280*, 130553.

Mehariya, S., Goswami, R., Verma, P., Lavecchia, R., & Zuorro, A. (2021c). Integrated approach for wastewater treatment and biofuel production in microalgae biorefineries *Energies* 2021, *14*(8), 2282.

Mehariya, S., Kumar, P., Marino, T., Casella, P., Lovine, A., Verma, P., & Musuma, A.M. (2021b). Aquatic Weeds: A Potential Pollutant Removing Agent from Wastewater and Polluted Soil and Valuable Biofuel Feedstock. In *Bioremediation Using Weeds* (pp. 59–77). Springer doi: https://doi.org/10.1007/978-981-33-6552-0_3

Mendoza, A., Aceña, J., Pérez, S., De Alda, M. L., Barceló, D., Gil, A., & Valcárcel, Y. (2015). Pharmaceuticals and iodinated contrast media in a hospital wastewater: a case study to analyse their presence and characterise their environmental risk and hazard. *Environmental Research*, *140*, 225–241.

Mofidi, Z., Norouzi, P., Larijani, B., Seidi, S., Ganjali, M. R., & Morshedi, M. (2018). Simultaneous determination and extraction of ultra-trace amounts of estradiol valerate from whole blood using FFT square wave voltammetry and low-voltage electrically enhanced microextraction techniques. *Journal of Electroanalytical Chemistry*, *813*, 83–91.

Mofidi, Z., Norouzi, P., Seidi, S., & Ganjali, M. R. (2017a). Determination of diclofenac using electromembrane extraction coupled with stripping FFT continuous cyclic voltammetry. *Analytica Chimica Acta*, *972*, 38–45.

Mofidi, Z., Norouzi, P., Seidi, S., & Ganjali, M. R. (2017b). Efficient design for in situ determination of amlodipine in whole blood samples using fast Fourier transform stripping square wave voltammetry after preconcentration by electromembrane extraction. *New Journal of Chemistry*, *41*(22), 13567–13575.

Mojiri, A., Vakili, M., Farraji, H., & Aziz, S. Q. (2019). Combined ozone oxidation process and adsorption methods for the removal of acetaminophen and amoxicillin from aqueous solution; kinetic and optimisation. *Environmental Technology & Innovation*, *15*, 100404.

Mojiri, A., Zhou, J., Vakili, M., & Van Le, H. (2020). Removal performance and optimisation of pharmaceutical micropollutants from synthetic domestic wastewater by hybrid treatment. *Journal of Contaminant Hydrology*, *235*, 103736.

Mompelat, S., Le Bot, B., & Thomas, O. (2009). Occurrence and fate of pharmaceutical products and by-products, from resource to drinking water. *Environment International*, *35*(5), 803–814.

Montforts, M. H. M. M. (1997). Environmental risk assessment for veterinary medicinal products. Part 1. Other than GMO-containing and immunological products. RIVM rapport 613310001.

Mus, F., Dubini, A., Seibert, M., Posewitz, M. C., & Grossman, A. R. (2007). Anaerobic acclimation in *Chlamydomonas reinhardtii*: anoxic gene expression, hydrogenase induction, and metabolic pathways. *Journal of Biological Chemistry*, *282*(35), 25475–25486.

Naghdi, M., Taheran, M., Brar, S. K., Kermanshahi-Pour, A., Verma, M., & Surampalli, R. Y. (2018). Removal of pharmaceutical compounds in water and wastewater using fungal oxidoreductase enzymes. *Environmental Pollution*, *234*, 190–213.

Nakajima, N., Teramoto, T., Kasai, F., Sano, T., Tamaoki, M., Aono, M., … & Saji, H. (2007). Glycosylation of bisphenol A by freshwater microalgae. *Chemosphere*, *69*(6), 934–941.

Nojavan, S., Shaghaghi, H., Rahmani, T., Shokri, A., & Nasiri-Aghdam, M. (2018). Combination of electromembrane extraction and electro-assisted liquid-liquid microextraction: a tandem sample preparation method. *Journal of Chromatography A*, *1563*, 20–27.

Norvill, Z. N., Toledo-Cervantes, A., Blanco, S., Shilton, A., Guieysse, B., & Muñoz, R. (2017). Photodegradation and sorption govern tetracycline removal during wastewater treatment in algal ponds. *Bioresource Technology*, *232*, 35–43.

Osundeko, O., Dean, A. P., Davies, H., & Pittman, J. K. (2014). Acclimation of microalgae to wastewater environments involves increased oxidative stress tolerance activity. *Plant and Cell Physiology*, *55*(10), 1848–1857.

Otto, B., & Schlosser, D. (2014). First laccase in green algae: purification and characterization of an extracellular phenol oxidase from Tetracystis aeria. *Planta, 240*(6), 1225–1236.

Overturf, M. D., Anderson, J. C., Pandelides, Z., Beyger, L., & Holdway, D. A. (2015). Pharmaceuticals and personal care products: a critical review of the impacts on fish reproduction. *Critical Reviews in Toxicology, 45*(6), 469–491.

OECD. Paris: OECD Publishing; 2017. Health at a Glance 2017: OECD Indicators. Available at: http://dx.doi.org/10.1787/health_glance-2017-en

Peng, F. Q., Ying, G. G., Yang, B., Liu, S., Lai, H. J., Liu, Y. S., ... & Zhou, G. J. (2014). Biotransformation of progesterone and norgestrel by two freshwater microalgae (*Scenedesmus obliquus* and *Chlorella pyrenoidosa*): transformation kinetics and products identification. *Chemosphere, 95*, 581–588.

Pereira, A., Silva, L., Laranjeiro, C., Lino, C., & Pena, A. (2020). Selected pharmaceuticals in different aquatic compartments: Part I—Source, fate and occurrence. *Molecules, 25*(5), 1026.

Petroutsos, D., Katapodis, P., Samiotaki, M., Panayotou, G., & Kekos, D. (2008). Detoxification of 2, 4-dichlorophenol by the marine microalga *Tetraselmis marina. Phytochemistry, 69*(3), 707–714.

Petrović, M., Gonzalez, S., & Barceló, D. (2003). Analysis and removal of emerging contaminants in wastewater and drinking water. *Trends in Analytical Chemistry, 22*(10), 685–696.

Prata, J. C., Lavorante, B. R., Maria da Conceição, B. S. M., & Guilhermino, L. (2018). Influence of microplastics on the toxicity of the pharmaceuticals procainamide and doxycycline on the marine microalgae *Tetraselmis chuii. Aquatic Toxicology, 197*, 143–152.

Quesada, H. B., Baptista, A. T. A., Cusioli, L. F., Seibert, D., de Oliveira Bezerra, C., & Bergamasco, R. (2019). Surface water pollution by pharmaceuticals and an alternative of removal by low-cost adsorbents: a review. *Chemosphere, 222*, 766–780.

Rahimi, A., & Nojavan, S. (2019). Electromembrane extraction of verapamil and riluzole from urine and wastewater samples using a mixture of organic solvents as a supported liquid membrane: study on electric current variations. *Journal of Separation Science, 42*(2), 566–573.

Raja, R., Hemaiswarya, S., Kumar, N. A., Sridhar, S., & Rengasamy, R. (2008). A perspective on the biotechnological potential of microalgae. *Critical Reviews in Microbiology, 34*(2), 77–88.

Rajapaksha, A. U., Vithanage, M., Lim, J. E., Ahmed, M. B. M., Zhang, M., Lee, S. S., & Ok, Y. S. (2014). Invasive plant-derived biochar inhibits sulfamethazine uptake by lettuce in soil. *Chemosphere, 111*, 500–504.

Restaino, S. M., Berger, A., & White, I. M. (2017). Inkjet-Printed Paper Fluidic Devices for Onsite Detection of Antibiotics Using Surface-Enhanced Raman Spectroscopy. In *Biosensors and Biodetection* (pp. 525–540). Humana Press, New York, NY.

Rizwan, M., Mujtaba, G., Memon, S. A., Lee, K., & Rashid, N. (2018). Exploring the potential of microalgae for new biotechnology applications and beyond: a review. *Renewable and Sustainable Energy Reviews, 92*, 394–404.

Rosenberg, J. N., Mathias, A., Korth, K., Betenbaugh, M. J., & Oyler, G. A. (2011). Microalgal biomass production and carbon dioxide sequestration from an integrated ethanol biorefinery in Iowa: A technical appraisal and economic feasibility evaluation. *Biomass and Bioenergy, 35*(9), 3865–3876.

Ryšavá, L., Dvořák, M., & Kubáň, P. (2019). The effect of membrane thickness on supported liquid membrane extractions in-line coupled to capillary electrophoresis for analyses of complex samples. *Journal of Chromatography A, 1596*, 226–232.

Santos, L. H., Araújo, A. N., Fachini, A., Pena, A., Delerue-Matos, C., & Montenegro, M. C. B. S. M. (2010). Ecotoxicological aspects related to the presence of pharmaceuticals in the aquatic environment. *Journal of Hazardous Materials, 175*(1–3), 45–95.

Seifrtová, M., Pena, A., Lino, C. M., & Solich, P. (2008). Determination of fluoroquinolone antibiotics in hospital and municipal wastewaters in Coimbra by liquid chromatography with a monolithic column and fluorescence detection. *Analytical and Bioanalytical Chemistry, 391*(3), 799–805.

Shah, A., & Shah, M. (2020). Characterisation and bioremediation of wastewater: a review exploring bioremediation as a sustainable technique for pharmaceutical wastewater. *Groundwater for Sustainable Development, 11*(3), 100383.

Shamsayei, M., Yamini, Y., Asiabi, H., Rezazadeh, M., & Seidi, S. (2017). Electromembrane surrounded solid-phase microextraction using a stainless-steel wire coated with a nanocomposite composed of polypyrrole and manganese dioxide. *Microchimica Acta, 184*(8), 2697–2705.

Shiroma, L. Y., Santhiago, M., Gobbi, A. L., & Kubota, L. T. (2012). Separation and electrochemical detection of paracetamol and 4-aminophenol in a paper-based microfluidic device. *Analytica Chimica Acta, 725*, 44–50.

Silva, A., Delerue-Matos, C., Figueiredo, S. A., & Freitas, O. M. (2019). The use of algae and fungi for removal of pharmaceuticals by bioremediation and biosorption processes: a review. *Water, 11*(8), 1555.

Sirés, I., & Brillas, E. (2012). Remediation of water pollution caused by pharmaceutical residues based on electrochemical separation and degradation technologies: a review. *Environment International, 40*, 212–229.

Sivaramakrishnan, R., & Incharoensakdi, A. (2018). Utilization of microalgae feedstock for concomitant production of bioethanol and biodiesel. *Fuel, 217*, 458–466.

Šlampová, A., & Kubáň, P. (2017). Direct analysis of free aqueous and organic operational solutions as a tool for understanding fundamental principles of electromembrane extraction. *Analytical Chemistry, 89*(23), 12960–12967.

Sofowora, A., Ogunbodede, E., & Onayade, A. (2013). The role and place of medicinal plants in the strategies for disease prevention. *African Journal of Traditional, Complementary and Alternative Medicines, 10*(5), 210–229.

Song, C., Wei, Y., Qiu, Y., Qi, Y., Li, Y., & Kitamura, Y. (2019). Biodegradability and mechanism of florfenicol via Chlorella sp. UTEX1602 and L38: experimental study. *Bioresource Technology, 272*, 529–534.

Subash Chandra bose, S. R., Ramakrishnan, B., Megharaj, M., Venkateswarlu, K., & Naidu, R. (2013). Mixotrophic cyanobacteria and microalgae as distinctive biological agents for organic pollutant degradation. *Environment International, 51*, 59–72.

Sun, M., Lin, H., Guo, W., Zhao, F., & Li, J. (2017). Bioaccumulation and biodegradation of sulfamethazine in *Chlorella pyrenoidosa. Journal of Ocean University of China, 16*(6), 1167–1174.

Sydney, E. B., Sturm, W., de Carvalho, J. C., Thomaz-Soccol, V., Larroche, C., Pandey, A., & Soccol, C. R. (2010). Potential carbon dioxide fixation by industrially important microalgae. *Bioresource Technology, 101*(15), 5892–5896.

Tabani, H., Shokri, A., Tizro, S., Nojavan, S., Varanusupakul, P., & Alexovič, M. (2019). Evaluation of dispersive liquid–liquid microextraction by coupling with green-based agarose gel-electromembrane extraction: an efficient method to the tandem extraction of basic drugs from biological fluids. *Talanta, 199*, 329–335.

Tahmasebi, Z., Davarani, S. S. H., & Asgharinezhad, A. A. (2018). Highly efficient electrochemical determination of propylthiouracil in urine samples after selective electromembrane extraction by copper nanoparticles-decorated hollow fibers. *Biosensors and Bioelectronics, 114*, 66–71.

Tanoue, R., Sato, Y., Motoyama, M., Nakagawa, S., Shinohara, R., & Nomiyama, K. (2012). Plant uptake of pharmaceutical chemicals detected in recycled organic manure and reclaimed wastewater. *Journal of Agricultural and Food Chemistry, 60*(41), 10203–10211.

Topp, E., Monteiro, S. C., Beck, A., Coelho, B. B., Boxall, A. B., Duenk, P. W., ... & Metcalfe, C. D. (2008). Runoff of pharmaceuticals and personal care products following application of biosolids to an agricultural field. *Science of the Total Environment, 396*(1), 52–59.

Vieno, N., Tuhkanen, T., & Kronberg, L. (2007). Elimination of pharmaceuticals in sewage treatment plants in Finland. *Water Research, 41*(5), 1001–1012.

Vithanage, M., Rajapaksha, A. U., Tang, X., Thiele-Bruhn, S., Kim, K. H., Lee, S. E., & Ok, Y. S. (2014). Sorption and transport of sulfamethazine in agricultural soils amended with invasive-plant-derived biochar. *Journal of Environmental Management, 141*, 95–103.

Vumazonke, S., Khamanga, S. M., & Ngqwala, N. P. (2020). Detection of pharmaceutical residues in surface waters of the Eastern Cape Province. *International Journal of Environmental Research and Public Health, 17*(11), 4067.

Wang, S. B., Chen, F., Sommerfeld, M., & Hu, Q. (2004). Proteomic analysis of molecular response to oxidative stress by the green alga *Haematococcus pluvialis* (Chlorophyceae). *Planta, 220*(1), 17–29.

Wijffels, R. H., Kruse, O., & Hellingwerf, K. J. (2013). Potential of industrial biotechnology with cyanobacteria and eukaryotic microalgae. *Current Opinion in Biotechnology, 24*(3), 405–413.

Xia, K., Bhandari, A., Das, K., & Pillar, G. (2005). Occurrence and fate of pharmaceuticals and personal care products (PPCPs) in biosolids. *Journal of Environmental Quality, 34*(1), 91–104.

Xiong, J. Q., Govindwar, S., Kurade, M. B., Paeng, K. J., Roh, H. S., Khan, M. A., & Jeon, B. H. (2019). Toxicity of sulfamethazine and sulfamethoxazole and their removal by a green microalga, *Scenedesmus obliquus. Chemosphere, 218*, 551–558.

Xiong, J. Q., Kurade, M. B., Abou-Shanab, R. A., Ji, M. K., Choi, J., Kim, J. O., & Jeon, B. H. (2016). Biodegradation of carbamazepine using freshwater microalgae *Chlamydomonas mexicana* and *Scenedesmus obliquus* and the determination of its metabolic fate. *Bioresource Technology, 205*, 183–190.

Xiong, J. Q., Kurade, M. B., & Jeon, B. H. (2017a). Biodegradation of levofloxacin by an acclimated freshwater microalga, *Chlorella vulgaris. Chemical Engineering Journal, 313*, 1251–1257.

Xiong, J. Q., Kurade, M. B., & Jeon, B. H. (2017b). Ecotoxicological effects of enrofloxacin and its removal by monoculture of microalgal species and their consortium. *Environmental Pollution, 226*, 486–493.

Xiong, J. Q., Kurade, M. B., & Jeon, B. H. (2018). Can microalgae remove pharmaceutical contaminants from water?. *Trends in Biotechnology, 36*(1), 30–44.

Xiong, J. Q., Kurade, M. B., Kim, J. R., Roh, H. S., & Jeon, B. H. (2017c). Ciprofloxacin toxicity and its co-metabolic removal by a freshwater microalga *Chlamydomonas mexicana. Journal of Hazardous Materials, 323*, 212–219.

Xiong, J. Q., Kurade, M. B., Patil, D. V., Jang, M., Paeng, K. J., & Jeon, B. H. (2017d). Biodegradation and metabolic fate of levofloxacin via a freshwater green alga, *Scenedesmus obliquus* in synthetic saline wastewater. *Algal Research, 25*, 54–61.

Xu, L., Cheng, X., & Wang, Q. (2018). Enhanced lipid production in Chlamydomonas reinhardtii by co-culturing with Azotobacter chroococcum. *Frontiers in plant science*, 9, 741.

Yang, G.C.C., Yen, C.-H., Wang, C.-L., 2014. Monitoring and removal of residual phthalate esters and pharmaceuticals in the drinking water of Kaohsiung City, Taiwan. *Journal of Hazardous Materials, 277*, 53–61.

Yang, K., Lu, J., Jiang, W., Jiang, C., Chen, J., Wang, Z., & Guo, R. (2017a). An integrated view of the intimate coupling UV irradiation and algal treatment on antibiotic: Compatibility, efficiency and microbic impact assessment. *Journal of Environmental Chemical Engineering, 5*(5), 4262–4268.

Yang, Y., Ok, Y. S., Kim, K. H., Kwon, E. E., & Tsang, Y. F. (2017b). Occurrences and removal of pharmaceuticals and personal care products (PPCPs) in drinking water and water/sewage treatment plants: A review. *Science of the Total Environment, 596*, 303–320.

Yaripour, S., Mohammadi, A., Esfanjani, I., Walker, R. B., & Nojavan, S. (2018). Quantitation of zolpidem in biological fluids by electro-driven microextraction combined with HPLC-UV analysis. *EXCLI Journal, 17*, 349.

Yoo, C., Jun, S. Y., Lee, J. Y., Ahn, C. Y., & Oh, H. M. (2010). Selection of microalgae for lipid production under high levels carbon dioxide. *Bioresource Technology, 101*(1), S71–S74.

Yu, Y., Zhou, Y., Wang, Z., Torres, O. L., Guo, R., & Chen, J. (2017). Investigation of the removal mechanism of antibiotic ceftazidime by green algae and subsequent microbic impact assessment. *Scientific Reports, 7*(1), 1–11.

Yuan, F., Hu, C., Hu, X., Qu, J., & Yang, M. (2009). Degradation of selected pharmaceuticals in aqueous solution with UV and UV/H2O2. *Water Research, 43*(6), 1766–1774.

Zaied, B. K., Rashid, M., Nasrullah, M., Zularisam, A. W., Pant, D., & Singh, L. (2020). A comprehensive review on contaminants removal from pharmaceutical wastewater by electrocoagulation process. *Science of the Total Environment, 726*, 138095.

Zhang, J., Fu, D., & Wu, J. (2012). Photodegradation of Norfloxacin in aqueous solution containing algae. *Journal of Environmental Sciences, 24*(4), 743–749.

Zhang, L., Cheng, J., Pei, H., Pan, J., Jiang, L., Hou, Q., & Han, F. (2018). Cultivation of microalgae using anaerobically digested effluent from kitchen waste as a nutrient source for biodiesel production. *Renewable Energy, 115*, 276–287.

Zhang, X., Yan, S., Tyagi, R. D., Surampalli, R. Y., & Valéro, J. R. (2016). Energy balance of biofuel production from biological conversion of crude glycerol. *Journal of Environmental Management, 170*, 169–176.

Zhang, Y., Habteselassie, M. Y., Resurreccion, E. P., Mantripragada, V., Peng, S., Bauer, S., & Colosi, L. M. (2014). Evaluating removal of steroid estrogens by a model alga as a possible sustainability benefit of hypothetical integrated algae cultivation and wastewater treatment systems. *ACS Sustainable Chemistry & Engineering, 2*(11), 2544–2553.

Zhao, B., Zhang, Y., Xiong, K., Zhang, Z., Hao, X., & Liu, T. (2011). Effect of cultivation mode on microalgal growth and CO2 fixation. *Chemical Engineering Research and Design, 89*(9), 1758–1762.

Zhou, J., & Broodbank, N. (2014). Sediment-water interactions of pharmaceutical residues in the river environment. *Water Research, 48*, 61–70.

Zhou, G. J., Peng, F. Q., Yang, B., & Ying, G. G. (2013). Cellular responses and bioremoval of nonylphenol and octylphenol in the freshwater green microalga *Scenedesmus obliquus*. *Ecotoxicology and Environmental Safety, 87*, 10–16.

Zhu, S., Liu, Y. G., Liu, S. B., Zeng, G. M., Jiang, L. H., Tan, X. F., ... & Yang, C. P. (2017). Adsorption of emerging contaminant metformin using graphene oxide. *Chemosphere, 179*, 20–28.

7 Green Nanotechnology: A Microalgal Approach to Remove Heavy Metals from Wastewater

Navonil Mal, Reecha Mohapatra, Trisha Bagchi, Sweta Singh, Yagya Sharma, Meenakshi Singh, Murthy Chavali, K. Chandrasekhar

CONTENTS

DOI: 10.1201/9781003155713-7

7.1 INTRODUCTION

Our planet is covered with 71% water, but humans can access only 1% of water for potable usage. As we know, water is a universal solvent, can dissolve more substances compared to many, and because of this property, water is peculiarly vulnerable to contamination. Further, deterioration of the aquatic ecosystem is caused by anthropogenic activities like mining, metal processing, and natural geochemical weathering of rocks and soil, which pose a great risk to water bodies (Ganeshkumar et al. 2018). Even at a minute scale, the concentration of heavy metals (HMs) can prove toxic to the biological systems as HM accumulation leads to bio-magnification. Therefore, bioremediation treatments have become a global concern in present times (Ahmad et al. 2020). In this regard, the phycoremediation approach proves promising as it exploits algae's natural ability to degrade organic constituents via symbiotic communication with aerobic bacteria (Pacheco et al. 2021; Ubando et al. 2021).

Microalgae's binding affinity with metals provides a great advantage of treating wastewater in an advanced and sustainable manner. It is possible because of its mass production, rapid metal uptake potential, an affinity for polyvalent metal ions, and high range of metal tolerance, asserting the superiority of microalgal biotechnology over traditional approaches for extraction of HM (García-Béjar et al. 2020). The metal ion uptake by microalgae can occur by two ways:

I. Bioaccumulation in living cells, where a continually self-sustaining system i.e., biofilm of metal-tolerant strain is used for extended periods,
II. Use of non-living biomass and other downstream products by algal sorption

The different strategies involved for algal uptake of metals include ion exchange, physical adsorption, complexation, etc. These techniques improve the metal-binding efficiency and bioaccumulation of toxic metallic species by introducing potential algal strains through genetic manipulations (Chaturvedi et al. 2021), which can

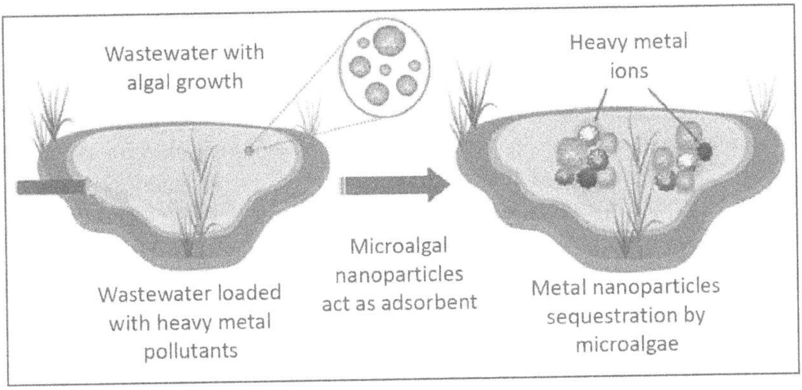

FIGURE 7.1 Biosynthesis of nanoparticles using algae.

be an eco-friendly, self-sustaining agent for remediation of metal-polluted water (Martín et al. 2015). The universal diversity of microalgae makes them a powerful nanofactory to biosynthesize nanoparticles (NPs) from wastewater compared to the rest of living organisms (Ramirez-Merida et al. 2015). Though both physical and chemical techniques can synthesize NPs, they cause harm to the environment by releasing toxic solvents and are more energy consuming and less efficient. To overcome this problem, the biological method of NP synthesis is carried out in an economical, non-toxic, and in a more efficient way. The living and dried biomass of algae is used as a raw material to biosynthesize metallic NPs to assemble metals and reduce metal ions (Sharma et al. 2019).

The following processes can perform biosynthesis of NPs using microalgae (See Figure 7.1):

i. boiling the algal extracts in the organic solvent for a specific time period,
ii. preparation of metal ions molar solution, and
iii. incubation of algal and molar solutions under controlled conditions (Rauwel et al. 2015).

Microalgal-associated nanotechnology is a unique and advanced approach that has been employed to eliminate HM from the polluted aquatic environment. Because of the several demerits and toxic effects of physical and chemical treatments, biology-mediated applications are best suited for HM remediation from wastewater. One of the most advanced technology for toxic HM elimination from aqueous solutions is biosorption because of its cost-effectiveness, higher uptake capacities, greater abundance, and renewability. Microalgal cells can accumulate metals and store them, and hence they serve as an ideal, economically viable, and non-toxic option as compared to other phytochemical techniques of metal exclusion and recovery from wastewaters (Martin-de-Lucia et al. 2018).

Silica-based nanomaterials are widely employed for purification and management of wastewater as they can be modified and engineered with various functional groups ($-NH_2$, $-SH$, etc.) on their surface and can act as assisting structures for nano-composites (Vincy et al. 2017). The remediation of HM depends on some factors such as:

i. The most common source of toxic metals is industrial effluents that gain entry into the food chain (because of HM poisoning), as they are not metabolized by the host body and accumulated in the cell organelles.
ii. The amount of metals removed from the water depends on the biomass dosage.
iii. The remediation of metals depends on the pH of the aqueous solution (wastewater) as it directly influences the biosorption efficiency to remove HM.
iv. The variation in temperature causes variation in biosorption behavior in dissimilar algal strains and for diverse metal ions.
v. The uptake capacities for HM depend on different algal strains.

Various freshwater algae have the capacity of degrading polycyclic hydrocarbons. Algal systems are involved in the tertiary wastewater treatment process. They are employed for various purposes such as removing coliform bacteria, lowering oxygen demand, removing N, P, and HM efficiently compared to chemical methods (Li et al. 1999). Anticancer, antifungal, antibacterial, and wound healing are some of the biomedical applications of NPs of algae. Metallic NPs like silver, gold, and copper oxide biosynthesized from *Galaxaura elongata*, *Padina pavonia*, *Caulerpa racemose*, *C. serruta*, and *Sargassum plagiophyllum*, etc. showed antibacterial properties (Hamdy, 2001; Jalali-Rad et al., 2004; Sheng et al., 2005). Also, improved biodiversity and aquatic life can be achieved by diatoms that take in CO_2 and discharge O_2, thus growing the concentration of dissolved O_2 in water bodies (Mal et al. 2021). Microalgae-derived NPs can also be used to counter multidrug-resistant bacteria by penetrating the extra-polysaccharide and cell membrane.

The chapter reviews algae's application as a biosorbent of HM or HM ions in industrial effluents/wastewater and debates on the available methods, prompting factors for commercial remediation. The role of microalgae acting as an elegant, powerful nanofactory for optimal elimination of hazardous metals from wastewater, due to their amazing metal-binding property, the corresponding components of their cell wall, and fast growth rate are discussed. The study of microalgal NPs, their efficiency, and physiological benefits over other NPs have also been explored briefly.

7.2 MICROALGAL NANOPARTICLES IN WASTEWATER TREATMENT

7.2.1 Classification of Algae as Sorbent

Microalgae serve as a potential bioremediating agent due to their exceptional metal-binding capability, extremely fast growth rate, and fewer growth requirements. The major difference in the biomineralization capacity in different algal groups comes from the variance in their cell wall architecture (Romera et al. 2007). Microalgal cell wall involves an array of functional groups including $-OH$, $-PO_3O_2$, $-NH_2$, $-COOH$, $-SH$, etc., which in combination impart a total negative charge to the cell wall and become a center of attraction to the cationic HM ions (Romero-Gonzalez et al. 2001). Every functional group has a precise detachment constant. It divides into an anionic counterpart and an H^+ to unmask the functional group's negative charge at a particular pH. In comparison with the other cell wall constituents, most of the metal-binding moieties remain resided on the polysaccharides and proteins.

The most primitive algal cell wall is found to be quite similar to that of the Gram-negative bacterial cell wall. The cyanobacterial cell wall's main component is peptidoglycan, which is the linear polymer of β-1, 4-N-acetylmuramic acid, and N-acetylglucosamine with pentapeptide chain, where carboxyl groups are the dominant counterparts for metal ion sequestration (Niu and Volesky, 2000). Outer to this peptidoglycan layer, the cyanobacterial cell envelope also involves an outer lipopolysaccharide layer. The phosphoryl group overwhelmingly dominates the metal-binding sites. In addition to this association, the presence of uronic acid and other negatively charged groups at the outermost layer also sequester a good number of metallic species. The cell wall of green and red algae mainly consists of hetero-polysaccharides and proteins, where predominantly carboxyl and sulphate groups provide the net negative charge, but these two groups are not that efficient for metal accumulation (de Philippis et al. 2001). According to Romera et al. 2007, the brown algae cell wall is predominated by "cellulose for the structural support, alginic acid complexed with light metals such as K, Na, Mg, Ca and sulphated polysaccharides".

The alginate content of brown algal cell walls ranges from 10–40% of their dry weight. Puranik et al. 1999 stated that polyglucuronic acid, a building block of alginate, possesses a great affinity for divalent metal ions. Another component, Fucoidan, α (1→2) linked and a branched polysaccharide sulphate ester with L-fucose building blocks show greater affinity towards trivalent cations (Figueira et al. 1999, 2000). All the algal members were compiled in a scheme of classification by Kumar et al. 2015, based on their metal-accumulating behavior and other related parameters, mentioned in the following section.

7.2.2 Molecular Mechanism of Action

Inexpensive growth requirements in association with novel technologies for innovative mass production, rapid metal uptake potential, an affinity for polyvalent metal ions, and a high range of metal tolerance assert the superiority of microalgal biotechnology to fill up the loopholes of the conventional approaches of metal ion removal.

Generally, these HM ions come under the class of metal elements with atomic numbers superior to twenty (Jing et al. 2018) and with five times superior atomic density than water (Herrera-Estrella et al., 2001). These HMs are generally any metal or metalloid that can't be biodegraded (so get bioaccumulated) and are responsible for environmental pollution. According to Wang and Chen, 2009, most of the HMs can be segregated into three major groups, poisonous metals (Hg, Cr, Pb, Cd, Co, As, Zn, Cu, Ni, Sn, etc.), radionuclides (U, Th, Ra, Am, etc.) and precious metals (Ag, Au, Pt, Pd, Ru, etc.). Microalgal metal acquisitive processes generally fall into two types:

I. bioaccumulation in living cells, where a continually self-sustaining system (biofilm of metal-tolerant strain) is used for extended periods, and
II. biosorption using dead biomass or other downstream products.

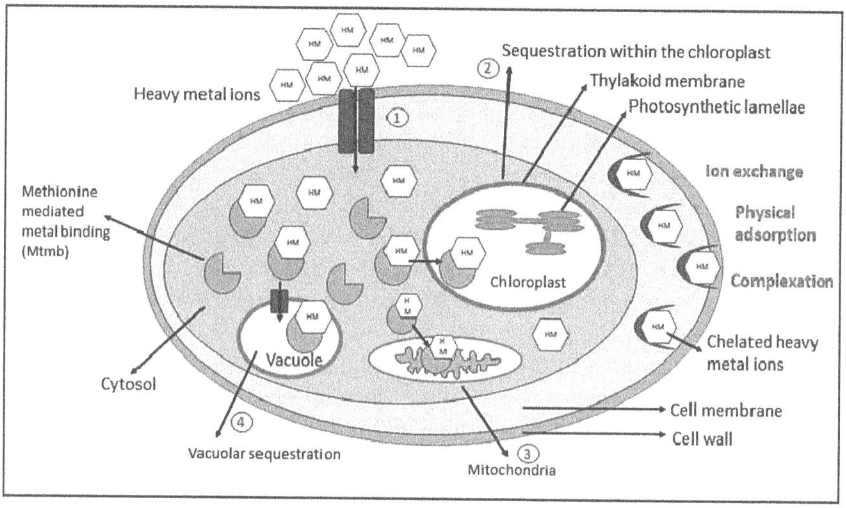

FIGURE 7.2 Different mechanisms of microalgal HM sequestration (1) passive adsorption via cell wall and cell membrane; (2) photocatalytic degradation of HM in the chloroplast where HM ions gain entry via thylakoid membrane; (3) active metabolic and biochemical sequestration of HM in the mitochondria; (4) vacuolar accumulation of HM ions through the cytosol.

The precise understanding of the use of biomass of different states, and the realizations of the state of art behind these, is summarized below (Figure 7.2).

Metal accumulation through living cell encounters a two-step route:

I. the first step includes an initial passive, reversible, non-metabolic adsorption of HM ions on the cell surface via contact by means of various cell-surface active groups, protruding out from the cell surface, followed by
II. a second step, which is comparatively slower, metabolism-dependent absorption within the cell.

The first phase happens in both alive and lifeless cells, where the metal ions get electrostatically attracted via the active groups of cell-wall materials through different strategies: ion exchange, coordination, chemisorption, chelation, complexation, entrapment, micro-precipitation, and many others. The second phase involves the metabolism-dependent inflow of HM ions within cytosol, crossways the cell membrane's barrier. As the HM ions are charged particles and hydrophilic in nature, the presence of several ion-specific plasma membrane transporter proteins is a prerequisite for this kind of ion uptake. For the sake of algal cells, these transporter proteins may also act as weapons for the first-line defense. Blaby-Haas and Merchant, 2012 have grouped the concerning transporter proteins into two distinct groups: Group A transporters include members from the Cu transporter (CTR), natural resistance-associated macrophage proteins (NRAMP), Fe transporter (FTR), and Zrt-, Irt-like proteins (ZIP) family of proteins. They are usually present on the plasma membrane as well as on the tonoplast membrane, enclosing the vacuole, and are involved in the enhancement of intracellular metal ion concentrations at the time of ion deficiency. On the contrary, Group B transporters are responsible for diminishing the intracellular metal-ion concentrations. They are so involved in the secretory pathways mediated by vesicle-mediated exocytosis of the redundant metal-ions. This group is represented by Ca (II)-sensitive cross-complementer 1 (Ccc1)/vacuolar iron transporter-1 (VIT1), FerroPortiN (FPN), P1B-type ATPases, and cation diffusion facilitator (CDF) family of proteins (Wan Maznah et al. 2012).

The second mechanism of algae-mediated HM removal relies on the passive adsorption of HM ions on the functional groups, each having a particular dissociation constant (pK_a) of the cell walls by virtue of electrostatic interactions. The intricate biochemical composition of the algal cell wall and its variety and variability among different groups contrasts with their absorption performance. The most promising candidates that interact with the metallic species are wall polysaccharides and proteins, existing on the algal cell wall's exterior surface. The polysaccharide skeleton of the algal cell wall provides the $-OH$, $-COOH$, $-PO_3O_2$, $-SH$, $-NH_2$, etc., that maintains a net negative charge over the surface of the cell and furnish the platform for adsorption of the cationic species. Besides, the amino acid constituents provide the amino, carboxyl, imidazole groups, etc. These functional groups form their corresponding anion through deprotonation of themselves. The extent of the proton displacement depends on specific pH. Carboxyl groups of N-acetylglucosamine and β-1, 4-N-acetylmuramic acid residues of peptidoglycan and phosphoryl groups of lipopolysaccharide layers are the major components of the cyanobacterial cell wall for metal adsorption (Chojnacka et al. 2005). The different metal-accumulative strategies, taken up by the algal cells, are discussed below:

7.2.2.1 Ion Exchange

Electron microscopy, as well as X-ray energy-dispersive analytical studies, suggest that algal cell wall encompasses dissimilar types of essentially anionic active groups on the cell surface and cationic metal-ions adsorb through counter-ionic interactions. Generally, all these functional groups remain protonated in normal seawater at low pH or remain masked by sodium or calcium ions at high pH. However, in the presence of HM ions like Mn^{2+}, Cu^{2+}, Ni^{2+}, Zn^{2+}, Cd^{2+}, Pb^{2+}, and Fe^{3+}, protons and the previously mentioned ions get released, liberating the negative charge of the surface-active groups, and then the latter metal ionic species get sorbed on to the cell surface of algae in exchange of the previous ones (Monteiro et al. 2012).

7.2.2.2 Physical Adsorption

According to Perpetuo et al. 2011, in this procedure, metallic ions attach to the microalgal cell wall's polyelectrolytic counterparts by virtue of covalent bonding, Van der Waals forces, and redox communication in order to be electroneutral. This process is reversible and metabolism-independent. Aksu et al. 1992 have shown that the copper adsorption process by *Chlorella vulgaris* occurs by virtue of Van der Waals interactions. Similarly, Cd, U, Cu, Co, and Zn sorption occur through electrostatic interactions on dead algal biomass (AB) (Kuyucak and Volesky, 1988).

7.2.2.3 Complexation or Coordination

Complexation or coordination refers to the precise process of the mixture of cations with any anions with lone pair electrons, either electrostatically or through covalent bonding. Here, the cationic species of HM stands as the central atom, whereas the anions or the combined molecules form the coordination shell around it, forming a complex or ligand. This process is further subdivided into two categories:

I. inner-sphere complexation, where the distance between the central cation and the other interacting molecules of the coordination complex is less, and
II. outer-sphere complexation, where the molecules of the coordination shell remain far apart from the central cationic species (Davis et al. 2003).

However, referring to Perpetuo et al. 2011, the capability of several organic acids like citric, fumaric, malic, and lactic acids to chelate HM ions, forming metallo-organic molecules, depends on the electrostatic interaction between chelating ligands and biopolymers excreted from the cells.

7.2.2.4 Metallothioneins

Metallothioneins are cysteine-rich metal-binding proteins by means of a molecular weight ranging between 6–7 kDa, accumulation of which within biological organisms may serve as a biomarker for HM poisonousness of the surrounding habitats. Perales-Vela et al. 2006 elucidated two major groups of metallothioneins gene-encoded proteins and enzymatically synthesized short-chain polypeptides named phytochelatins. In algae, phytochelatin synthesis's major enzymatic counterpart is the enzyme phytochelatin synthase, which is post-translationally activated by HM ions. Several reports show the accumulation of phytochelatins in algae as a consequence of HM harmfulness. For example, phytochelatin production by a freshwater alga *Stigeoclonium sp.* grown in metal-contaminated mining water has been reported by Pawlik-Skowrońska, 2001.

7.2.2.5 Vacuolar Sequestration of Heavy Metals

Based on the data from microscopy and X-ray analytical studies, it is confirmed that the metal-MtIII complex (cadmium-containing electron-dense particles) accumulates within the vacuole of a cell of marine microalga *Tetraselmis suecica* (Pérez-Rama et al. 2002; Shanab et al. 2012). According to them, the metal ions along with their chelating partners move into the vacuole, thereby reducing the cytotoxic metal concentration. Within vacuole, the acidic pH releases chelating peptides into the cytoplasm, while the toxic metal ions get sequestered by the organic acids present in high concentration within the vacuole.

7.2.2.6 Chloroplast and Mitochondrial Sequestration

Chloroplast and mitochondria may serve as a great reservoir of poisonous metal ions in the algal cell. Accumulation of Cu^{2+} ions in the region of thylakoid and pyrenoid in Oocystis nephrocytioides has been reported when exposed to a cytotoxic copper concentration (Soldo et al. 2005). Similarly, Nagel et al. 1996 reported cadmium accumulation in the chloroplast of *Chlamydomonas reinhardtii*.

Accumulation of cadmium and mercury within the mitochondria of the heterotrophic cells of *Euglena gracilis* has been describead by Avilés et al. 2003. The occurrence of class III metallothioneins within the mitochondria and chloroplast of *Euglena gracilis* has been elucidated by Mendoza-Cózatl and Moreno-Sánchez, 2005a; Mendoza-Cózatl et al. 2005b. According to them, these chelating peptides may either be synthesized within the cytosol or by the reservoir organelle itself. However, Perales-Vela et al. 2006 reported the biosynthesis of class III metallothioneins in both places.

7.2.2.7 Polyphosphate Bodies

Dwivedi, 2012 elucidated the role of polyphosphate bodies, which are electron-dense materials present within the cytoplasm and are of huge taxonomic implications for microalgal classification. Polyphosphate bodies or acidocalcisomes (Ruiz et al. 2001) serve as a storage pool for several other ionic species, including Pb, Ti, Cd, Sr, Hg, Ni, Mg, Zn, Co, and Cu in addition to polyphosphate compounds, thereby facilitating metal detoxification.

7.2.2.8 Other Responses

Interestingly enough, Abd-El-Monem et al. 1998 reported sexual reproduction in *Scenedesmus sp.* upon exposure to HM shock. Peña-Castro et al. 2004 studied the phenotypic plasticity adopted by *Scenedesmus incrassatulus* upon exposure to cytotoxic cadmium and copper concentration in their growth environment. The role of glutathione as a HM scavenger has also been established by Satoh et al. 1999 in *Tetraselmis tetrathele* and *Chlamydomonas reinhardtii* by Perales-Vela et al. 2006.

7.2.3 Genetic Manipulation for Efficient Metal Binding

Recent research focuses on enhancing the efficiency of metal-binding and bioaccumulation of hazardous metallic species by potential algal strains through genetic manipulations, which can be used as an eco-friendly, self-sustaining agent for remediation of metal-polluted water from different sources. This is achieved through transgenic technology, which provides a great deal to manipulate an organism's genetic constitution. More precisely, the strategy of transgenesis, in this respect, includes the valorization of the synthesis of those metabolites, synthesized as a reply to HM stress, over-expression genes of the cell surface as well as cytosolic metal-binding proteins and other chelating agents (Rajamani et al. 2007).

Due to the availability of the entire genome sequence, *Chlamydomonas sp.* serves as the model organism in this regard. Eleven novel gene families have been reported in the *Chlamydomonas* genome, which encodes proteins responsible for HM sequestration (Hanikenne et al. 2005). The major metal-binding protein in algae appears to be the metallothionein family of proteins. The sequences encoding the metal-binding domains of metallothionein have been introduced along with the sequences encoding a low CO_2-induced plasma membrane protein in transgenic *Chlamydomonas* (Siripornadulsil et al. 2002). Interestingly, the transgenic cells show growth rates similar to that of the wild-type cells when exposed to a lethal dose of cadmium (120 µm for wall-less cells), but shows five times greater accumulation of cadmium than the wild-type cells, by virtue of the overexpression of the extracellular metal-binding domains (Sayre et al. 2005). This approach has often been used to sequester HMs released by in-situ sonication of the contaminated sediment (He et al. 2011).

HM can also be sequestered from the metal-polluted sediments by *in situ* solubilization using dilute acetic acid within the injection pump grid. The metallic solution can then be extracted within recovery wells, attached to the injection pumps. This metal-containing acid solution can be used in the culture of *Chlamydomonas* as a source of trace metals. But to mitigate the biological and societal considerations, the escape of transgenic cells in the environment should be precluded. To accomplish this, several mutations have been introduced into the metal-tolerant host strain to diminish the viability and growth potential of the transgenic cells and to preclude the exchange of genetic materials with non-transgenic cells. Mutant cells like *nit1-30* and *pf-14* prevent mating and genetic exchange due to deformity in the flagellar apparatus. In this context, maternally inherited genes are of special interest, as they constitutively segregate onto the next generations. In order to preclude photosynthesis and to include heterotrophism, mutation of one chloroplast gene, namely *psbA*, can serve this purpose. Therefore, triple mutant cells containing lesions like *psbA*, *nit1-30*, and *pf-14* deletions cannot mate and require acetate and ammonia for their survival (Diener et al. 1990).

In addition to these deletions, several reports also show that introduction and overexpression of genes or regulatory sequences help in the increment in metal-accumulation efficiency. It has been proved that the genetics of phytochelatin synthesis is activated by the presence of glutathione-HM adducts (Vatamaniuk et al. 2000). Siripornadulsil et al. 2002 engineered *Chlamydomonas* strains through the introduction of two genes: *P5CS*, which codes for the enzyme pyrroline-5 carboxylate synthase, a rate-limiting enzyme, involved in proline biosynthesis (Hu et al. 1992), and *HAL2* gene, involved in cysteine biosynthesis. Transgenic cells showed two-fold elevated proline levels in association with four times greater metal-accumulating capacity than wild-type cells, when cultured in lethal cadmium concentration. This enhancement was accompanied by a four times increase of reduced glutathione levels compared to their oxidized counterparts. The enhanced proline level was also convenient to detox several ROS-mediated cytotoxicities, including lipid peroxidation. In continuation with the previous result, transgenic microalgal cells expressing *HAL2* gene have expressed a 2.5 times increase in cadmium accumulation along with the enhancement of reduced glutathione level as the gene product involves cysteine biosynthesis.

7.2.4 Commercial Feasibility

To establish algae-based metal remediation technology at the industrial level and make it commercially feasible, considering some recommendations of its metal-binding capacity, the same biomass can be utilized for many sorption-desorption cycles. Appropriate care should be taken in the choice of the eluent consumed for the desorption of metal ions from the biomass studied. The eluent should work very fast and remove all the bound metal ions and not hamper the biomass's metal absorbing capacity after multiple uses (Borowitzka & Borowitzka, 1989). An electrolyte is better, so removal can be achieved through electrolysis. In addition, optimization of the sorption process and the option of the eluent should also be subject to the nature of the biomass used. For example, $CaCl_2$ in combination with hydrochloric acid is a widely used eluent for metal cations. Interestingly, HCl desorbed 100% Cd from *Sargassum sp.* (Esteves et al. 2000) but only 1.2% Cd from *Chlorella minutissima* (Roy et al. 1993), which suggests the difference in the metal-sorption strategy among different groups mainly due to their difference in cell wall architecture. Therefore, it is unconditionally required to standardize the entire process by manipulating different parameters to achieve the best result.

The use of hydrochloric acid as an ideal eluting agent is not recommended for the regeneration of ABs despite showing the metal desorption efficacy of the biomass after successive rounds of elution. The probable reason may be because of damage to metal-binding sites and hydrolysis of the cell wall polysaccharides (Chu et al. 1997). Similarly, Kuyucak and Volesky, 1989 reported a 74% reduction in cobalt accumulation using the HCl-washed *Ascophyllum nodosum* that reveal reversible mechanism of cobalt binding in the cell wall of the seaweed.

7.3 MICROALGAE-MEDIATED NANOTECHNOLOGY TECHNIQUES TO REMOVE HEAVY METALS

In recent times, the concern for wastewater treatment is being raised globally and is the critical debate of the hour. As a result, it has become an essential issue to safeguard and maintain energy resources and implement innovative and unconventional wastewater treatment technologies (Kumar et al. 2018; Agrawal et al. 2020a; Goswami et al. 2021a). Several investigations and explorations are in progress to furnish efficient, beneficial, and feasible solutions for managing wastewater and reducing pollutant levels. Therefore, serious attention is being given to the natural ways of wastewater treatment that have re-emerged in this decade. Recently, the use of engineered aquaculture systems for wastewater (either for domestic or large-scale industrial purposes) treatment and water reprocessing and reuse has been leveled up vastly around the globe (Agarwal et al. 2019).

Amidst all the biological systems, microalgae have won a tremendous amount of attention and interest since they can be used in the remediation of toxic metals to more pliable forms from wastewater. In nature, microalgae are ubiquitous organisms and have great diversity among themselves. Furthermore, they are easy to handle, convenient to use, effective in energy utilization, have lower risk factor to the environment, and can gather toxic HMs and harmful contaminants from wastewater for processing, proposing them to be an excellent and potential nanofactory (Hwang et al. 2016). Microalgae culture systems provide cost-effective advanced approaches for treating toxic contaminants and hazardous chemicals from wastewater effectively (Mehariya et al. 2021a, 2021b). The biotreatment of wastewater with microalgae is fascinating particularly due to its photosynthetic capabilities, in which microalgae convert carbon dioxide and solar power into beneficial biomass enriched in N and P, causing eutrophication in water bodies. This is also known as phycoremediation (Larsdotter, 2006; Ahmad et al. 2020; Goswami et al. 2020). Hence, microalgae culture offers a combined benefit of treating and managing polluted water, acting as a growth medium as well as producing a large amount of valuable biomass (Agrawal et al. 2020b) for several other applications such as livestock feed and composting.

The amalgamation of nanotechnology with biotechnology has crafted a new branch of science that is nano-biotechnology. Utilization of these techniques in combination helps in providing efficient outcomes. In comparison to other microorganisms like bacteria and fungi, microalgae are equivalently efficient in manufacturing nano-bioparticles; thus, the investigation and analysis of bio-production of nanomaterials by microalgae can be termed as phyco-nanotechnology and further findings can be approached and employed in unique and innovative ways for better results (Agarwal et al. 2019). Several advanced approaches and resulting research outcomes have been worked upon to develop NPs composed of metals from their heterogeneous salt solutions by utilizing microalgae (Oscar et al. 2017). Exploitation of NPs' bio-production by implementing several microalgae of varying strains and species has been used immensely. In addition, the biomolecules extracted by microalgae are used for the practice of nano-bioparticle synthesis (Patel et al. 2015; El-Sheekh and El-Kassas, 2016). To avoid the tedious and time-consuming methods of microalgae culture systems, microalgal-associated nanotechnologies have been employed as a unique and enhanced procedure.

7.3.1 Biosorption

Numerous physiochemical processes, such as surface adsorption and absorption, precipitation of chemical molecules, extraction and isolation of solutes from their respective solvents, separation through selectively permeable membranes, and interaction and exchange of ionic forms of metals, etc., are widely practiced for toxic metal exclusion from wastewater, to treat and recycle for its further utilization (Eccles, 1999). However, due to various demerits of physio-chemical methods, such as inefficiency in metal removal, high-cost maintenance of equipment, greater energy requirements, need high reagent supplies and generate a lot of toxic sludge or other wastewater that needed proper dumping. Therefore, the application of biological mediators and procedures can help overcome certain drawbacks of physical and chemical practices and can be considered a potent contender for the management of wastewater concentrated with hazardous HMs.

The process of biosorption involves the absorption of materials in contact with the use of biological agents or their derivatives, which act as sorbents. This occurrence was witnessed in the mid-19th century when wastewater (containing radioactive elements) discharged from a nuclear power plant was observed to be the habitat for several microalgal colonies. Earlier laboratory findings revealed that biosorption is cost-effective and a favorable area of research. It is considered as one among the very few promising technologies for toxic HM exclusion from aqueous solutions because of its higher uptake capacities, greater abundance, and renewability. In addition to this, it is also useful in recovering valuable metal ions like gold and silver (Kumar and Oommen, 2012; He and Chen, 2014). To exclude unsafe HM ion concentrations from contaminated waters, a huge amount of concern and interest is devoted towards a variety of cheaper biomass availability, their capability in metal adsorption and uptake, and removal of unhealthy and undesired concentrations of metallic forms. These approaches have received better responsiveness since past few years because of their potential application in the safety of the environment and retrieval of valued or premeditated metals (Malik, 2004).

Algal biomass is highly effective in nature with no nutrient supply requirement, lower operating expenses in cleaning HMs even in their minute levels, and reduced proportion of spent biosorbent for the final disposal, leading to ecological safety (Sheng et al. 2007). Moreover, several investigate works have revealed that the dried (inactive) biomass of microalgae can be more effective in comparison to living (active) algal cells for HM detoxification, as the inactive biomass does not require any nutrition for its growth and development (Gautam et al. 2015). At the same time, active microalgal-based sorbents for HM segregation and elimination depend largely on the efficacy and growth rate of AB in heterogeneous aqueous solutions concentrated with HM, as HM ions may pose detrimental effects on microalgal cells, leading to a reduction in their potential to remove HM. The mechanism of uptake and segregation of HMs and their elimination from contaminated wastewater by living microAB is more intricate and elaborate than in the case of inactive or non-living biomass of microalgae. The active biomasses of microalgae absorb and utilize these ionic forms of HMs in their intracellular metabolic pathways, whereas, unlike the active AB, the inactive or dead biomass of microalgae can adsorb and uptake HM ions onto the external surfaces of their cell walls. Passive AB occurs in the form of aggregation of a diverse range of polymers; they may be composed of carbohydrates, cellulose, pectin, glycoproteins, etc., which enables them to hold onto and bind effectively to the HM ions as bioadsorbents. They are also considered economically feasible means of wastewater treatment relative to other chemical technologies (Misbah et al. 2014; Ahmad et al. 2020).

Depending upon the HM dosage in microalgae's external environment, they accumulate a huge amount of HM ions. The concentration of HMs has a great effect on sorption capacity, which fluctuates significantly in different species of algal cells; however, it escalates if the wastewater tends to have a lesser concentration of metal ions (Kelly, 1988). Because the concentration of metal accumulation by microalgae is affected directly by the accumulation of metals in the wastewater, there is a possibility that native microalgal species can be employed for keeping a check on metal contamination in polluted water bodies and their proper biomonitoring to safeguard the availability of re-usable water (De Filippis and Pallaghy, 1994).

Moreover, microalgae possess the in-built ability of segregation and elimination of HMs, which could be exploited for recovering useful metals like gold and silver, meanwhile discarding the burden of harmful metals. This will lead to wastewater management without causing damage to the environment and will even be beneficial to the economy. The process of adsorption and accumulation of HMs by the AB predominantly depends on their cell walls' structure and composition and other layers of membranes. The existence of dissimilar cellular molecules on the surface like polysaccharides, proteins or intracellular agents like cytoplasmic ligands, phytochelatins, and metallothioneins result in varying adsorption capacities (Aksu, 1998). Thus, it encloses metal ion binding sites on the extracellular surfaces in addition to intracellular ligands.

Experimentally, it was found that microalgae-based sorbents are mostly loaded with organic ligands or different functional groups incorporated into their structural components that serve a significant role in the elimination of HM toxicants. They have a great deal of metal-binding capacity, which makes them a potential contender of biosorption and have been widely acknowledged (Ahalya et al. 2003). Fascinatingly, the ability to stand a difficult dosage of toxic HMs by the microalgae is said to be immensely dependent on the mechanism of surface adsorption (Lombardi et al. 2002). The relative magnitude of surface adsorption of metal ions and accumulation may differ from species to species (Mehta and Gaur, 2005). This occurs because of the variation in cell wall assembly and configuration, the existence of dissimilar functional groups and their number, and the distribution of different groups of algae with dissimilar requirements. The cell walls consist of chemically active functional groups mostly conferring negative charge on the external cell surface, which in turn attracts and binds to commonly found positively charged metal ions present in the aqueous solutions (Xue et al. 1988; Crist et al. 1991, 1994; Skowronski and Ska, 2000; Romero-Gonzalez et al. 2001).

Even though diverse potential functional groups attached to the structural backbone of the cell wall of algae have been studied for wastewater recovery, just their presence does not guarantee the participation in the sorption of HM ions. There are several factors such as steric hindrance (due to bulkier functional groups), variation in conformations or crosslinking, and the formation of branched structure that may fluctuate according to the ecological circumstances, and may hinder some of the active groups on the cell exterior from binding to the metallic ions (Adhya et al. 2002).

Among the variety of biomolecules present on algae's cell surfaces, polysaccharides and proteins prove to be having countless metal-binding sites. They serve as great constituents for metal sorption. According to findings of Chojnacka et al. 2005 in Cyanobacteria, several biomolecules on present on cell surfaces, like lipopolysaccharides, lipids, and lipoproteins, that possess metal-binding properties. Carboxyl groups have also exhibited a great deal of metal-binding capabilities and are thus considered as dominant active sites in the cell walls. Therefore, numerous studies related to active sites for HM sorption suggest that carboxyl groups on cell walls of both algae and cyanobacteria play a vital role (Chaturvedi et al. 2013).

Microalgae tend to accumulate HM ions in various ways, broadly categorized into two: active and passive methods. In the initial phase of adsorption, algal cells rapidly take up the metal ions from their aqueous solutions (passively), but in the later phase, the rate of uptake becomes slower (actively). The passive uptake of metals is very frequent (short timespan) as is independent of cell metabolism, whereas the active process involves cell metabolic activities and consumes relatively more time. The process of active sorption includes energy expenditure to transport the metal ions from the outside to the inside of the cells. Metal ions need to cross the cell wall, plasma membrane and then accumulate in the cytoplasm of the algal cells. In certain environments with large dosage of HMs, the permeability of the membranes tends to change, leading to greater sorption through a passive mode of diffusion (Gadd, 1988). The sorption of HMs by numerous processes, for instance, ion exchange, complex formation, electrostatic interactions, and micro-precipitation, occurs at the micro-scale. (Kumar and Oommen, 2012).

Investigations using an electron microscope and X-ray analysis resulted in determining the metal ions' subcellular localization and position. Experimental studies with these tools and techniques revealed that the metal ion-binding sites are located on microalgae's cell surfaces (Klimmek et al. 2001). Several spectroscopy techniques like Fourier Transform Infrared Spectroscopy (FTIR) and titration methods are conducted to recognize the active working groups present on the algal cells' external surfaces, which prove to adsorb metals remarkably. It is even observed that treating algal cells with certain chemicals or reagents, which block these active functional groups, making them inactive or dysfunctional, eventually decreasing the rate of metal sorption (Mehta and Gaur, 2005).

Several investigations are conducted on the elimination of HM ions using biosorption capacity of brown algae (Davis et al. 2000, 2003; Lodeiro et al. 2005; Luna et al. 2010; Kleinübing et al. 2011), green algae (Deng et al. 2009; Zakhama et al. 2011; Wan Maznah et al. 2012), and red algae (Ibrahim, 2011). Although there is wide acceptance of algal technology in metal exclusion and wastewater management, for large-scale commercial usage, we need to focus on several essential factors, such as:

i. selection and isolation of beneficial strain types possessing superior metal sorption ability,
ii. in-detail knowledge about the modes and mechanisms of metal sorption,
iii. employment of advanced cell immobilization methods,
iv. use of highly efficient yet low-cost maintenance procedures for maximum profit, and
v. the utilization of bioengineered microalgae for their better performance in metal binding and adsorption processes (Chen et al. 2012).

7.3.2 Biogenic Silica-Based Filtration

Marine algae such as diatoms are the only organism on earth with cell walls made up of silica. The silica composition is widely used to purify and manage wastewater treatment along with nanotechnological methods. Nanomaterials that are composed of silica and its derivatives are considered crucial nanomaterial for the exclusion of HMs because of their vital features, like non-toxic property and excellent surface characteristics (Vincy et al. 2017). Silica-based nanomaterials can be altered and engineered with various functional groups ($-NH_2$, $-SH$, etc.) on their surface or act as supporting structures for nano-composites.

Accelerating the growth of diatoms is a difficult task, for which an ample amount of silica is required in the process. Nualgi is a novel product that uses silica-based nanotechnology material (nanosilica) serving as a chief micronutrient along with several other trace elements such as Mg, Co, B, Ca, Mn, Zn, Mo, etc., for inducing a high rate of growth of diatoms (Kumar, 2008). Referring to the research done by Agarwal et al. 2019, Nualgi triggers the rapid growth of diatoms in the aqueous environment by removing carbon content and elevates the availability of nutrient requirements to the AB. The nano-size of Nualgi allows it to explore even the smallest of spaces in the subsurface. It is advantageous over other methods because of its non-requirements of electricity and risk-free procedures.

In an experimental analysis, Kotsyuda et al. 2017 artificially developed silica-based nanospheres composed of functional groups like 3-aminopropyl and phenyl groups that were bioengineered with the help of advanced tools and techniques. They were further bio-functionalized and have undergone biocompatibility tests by examining them for their capacity to segregate and exclude divalent copper ions and cationic thiazine dye in its static mode. The research carried out resulted in fruitful consequences. It was observed that the silica-based biofunctionalized nanospheres acquired higher capability for divalent copper metal adsorption and methylene blue, in contrast to the functionalized amino-based nanosilica. The findings indicated that functionalized silica nanospheres had twice the methylene blue adsorption capacity compared with amino-based silica NPs. This bioengineered nanomaterial also displayed reasonable antibacterial properties.

In the research performed by Najafi et al. 2012, on using three varieties of silica-based nanomaterials phase by phase for divalent cadmium, nickel, and lead ions, the proficiency of metal adsorption and exclusion potential was found to be adhering to a pattern of increase as follows: (i) (Ni II) < (Cd II) < (Pb II); (ii) non-functionalized silica nano hollow sphere (SNHS) < amino-functionalized silica gel (NH_2-SG) < amino-functionalized silica nano-hollow sphere (NH_2-SNHS) respectively. In addition to the property of surface modifications, the synthesis of nanocomposites from silica has been widely reported. Silica-based magnetic materials have also been a matter of attraction for researchers. From the findings of Pogorilyi et al. 2014, silica layers could be coated on the surface of magnetite particles, which leads to large-scale applications for industrial purposes. There are pieces of evidence from work carried out by Mahmoud et al. 2016 that linear chain and crosslinked nano-polyaniline, when introduced into the nanosilica-based nanocomposites, exhibited tremendous adsorption effects for divalent metal ions such as Cu (II), Cd (II), Hg (II), and Pb (II). Applying the Langmuir isotherm to the results, they concluded that crosslinked nano-polyaniline silica particles could be economically feasible as well as an excellent adsorbent for HM extraction (Yang et al. 2019).

7.3.3 Bioreactors

Most of the biotechnological processes are carried out in an apparatus known as bioreactor, where new useful products are synthesized or harmful substances are degraded, with the help of microbes, maintained under controlled conditions (optimum temperature, pH, nutrient supply, pressure, aeration, sterile environment, etc.), so that essential functions for the process could be regularly measured and checked. Several biological reactions occur during this process while conducted in a closed system for bioprocessing purposes (Enamala et al. 2019). There are several diverse kinds of bioreactors, for example, fixed-bead, air-lift, and stirred tank. Among these, the most widely used and acknowledged bioreactors are the stirred-tank. An assortment of organisms like bacteria, fungi, yeast, and algae are used to produce a large number of desired products. In recent years, AB is employed widely in these bioreactors to treat wastewater and recover HMs from aqueous solutions. Generally, tubular photobioreactors are made up of a closed system of clear tubes. These tubes serve as growth chambers for algal cells to flourish and contain gas exchanging units where CO_2 can be supplied. In turn, O_2 generated can be stripped out from the medium. For the maintenance of temperature, heat exchangers can be added to the reactors. Several factors, such as gaseous concentrations, their circulation speed, nutrient supply for growth, and development of AB, are taken care of at regular intervals without damaging fragile algal cells (Alaswad et al. 2015; Yadavalli et al. 2020, 2021).

7.3.4 The Hybrid Nano-Membrane Technique to Treat Wastewater

At present, several attempts and investigations are constantly carried out to improve the efficiency of the already existing membranes in terms of better separation capabilities, anti-compaction, and antifouling properties (Table 7.1). These membranes need to possess good thermal stability, anti-corrosion property and should be highly hydrophobic. Hence, the use of nanobiotechnology could be a solution for producing these hybrid membrane systems. Therefore, hybrid membrane preparation with the above required qualities is still elusive and a great challenge for researchers (Giagnorio et al. 2019, Fabris et al. 2020).

7.4 FACTORS AFFECTING HEAVY METAL REMEDIATION

Several factors are responsible for HM infiltration in the water resources; it can be caused naturally by soil erosion, weathering of rocks, volcanic eruptions, and other geological accidents. Anthropogenic factors leading to the release of toxic metallic ions are urban drainage, agricultural runoff, air pollutants, and polluted wastewater for agricultural activities (Kinuthia et al. 2020).

7.4.1 Metal Toxicity

HM toxicity can severely affect the life of inhabitant organisms such as phytoplanktons, zooplanktons, fishes, arthropods, and other animals (De Filippis and Pallaghy, 1994). These toxic metal ions can affect the cellular organelles and interact with cell membranes, mitochondria, endoplasmic reticulum, and even enzymes involved in metabolic activities. The level of toxicity in the exposed individuals can vary according to various physicochemical properties, dose, exposure time, mode of adsorption, etc. The small amount of highly toxic HMs, such as As, Pb, Hg, Cd, Fe, and Al, can severely impact human health and even cause death (Baby et al. 2010; Blaby-Haas and Merchant, 2012). These HMs gain entry to the wastewater by various industrial effluents, agricultural runoff, mining, smelting activities and present naturally in the earth's crust.

7.4.2 Biomass Concentration

The biomass concentration of algal cells has a crucial role in the efficient exclusion of HM ions from contaminated water. Roy et al. 1993 studied dried biomass of *Chlorella sp.* and found a 91% reduction in the metal cadmium. Mehta and Gaur, 2001a, 2001b tested the affinity of biomass for metal attaching on copper

TABLE 7.1
Different Nano Techniques Employed to Study the Efficiency of HM Removal in Various Wastewater

Technique	Contaminant	Efficiency	Findings	References
Photocatalysis via nanofiber	Dairy effluent water	75–95%	Nanomaterials ($AgTiO_2$) within the nanofiber membrane show a critical function in the photocatalytic breakdown of dairy effluents	Kanjwal et al. 2010
Nano-membrane with poly-gamma-glutamic acid (γ-PGA)	Lead (Pb) ions	99.8%	γ-PGA binds with HM ions and NF membrane remove optimal Pb^{2+} at fast rate	Hajdu et al. 2012
Nano-filtration membranes	Reactive dye Black 5 removal from textile effluent	99%	The composite leaves low electrolytes in the effluent with high cleaning efficiency and antifouling property	Zhu et al. 2013
NF membrane using γ-alumina and titania nano-crystallites	Microorganisms and ions rejection from wastewater	Microbes (100%), ions (25%)	Bar the entry of microorganisms and reject ions by adjusting the pH	Shayesteh et al. 2016
NP membrane filtration	Total suspended solids (TSS), Total dissolved solids (TDS), oil, grease, Chemical oxygen demand (COD), Biological oxygen demand (BOD) from oil wastewater	TSS (100%), TDS (44%), oil (99%), grease (80%), BOD (76%)	The optimal circumstances for effective water purification were “temperature-45°C, the cross-flow velocity-1.3 m/s, trans-membrane pressure-4 bar, salt concentration-11.2 g/L and pH-10”. Treated water can be used for agricultural purpose	Salahi et al. 2015
ZrO_2 microfiltration membrane	Pre-treatment of dimethylformamide (DMF) wastewater	Turbidity removal (99.6%), suspended solids (99.9%)	Active elimination of suspended fine particles and flux recovery from DMF wastewater	Zhang et al. 2014

and nickel sorption using *Chlorella* species. The Langmuir constant Q_{max} of copper and nickel showed a noticeable reduction in metal sorption measurements by 2.6 and 3-fold, respectively. However, the association amid a concentration of AB and sorption may be restricted to metal availability, increased electrostatic connections, meddling among binding sites, and insufficient mixing at raised biomass concentrations of different biosorbents. The biosorbent concentration at a definite preliminary metal ion dosage regulates the biosorption volume and the elimination efficacy of the biomass in the direction of the HM elimination. However, it can be predicted that the higher biomass concentration depicted higher and effective elimination of metal ions from the wastewater/source. Meanwhile, it is also possible that lower biomass concentrations will be depicted as lower and low metal elimination efficacy of HM ions from the wastewater/source was due to the biosorbent saturation with excess metal pollutants.

7.4.3 pH

Among the several physical factors, pH is one of the leading domineering factors that directly impact the biosorption procedure to eliminate metallic ions from wastewater. Ajayan, 2015 studied metallic ion removal from tannery wastewater and found an essential role of pH in the sorption of metal ions in the aqueous solution. Consequently, several researchers seriously investigated the optimum pH conditions for algae-mediated enhanced metal removal, mentioned in Table 7.2. It is well known that several industrial effluents are rich in inorganic ligands such as bicarbonate, carbonate, Cl^-, SO_4^-, HS^-, and several organic acids for metal binding. The lower pH solution usually intensify the number of free metal ions whereas rise in pH medium usually upsurges the free HM ions' concentration (Yu and Kaewsarn, 1999). The aqueous solution pH can significantly influence the solubility of the HM, the quantity

TABLE 7.2
The Optimum pH for Sorption of Heavy Metals by Different Microalgae

Metal ions	Algae	pH	Reference
Ag	*Cladophora prolifera*	5	Zhao et al. 1994
	Oedogonium sp.	7	Crist et al. 1988
Al	*Cladophora prolifera*	4–5	Zhao et al. 1994
	Padina pavonica		
Au	*Cladophora prolifera*	2–4	Zhao et al. 1994
	Padina pavonica		Kaewsarn, 2002
Cd	*Cladophora prolifera*	6	Zhao et al. 1994
Cu & Cd	*Fucus vesiculosus*	3.54–6.8	Sandau et al. 1996
Cr (III)	*Laminaria japonica* and *Sargassum kjellmanianum Sargassum sp.*	4–5	Zhou, Huang, and Lin, 1998
Cu		4	Cossich et al., 2002
	Microcystis aeruginosa and Spirulina platensis	>2	Mehta and Gaur, 2001a

TABLE 7.3
Optimal Metal Removal by Passive AB at Diverse Temperatures

Algal Species	Metal Ions	Temp (°C)	% Removal	References
Cystoseira barbata	Cd^{2+}	20	68	Yalcin et al. 2012
Cystoseira barbata	Ni^{2+}	20	65	Yalcin et al. 2012
Cystoseira barbata	Pb^{2+}	20	52	Yalcin et al. 2012
Lessonia nigrescens	As^{+}	20	77	Hansen et al. 2006
Sargassum muticum	Sb^{2+}	23	50	Ungureanu et al. 2015a, 2015b
Spirogyra sp.	Pb^{2+}	25	30	Gupta and Rastogi, 2008

of the counterions on the functional group of the biosorbent, the HM ion linking site on the AB, and the HM ion chemistry in the aqueous solutions. At lower pH values, H^+ ions contest with these HM ions in the industrial effluents for the active binding sites on the AB surface.

7.4.4 Temperature

Temperature variations can influence microalgae's sorbent capacity; the metal adsorption rises with increasing temperature and causes diverse biosorption behaviors in numerous algal strains with dissimilar HM ions. Algae-mediated biosorption of HMs is unaffected in the temperature range of 20–35°C. While in the 40–50°C range, a visible increment in biosorption will occur. However, high temperatures may lead to structural changes to the algal cells (Sharma and Azeez, 1988). As a result, metal uptake declines. Biosorption, which is an exothermic process, is mainly based on an adsorption reaction. With a drop in temperature, the degree of HM adsorption by algae rises. Ali et al. 2012 documented the absorption of metal by low cost adsorbent, *Spirulina platensis* to remove organic pollutants. With rising temperature, it gradually increased, and maximum Cu uptake (90.61%) was found at a temperature of 37°C. Therefore, by AB, temperature plays a critical effect on metal uptake (Ahmad et al. 2020). Metal removal by different passive AB at different temperatures is given in Table 7.3.

7.5 PHYSIOLOGICAL BENEFIT OF MICROALGAL NANOPARTICLES OVER OTHER NANOMATERIALS

Nanomaterials have their structural composition between 1–100 nm. Many nanomaterials possess properties such as catalysis, high reactivity, and adsorption due to their different mechanical, optical, magnetic, and optical properties. Till now, NPs have been employed in biology, sensing, catalysis, medicine, and others, including wastewater treatment. Nanomaterials have a small size and high specific surface area, which confer them with high mobility, high adsorption, and reactivity (Lu et al. 2016).

When compared to conventional treatments of wastewater, microalgal techniques in the purification of wastewater prove to be more eco-friendly and require low cost.

Moreover, integration of industrial and municipal algal systems provides many benefits such as the production of biofuel, utilization of CO_2 for algal growth, removal of nitrogen and phosphorus from wastewater, and generation of other value-added products (Kumar et al. 2019; Bhardwaj et al. 2020; Goswami et al. 2020, 2021b, 2021c; Mehariya et al. 2021c), for example, carbohydrates, pigments, and proteins for various uses in the pharmaceutical industry (Hwang et al. 2016).

Nanotechnology revolves around nano-sized objects with novel properties. Modern-day human activities lead to the release of contaminants in the environment, causing environmental pollution. The pollution of water generates wastewater, treatment of which is very crucial. Efficient advanced techniques are continually being adopted for wastewater treatment, and one such technique is microalgal culture. To counter the limitations associated with microalgal culture systems, such as time-consumption, requirement of arable land, etc., nanotechnology is being looked up to for increasing removal efficiency. NPs have some unique and efficient crystallographic, physiochemical, thermal, optical, electronic, and biological properties.

NPs find their applications in the fields of medical treatment, solar and fuel cells, water and air filters, and catalysis in the manufacturing process for replacing toxic substances. Wet methods of physical and chemical nature are conventional means for NP production. But these methods pose a negative environmental impact, along with high cost and laborious production. NPs have high surface energies and therefore interact with other atoms and molecules, resulting in a change in their surface properties.

The biosynthetic route of NP synthesis is simpler than conventional techniques. Fungi, yeast, and bacteria help in remediating metal ions, whereas microalgae can neutralize toxic metals. Microalgae develop metallic NPs from their corresponding salt solutions. NPs like gold and silver can be extracted from microAB. Many species of algae, such as *Spirulina platensis, Lyngbya sp.*, and *Chlorella vulgaris* are capable of the synthesis of silver NPs. Microalgae have been testified to be an efficient nanofactory as NPs' biosynthesis takes less time than other microorganisms. Till now, NPs of gold, silver, and cadmium have been biosynthesized from microalgae. The reduction in metal ions takes place by polysaccharides, peptides, and pigments. In contrast, capping and stabilization of the metal NPs are achieved by proteins and sulphated polysaccharides (Agarwal et al. 2019).

7.5.1 Activated Carbon-Based Nanomaterial

Carbonaceous nanomaterials include carbon nanotubes (single-walled carbon nanotubes, SWCNT, or multi-walled carbon nanotubes, MWCNT), carbon nanofibers, graphene, fullerene, and amorphous carbonaceous composites. These have a high surface area and the surfaces are hydrophobic, which helps in targeting specific pollutants. These may be used as filters or membranes to remove pollutants (Adeleye et al. 2015).

7.5.2 Zero Valent Metal-Based Nanomaterial

Silver nanoparticles (AgNPs) have antimicrobial properties, capable of killing many viruses, bacteria, and fungi. AgNPs have proved to be active in wastewater treatment during the last few years, with some drawbacks. These include aggregation of NPs in an aqueous medium, reducing their efficiency. However, AgNPs with filter materials have advantages like low cost and high antibacterial activity during disinfection of water (Lu et al. 2016).

7.5.3 Metal Oxide-Based Nanomaterial

Metal oxide NPs (MNPs) are regarded as environment-friendly, sustainable, and low-cost technique for remediation of wastewater. These NPs have a short intraparticle diffusion distance and are reusable as well as can be easily regenerated. Their adsorption efficiency is higher than activated carbon because of their superparamagnetic nature. Adsorption of HMs along the walls of the surface of MNPs is affected by intraparticle diffusion. MNPs have removed HMs like nickel, copper, lead, chromium, arsenic, cadmium, and mercury. Metal oxides of aluminum, titanium, zinc, and iron have been stated to be active, low-cost adsorbents for the removal of HMs as well as radionuclides (Singh et al. 2019).

7.5.4 Nanocomposites

Nanocomposites have gained importance over the last few years because of their capability to absorb contaminants like dyes, pesticides, and HMs. The various characteristics of nanocomposites like high surface area and the addition of photoactive and antimicrobial NPs upsurge the efficacy of removal of pollutants and minimize foul smell effectively (Liu et al. 2009; Ge et al. 2015; Anjum et al. 2016; Surendhiran et al. 2017; Akharame et al. 2018).

7.5.5 Hybrid Nanoparticles

A variety of NPs, including bio-inspired nanofibers, porous nanofibers, 2D- nanofibers/nets, and cable-like nanofibers, are synthesized through electrospinning methods. The advantages of these materials are high surface-to-volume ratio, adjustable functionality, and large porosity, allowing them to perform better at liquid filtration and particulate separation than polymeric and nonwoven membranes. Treatment of wastewater is done by hybridizing bio-macromolecules with nanofiber membranes. Electrospun nanofiber membranes can separate particulates from wastewater. The combined effect of TiO_2/Ag nanomaterial has been studied by Pan et al. 2017 in photo removal of Cr(IV) under UV irradiation. Reactive dyes were removed from water with the help of immobilization of microalgae *Chlamydomonas reinhardtii*, and polysulphone nanofibrous web was developed from this system by electrospinning (Hu et al. 2001). In the same way, Eroglu et al. 2012 removed nitrate from effluents through immobilization of microalgal cells on electrospun nanofiber mats. For removing HM ions from wastewater, Liu et al. 2009 made magnetic chitosan nanocomposites, which can be removed easily from wastewater by an external magnet.

7.6 STRATEGIES IN ALGAL NANOFACTORY FOR OPTIMAL ELIMINATION OF HAZARDOUS METALS

Approximately two decades ago, the field of green chemistry had drawn the Environmental Protection Agency's attention, aiming to reduce pollution caused by urbanization, industrialization, and modern agricultural trends (Ramirez-Merida et al. 2015). Wastewater from industrial or agricultural sewage releases plenty of metal pollutants into the environment. Changes in the HM profile in the environment may alter the life of the concerning flora and fauna through genetic means, which ultimately establish the stepping stones for various health disorders in animals and plants. Material scientists treated the potentiality of microorganisms in HM remediation with additional interest for eco-friendly nanofabrication of toxic metal ions into nanometals with highly optimized properties (Pan et al, 2017). Microalgae can serve as an elegant, powerful nanofactory for optimal elimination of hazardous metals due to their amazing metal-binding property, the corresponding components of their cell wall, and fast growth rate along with their nominal growth requirements. Various strategies have been taken into consideration to elicit the microalgae-based bio-mineralization process (Rajendran and Gunasekaran, 2007).

7.6.1 Use of Immobilized Algal Beads

The use of metabolically active immobilized algal beads is precisely studied by Hameed and Ebrahim, 2007. They recognized many microalgal species eliminating HM ions effectively by employing immobilized beads, prepared with alginate as a suitable system for metal recovery. De-Bashan and Bashan, 2010 had summarized the algal strains and their corresponding immobilized bead materials, for example, carrageenan, alginate, and poly-urethane foam for the surface of *Sargassum sp.* and *Ulva sp.*, silica gel, epoxy resin, milk casein+ glutaraldehyde, polyacrylamide, and polysulphone. *Chlorella sorokiniana* cells were immobilized in the matrix made up of *Luffa cylindrica*, a cylindrical sponge as a sorption bead to remove cadmium and copper (Akhtar et al. 2003, 2008).

Besides, the immobilization bead's nature should be taken into consideration for enhancement of the rate of metal removal. The shape and size of the immobilized beads are critical determinants, and according to Volesky, 2001, "bead size should range from 0.7 to 1.5 mm, related to the size of commercial resins worthy for metal removal".

7.6.2 Desorption of Metal Ions

The metal molecules sorbed on the biomass should be adequately desorbed to use the same biomass for multiple sorption-desorption cycles (Kumar et al. 2018). In addition to high desorption efficiency, the desorbing eluent should not hamper the AB's metal-binding efficacy. Reduction in pH along with the use of dilute organic acid solution may effectively serve as a strong eluent (Vannela and Verma, 2006). The advantage of nitric acid as an eluent compared with 0.1M ethylenediamine tetra acetic acid (EDTA), 0.1M HNO_3, and deionized water for removing cadmium and copper using *Spirulina sp.* has been demonstrated by Chojnacka et al. 2004, 2005. However, the use of acid has several limitations. It damages the metal-binding cell-surface proteins and triggers the hydrolysis of wall polysaccharides (Monteiro et al. 2012).

7.6.3 Use of Transgenic Technology

Recent research trends to the inclusion of transgenic techniques to enhance the microalgal bio-mineralization process at a precise molecular level, which encompasses the overexpression of the metal-detoxifying enzymes or metal-binding proteins to valorize the metal-remediation process using the biological means (Rajamani et al. 2007). According to Huang et al. 2006, "the eukaryotic microalga *Chlorella sp.* DT, transformed with the *Bacillus megaterium* strain MB1 *merA* gene that encodes mercuric reductase (*merA*), is revealed to facilitate the reduction of Hg^{2+} to volatile elemental mercury; here the *merA* gene was effectively combined into the genome of transgenic strains and functionally stated to endorse mercury removal". However, even though there is a potential market for the betterment of microalgal bioremediation, extreme care should be taken to prevent the chance of escape of these transgenic microalgal cells into the environment.

7.7 CONCLUSION

The chapter discussed the various extraction techniques to remove toxic HMs found in wastewater using algal NPs. The molecular mechanism involved in manipulating algae-mediated nanotechnology for detoxification of wastewater concerning the process's physiological factors is evaluated. A brief comparison of algal NPs with other NPs is made, and various strategies are discussed for a better remediation tool. We can summarize the essential points:

I. Microalgae serve as potential bioremediating agents due to their exceptional metal-binding capability, rigorous growth rate, and minimal growth requirements.
II. The amalgamation of nanotechnology and biotechnology has led to the production of algal NPs that have provided efficient outcomes for treating HMs from wastewater.
III. Among HMs, arsenic is the most toxic to the environment and can prove hazardous if it reaches the food chain. The lower biomass is effective in metal removal as the ambient temperature moves microbes' ability to adsorb HMs. pH and temperature variations can reduce the sorption ability of microalgae and vary from metal to metal.
IV. Microalgal-mediated NPs are most efficient over other NPs as they take lesser time for synthesis.
V. Various strategies are followed for efficient removal of the hazardous HMs from wastewater, such as by means of the metabolically active immobilized algal bead for the enhanced rate of HM removal, use of the same biomass for multiple sorption-desorption cycles without disturbing the metal-binding efficiency of AB, and application of the transgenic technique at a molecular level to improve the microalgal biomineralization process for enhancement in metal remediation.

Biosynthesis of NPs can curb ecological damage and human health risks. Microalgae are constantly being researched for the syntheses of NPs, as the field of nanobiotechnology is quite new. Treatment of wastewater by microalgae can be considered as an addition to the existing biological treatment.

ABBREVIATIONS

AB, algal biomass; AgNPs, silver nanoparticles; BOD, Biological oxygen demand; COD, Chemical oxygen demand; COOH, carboxyl group; CO_3^{2-}, carbonate ion; CO_2, carbon dioxide; CTR, Cu transporter; DMF, dimethyl formamide; EDTA, ethylenediamine tetra acetic acid; FTR, Fe transporter; FTIR, Fourier Transform Infrared Spectroscopy; HAL2, yeast halotolerant gene; HM, heavy metal; HS^-, sulfhydryl ion; kDa, kilodalton unit; MerA, gene mercuric reductase from *Bacillus megaterium*; MNPs, metal-oxide nanoparticles; MWCNT, multi-walled carbon nanotubes; NRAMP, natural resistance-associated macrophage proteins; NPs, nanoparticles; NP, nanoporous; Nit1 30, nitrilase-like protein 1; P, phosphate; P5CS, pyrroline–5-carboxylate synthase; PO_3O_2, phosphoryl group; pf14, mutant strain; psbA, gene–photosystem II protein; ROS, reactive oxygen species; SG, silica gel; SH, sulphydryl group; SNHS, silica nano-hollow sphere; SWCNT, single-walled carbon nanotubes; TSS, Total suspended solids; TDS, Total dissolved solids; TiO_2, titanium dioxide; TISTR 8217, immobilized cells of *Spirulina platensis*; UV, ultraviolet; Qmax, Langmuir constant; ZrO_2, zirconium dioxide; ZIP, Zrt-, Irt-like proteins

ACKNOWLEDGEMENT

The authors wish to thank Dr. Sandeep B. Gaudana, Senior Research Scientist, Reliance Industries Limited, Vadodara, Gujarat, India for proofreading the chapter.

Funding Statement: This research did not receive any specific grant from funding agencies in the public, commercial, or not-for-profit sectors.

CONFLICT OF INTEREST

The authors declare no conflict of interest.

REFERENCES

Abd-el-Monem, H. M., Corradi, M. G., and Gorbi, G. 1998. Toxicity of copper and zinc to two strains of *Scenedesmus acutus* having different sensitivity to chromium. Environ. Exp. Bot. 40(1): 59–66.

Adhya, J., Cai, X., Sayre, R. T., and Traina, S. J. 2002. Binding of aqueous cadmium by the lyophilized biomass of *Chlamydomonas reinhardtti*. Colloids Surf. A Physicochem. Eng. Aspects. 210: 1–11.

Adeleye, A.S., Wang, X., Wang, F., Hao, R., Song, W and Li, Y. 2018. Photoreactivity of graphene oxide in aqueous system: Reactive oxygen species formation and bisphenol A degradation, Chemosphere, 195, 344–350, https://doi.org/10.1016/j.chemosphere.2017.12.095.

Agarwal, P., Gupta, R., and Agarwal, N. 2019. Advances in synthesis and applications of microalgal nanoparticles for wastewater treatment. J. Nanotechnology. 1–9.

Agrawal, K., Bhatt, A., Bhardwaj, N., Kumar, B., and Verma, P. 2020a. Integrated Approach for the Treatment of Industrial Effluent by Physico-chemical and Microbiological Process for Sustainable Environment. In Combined Application of Physico-Chemical & Microbiological Processes for Industrial Effluent Treatment Plant, pp. 119–143. Springer, Singapore.

Agrawal, K., Bhatt, A., Bhardwaj, N., Kumar, B., and Verma, P. 2020b. Algal Biomass: Potential Renewable Feedstock for Biofuels Production–Part I. In Biofuel Production Technologies: Critical Analysis for Sustainability, pp. 203–237. Springer, Singapore.

Ahalya, N., Ramachandra, T. V., and Kanamadi, R. D. 2003. Biosorption of heavy metals. Res. J. Chem. Environ. 7: 71–78.

Ahmad, S., Pandey, A., Pathak, V. V., Tyagi, V. V., and Kothari, R. 2020. Phycoremediation: Algae as eco-friendly tools for the removal of heavy metals from wastewaters. In Bioremediation of Industrial Waste for Environmental Safety, pp. 53–76. Springer.

Ajayan, K. V. 2015. Phycoremediation of tannery wastewater using microalgae *Scenedesmus* species. Int. J. Phytoremediation. 17(10): 907–916.

Akhtar, N., Saeed, A., and Iqbal, M. 2003. *Chlorella sorokiniana* immobilized on the biomatrix of vegetable sponge of *Luffa cylindrica:* a new system to remove cadmium from contaminated aqueous medium. Bioresour. Technol. 88(2): 163–165.

Akhtar, N., Iqbal, M., Zafar, S. I., and Iqbal, J. 2008. Biosorption characteristics of unicellular green alga *Chlorella sorokiniana* immobilized in loofa sponge for removal of Cr (III). J. Environ. Sci. 20(2): 231–239.

Aksu, Z., Sai Y, and Kutsal T. 1992. The biosorption of copper (ii) by *C. vulgaris* and *Z. ramigera*. Environ. Technol. 13: 579–586.

Aksu, Z. 1998. Biosorption of heavy metals by microalgae in batch and continuous systems. In: Wastewater Treatment with Algae, pp. 37–52. Wong, Y.S., and Tam, N.F.Y., Eds., Springer-Verlagand Landes Bioscience.

Alaswad, A., Dassisti, M., Prescott, T., and Olabi, A. G. 2015. Technologies and developments of third generation biofuel production. Renew. Sustain. Energy Rev. 51: 1441–1460.

Ali, I., Asim, M., and Khan, T. A. 2012. Low cost adsorbents for removal of organic pollutants from wastewater. J. Environ. Manag. 113: 170–183. DOI: 10.1016/j.jenvman.2012.08.028

Anjum, M., Miandad, R., Waqas, M., Gehany, F., and Barakat, M. A. 2016. Remediation of wastewater using various nano-materials. Arab. J. Chem. 12(8): 4897–4919.

Akharame, M. O., Fatoki, O. S., Opeolu, B. O., Olorunfemi, D. I., and Oputu, O. U. 2018. Polymeric nanocomposites (PNCs) for wastewater remediation: an overview. Polym. Plast. Technol. Eng. 57(17): 1801–1827.

Avilés, C., Loza-Tavera, H., Terry, N., and Moreno-Sánchez, R. 2003. Mercury pretreatment selects an enhanced cadmium-accumulating phenotype in *Euglena gracilis*. Arch. Microbiol. 180(1): 1–10.

Baby, J., Raj, J. S., Biby, E. T., Sankarganesh, P., Jeevitha, M. V., Ajisha, S. U., and Rajan, S. S. 2010. Toxic effect of heavy metals on aquatic environment. Int. J. Biol. Chem. Sci. 4(4).

Bhardwaj, N., Agrawal, K., and Verma, P. (2020). Algal Biofuels: An Economic and Effective Alternative of Fossil Fuels. In Microbial Strategies for Techno-Economic Biofuel Production, pp. 59–83. Springer.

Blaby-Haas, C. E., and Merchant, S. S. 2012. The ins and outs of algal metal transport. Biochim. Biophys. Acta. Mol. Cell Res. 1823(9): 1531–1552.

Borowitzka, L.J., and Borowitzka, M.A. 1989. Industrial production: methods and economics. In: Algal and Cyanobacterial Biotechnology, pp. 244–316. Cresswell, R.C., Rees, T.A.V., and Shah, N. Eds., Longman Scientific, London.

Chaturvedi, V., Chandravanshi, M., Rahangdale, M., and Verma, P. 2013. An integrated approach of using polystyrene foam as an attachment system for growth of mixed culture of cyanobacteria with concomitant treatment of copper mine waste water. J. Waste Manag. (1): 1–7.

Chaturvedi, V., Goswami, R. K., and Verma, P. 2021. Genetic Engineering for Enhancement of Biofuel Production in Microalgae. In Biorefineries: A Step Towards Renewable and Clean Energy, pp. 539–559. Springer.

Chen, C., Chang, H., Kao, P., Pan, J., and Chang, J. 2012. Biosorption of cadmium by CO2 -fixing microalga *Scenedesmus obliquus* CNW-N. Bioresour. Technol. 105: 74–80.

Chojnacka, K., Chojnacki, A., and Górecka, H. 2004. Trace element removal by *Spirulina sp.* from copper smelter and refinery effluents. Hydrometallurgy. 73(1–2): 147–153.

Chojnacka, K., Chojnacki, A., and Gorecka, H. 2005. Biosorption of Cr3+, Cd2+ and Cu2+ ions by blue–green algae *Spirulina sp.*: kinetics, equilibrium and the mechanism of the process. Chemosphere. 59(1): 75–84.

Chu, K.H., Hashim, M.A., Phang, S.M. and Samuel, V.B. 1997. Biosorption of cadmium by algal biomass: Adsorption and desorption characteristics, Water Science and Technology, 35 (7), 115–122. https://doi.org/10.1016/S0273-1223(97)00121-2.

Cossich, E., Tavares, C., and Ravagnani, T. 2002. Biosorption of chromium (III) by Sargassum sp. biomass. Electronic Journal of Biotechnology, 5. 6–7. DOI: 10.4067/S0717-34582002000200008.

Crist, R. H., Martin, J. R., and Crist, D. R. 1991. Interaction of metals and protons with algae. Equilibrium constants and ionic mechanisms for heavy metal removal as sulfides and hydroxides. In: Mineral Bioprocessing, pp. 275–287. Smith, R. W., and Misra, M., Eds., The Mineral, Metals, and Materials Society, Warrendale, PA.

Crist, R. H., Martin, J. R., Carr, D., Watson, J. R., Clarke, H. J., and Carr, D. 1994. Interaction of metals and protons with algae. 4. Ion exchange vs adsorption models and a reassessment of Scatchard plots; ion-exchange rates and equilibria compared with calcium alginate. Environ. Sci. Technol. 28(11): 1859–1866.

Crist, R. H., Oberholser, K., Schwartz, D., Marzoff, J., Ryder, D., and Crist, D. R. 1988. Interactions of metals and protons with algae. Environ. Sci. Technol. 22(7): 755–760.

Davis, T. A., Volesky, B., and Vieira, R. H. S. F. 2000. Sargassum seaweed as biosorbent for heavy metals. Water Res. 34(17): 4270–4278.

Davis, T. A., Volesky, B., and Mucci, A. 2003. A review of the biochemistry of heavy metal biosorption by brown algae. Water Res. 37(18): 4311–4330.

De-Bashan, L. E. and Bashan, Y. 2010. Immobilized microalgae for removing pollutants: review of practical aspects. Bioresour. Technol. 101(6): 1611–1627.

De Filippis, L. F., and Pallaghy, C. K. 1994. Heavy metals: Sources and biological effects. In: Algae and Water Pollution, pp. 31–77. Rai, L. C., Gaur, J. P., and Soeder, C. J., Eds., E.Schweizerbart'sche Verlagsbuchhandlung (Nägele u. Obermiller), Stuttgart.

de Philippis, R., Sili, C., Paperi, R., and Vincenzini, M. 2001. Exopolysaccharide-producing cyanobacteria and their possible exploitation: a review. J. Appl. Phycol. 13(4): 293–299.

Deng, L., Zhang, Y., Qin, J., Wang, X., and Zhu, X., 2009. Biosorption of Cr(VI) from aqueous solutions by nonliving green algae *Cladophora albida*. Miner. Eng. 22(4), 372–377.

Diener, D. R., Curry, A. M., Johnson, K. A., Williams, B. D., Lefebvre, P. A., Kindle, K. L., and Rosenbaum, J. L. 1990. Rescue of a paralyzed-flagella mutant of *Chlamydomonas* by transformation. Proc. Natl. Acad. Sci. USA. 87(15): 5739–5743.

Dwivedi, S. 2012. Bioremediation of heavy metal by algae: current and future perspective. J. Adv. Lab. Res. Biol. 3(3): 195–199.

Eccles, H. 1999. Treatment of metal-contaminated wastes: why select a biological process? Trends Biotechnol. 17(12): 462–465.

El-Sheekh, M. M., and El-Kassas, H. Y. 2016. Algae mediated biosynthesis of inorganic nanomaterials as a promising route in nanobiotechnology- a review. Green Chem. 19(3): 552–587.

Enamala, M. K., Enamala, S., Chavali, M., Donepudi, J., Yadavalli, R., Kollapalli, B., Aradhyula, T. V., Velpuri, J., and Chandrashekar, K. 2019. Production of biofuels from microalgae-a review on cultivation, harvesting, lipid extraction, and numerous applications of microalgae. Renew. Sustain. Energy Rev. 94, 49–68.

Eroglu, E., Agarwal, V., Bradshaw, M., Chen, X., Smith, S. M., Raston, C. L., and Swaminathan Iyer, K. 2012. Nitrate removal from liquid effluents using microalgae immobilized on chitosan nanofiber mats. Green Chem. 14(10): 2682.

Esteves, A.J.P., Valdman, E. and Leite, S. 2000. Repeated Removal of Cadmium and Zinc from an Industrial Effluent by Waste Biomass Sargassum sp. Biotechnology Letters. 22. 499–502. 10.1023/A:1005608701510.

Fabris, M., Abbriano, R. M., Pernice, M., Sutherland, D.L., Commault, A. S., Hall, C. C., Labeeuw, L., McCauley, J. I., Kuzhiuparambil, U., Ray, P., Kahlke, T., and Ralph, P.J. 2020. Emerging Technologies in Algal Biotechnology: Toward the Establishment of a Sustainable, Algae-Based Bioeconomy, Frontiers in Plant Science, 11, 279. https://www.frontiersin.org/article/10.3389/fpls.2020.00279

Figueira, M. M., Volesky, B., and Mathieu, H. J. 1999. Instrumental analysis study of iron species biosorption by *Sargassum* biomass. Environ. Sci. Technol. 33(11): 1840–1846.

Figueira, M. M., Volesky, B., Ciminelli, V. S. T., and Roddick, F. A. 2000. Biosorption of metals in brown seaweed biomass. Water Res. 34: 196–204.

Gadd, G. M. 1988. Accumulation of metals by microorganisms and algae. In: Biotechnology, pp. 401–434. Rehm, H. J., Ed., VCH, Weinheim.

García-Béjar, B., Arévalo-Villena, M., Guisantes-Batan, E. Rodríguez-Flores, J and Briones, A. 2020. Study of the bioremediatory capacity of wild yeasts. Sci Rep 10, 11265. https://doi.org/10.1038/s41598-020-68154-4

Ganeshkumar, V., Subashchandrabose, S. R., Dharmarajan, R., Venkateswarlu, K., Naidu, R., and Megharaj, M. 2018. Use of mixed wastewaters from piggery and winery for nutrient removal and lipid production by *Chlorella sp.* MM3. Bioresour. Technol. 256, 254–258.

Gautam, R. K., Sharma, S. K., Mahiya, S., and Chattopadhyaya M. C. 2015. Contamination of heavy metals in aquatic media: transport, toxicity and technologies for remediation. In: Heavy Metals in Water: Presence, Removal and Safety, pp. 1–24. Sharma, S.K., Ed., The Royal Society of Chemistry, Cambridge, UK.

Ge, S., Agbakpe, M., Zhang, W., Kuang, L., Wu, Z., and Wang, X. 2015. Recovering magnetic Fe3O4–ZnO nanocomposites from algal biomass based on hydrophobicity shift under UV irradiation. ACS Appl. Mater. Interfaces. 7(21), 11677–11682.

Giagnorio, M., Ricceri, F., Tagliabue, M., Zaninetta, L., and Tiraferri, A. 2019. Hybrid Forward Osmosis-Nanofiltration for Wastewater Reuse: System Design. Membranes, 9(5), 61. https://doi.org/10.3390/membranes9050061

Goswami, R. K., Agrawal, K., Mehariya, S., Molino, A., Musmarra, D., and Verma, P. 2020. Microalgae-Based Biorefinery for Utilization of Carbon Dioxide for Production of Valuable Bioproducts. In Chemo-Biological Systems for CO_2 Utilization, pp. 199–224. Taylor & Francis.

Goswami, R. K., Mehariya, S., Verma, P, Lavecchia, R., and Zuorro, A. 2021a. Microalgae-based biorefineries for sustainable resource recovery from wastewater. J. Water Process Eng. 40: 101747.

Goswami, R. K., Mehariya, S., Karthikeyan, O. P. K., and Verma, P. 2021b. Advanced micro-algae-based renewable biohydrogen production systems: a review. Bioresour. Technol. 320(A): 124301.

Goswami, R. K., Agrawal, K., and Verma, P. 2021c. An Overview of Microalgal Carotenoids: Advances in the Production and Its Impact on Sustainable Development. In Bioenergy Research: Evaluating Strategies for Commercialization and Sustainability, pp. 105–128. John Wiley & Sons.

Gupta, V. K., and Rastogi, A. 2008. Biosorption of lead (II) from aqueous solutions by non-living algal biomass Oedogonium sp. and Nostoc sp.- a comparative study. Colloids Surf. B Biointerfaces. 64(2): 170–178.

Hajdu, I., Bodnar, M., Csikos, Z., Wei, S., Daroczi, L., Kovacs, B., and Borbely, J. 2012. Combined nano-membrane technology for removal of lead ions. J. Membr. Sci. 409: 44–53.

Hamdy, A. A. 2000. Biosorption of heavy metals by marine algae. Curr. Microbiol. 41(4): 232–238.
Hameed, M. A., and Ebrahim, O. H. 2007. Biotechnological potential uses of immobilized algae. J. Agric. Biol. 9(1): 183–192.
Hanikenne, M., Krämer, U., Demoulin, V., and Baurain, D. 2005. A comparative inventory of metal transporters in the green alga *Chlamydomonas reinhardtii* and the red alga *Cyanidioschizon merolae*. Plant Physiol. 137(2): 428–446.
Hansen, H., Ribeiro, A. B., and Mateus, E. P. 2006. Biosorption of Arsenic(V) with *Lessonia nigrescens*. Miner. Eng. 19(5):486–490.
He, J. and Chen, J. P. 2014. A comprehensive review on biosorption of heavy metals by algal biomass: materials, performances, chemistry, and modeling simulation tools. Bioresour. Technol. 160: 67–78.
He, Z., Siripornadulsil, S., Sayre, R. T., Traina, S. J., & Weavers, L. K. (2011). Removal of mercury from sediment by ultrasound combined with biomass (transgenic *Chlamydomonas reinhardtii*). Chemosphere. 83: 1249–1254.
Herrera-Estrella, L.R., Guevara-García, A.A. and López-Bucio, J. (2001). Heavy Metal Adaptation. In eLS, (Ed.). https://doi.org/10.1038/npg.els.0001318
Hu, C. A., Delauney, A. J., and Verma, D. P. 1992. A bifunctional enzyme (delta 1-pyrroline-5-carboxylate synthetase) catalyzes the first two steps in proline biosynthesis in plants. Proc. Natl. Acad. Sci. USA. 89(19): 9354–9358.
Hu, S., Lau, K. W., and Wu, M. 2001. Cadmium sequestration in *Chlamydomonas reinhardtii*. Plant Sci. 161(5): 987–996.
Huang, C. C., Chen, M. W., Hsieh, J. L., Lin, W. H., Chen, P. C., and Chien, L. F. 2006. Expression of mercuric reductase from *Bacillus megaterium* MB1 in eukaryotic microalga *Chlorella sp.* DT: an approach for mercury phytoremediation. Appl. Microbiol. Biotechnol. 72(1): 197–205.
Hwang, J. H., Church, J., Lee, S. J., Park, J., and Lee, W. H. 2016. Use of microalgae for advanced wastewater treatment and sustainable bioenergy generation. Environ. Eng. Sci. 33(11): 882–897.
Ibrahim, W. M. 2011. Biosorption of heavy metal ions from aqueous solution by red macroalgae. J. Hazard. Mater. 192(3): 1827–1835.
Ince, M., and Ince, O. K. 2017. An overview of adsorption technique for heavy metal removal from water/wastewater: A critical review. International Journal of Pure and Applied Sciences. 10–19. DOI: 10.29132/ijpas.372335.
Jalali-Rad, R., Ghafourian, H., Asef, Y., Dalir, S. T., Sahafipour, M. H., Gharanjik, B. M. 2004. Biosorption of cesium by native and chemically modified biomass of marine algae: introduce the new biosorbents for biotechnology applications. J. Hazard. Mater. 116(1–2): 125–134.
Jing, F., Chen, X., Yang, Z., & Guo, B. (2018). Heavy metals status, transport mechanisms, sources, and factors affecting their mobility in Chinese agricultural soils. Environmental Earth Sciences, 77, 104. DOI:10.1007/s12665-018-7299-4
Kaewsarn, P. 2002. Biosorption of copper(II) from aqueous solutions by pre-treated biomass of marine algae padina sp. Chemosphere. 47(10): 1081–1085.
Kanjwal, M. A., Barakat, N. A., Sheikh, F. A., Khil, M. S., and Kim, H. Y. 2010. Functionalization of electrospun titanium oxide nanofibers with silver nanoparticles: strongly effective photocatalyst. Int. J. App. Ceramic Technol. 7: E54–E63. https://doi.org/10.1111/j.1744-7402.2009.02397.x.
Kelly, M. 1988. Mining and the Freshwater Environment. BP Elsevier Applied Science, London.
Kinuthia, G. K., Ngure, V., and Kamau, L. 2020. Levels of heavy metals in wastewater and soil samples from open drainage channels in Nairobi, Kenya: community health implication. Sci. Rep. 10: 8434.

Kleinübing, S. J., da Silva, E. A., da Silva, M. G. C., and Guibal, E. 2011. Equilibrium of Cu(II) and Ni(II) biosorption by marine alga *Sargassum filipendula* in a dynamic system: competitiveness and selectivity. Bioresour. Technol. 102(7): 4610–4617.

Klimmek, S., Stan, H. J., Wilke, A., Bunke, G., and Buchholz, R. 2001. Comparative analysis of the biosorption of cadmium, lead, nickel, and zinc by algae. Environ. Sci. Technol. 35: 4283–4288.

Kotsyuda, S. S., Tomina, V. V., Zub, Y. L., Furtat, I. M., and Melnyk, I. V. 2017. Bifunctional silica nanospheres with 3-aminopropyl and phenyl groups. Synthesis approach and prospects of their applications. Appl. Surf. Sci. 420: 782–791.

Kumar, J. I. N., and Oommen, C. 2012. Removal of heavy metals by biosorption using freshwater alga *Spirogyra hyaline*. J. Environ. Biol. 33: 27–31.

Kumar, K. S., Dahms, H. U., Won, E. J., Lee, J. S., and Shin, K. H. 2015. Microalgae–a promising tool for heavy metal remediation. Ecotoxicol. Environ. Saf. 113: 329–352.

Kumar, T. S. 2008. Nualgi nanobiotech Bangalore-41. Sewage/effluent treatment by growth of diatom algae. http://www.indiawaterportal.org/sites/indiawaterportal.org/ files/uploads/2008/07/nualgi_final1udaipur-conference.pdf.

Kumar, M., Singh, A.K. and Sikandar, M. 2018. Study of sorption and desorption of Cd (II) from aqueous solution using isolated green algae Chlorella vulgaris. Appl Water Sci 8, 225. https://doi.org/10.1007/s13201-018-0871-y

Kumar, B., Agrawal, K., Bhardwaj, N., Chaturvedi, V., and Verma, P. 2019. Techno-Economic Assessment of Microbe-Assisted Wastewater Treatment Strategies for Energy and Value-Added Product Recovery. In Microbial Technology for the Welfare of Society, pp. 147–181. Springer.

Kuyucak, N., and Volesky, B. 1988. Biosorbents for recovery of metals from industrial solutions. Biotechnol. Lett. 10(2): 137–142.

Kuyucak, N., and Volesky, B. 1989. The mechanism of cobalt biosorption. Biotech. Bioeng. 33(7): 823–831.

Larsdotter, K. 2006. Wastewater treatment with microalgae – a literature review. Vatten. 62: 31–38.

Li, Y., Duan, X., Qian, Y., Li, Y., and Liao, H. 1999. Nanocrystalline silver particles: synthesis, agglomeration, and sputtering induced by electron beam. J. Colloid Interface Sci. 209 (2): 347–349.

Liu, X., Hu, Q., Fang, Z., Zhang, X., and Zhang, B. (2009). Magnetic chitosan nanocomposites: a useful recyclable tool for heavy metal ion removal. Langmuir. 25(1), 3–8.

Lodeiro, P., Cordero, B., Barriada, J. L., Herrero, R., and Sastre de Vicente, M. E. 2005. Biosorption of cadmium by biomass of brown marine macroalgae. Bioresour. Technol. 96(16): 1796–1803.

Lombardi, A. T., Vieira, A. V. H., and Sartori, L. A. 2002. Mucilaginous capsule adsorption and intracellular uptake of copper by Kirchneriella aperta (Chlorococcales). J. Phycol. 38: 332–337.

Lu, H., Wang, J., Stoller, M., Wang, T., Bao, Y., and Hao, H. 2016. An overview of Nanomaterials for Water and Wastewater Treatment. Advances in Material Science and Engineering. 2016, 4964828. https://doi.org/10.1155/2016/4964828

Luna, A. S., Costa, A. L. H., da Costa, A. C. A., and Henriques, C. A. 2010. Competitive biosorption of cadmium(II) and zinc(II) ions from binary systems by *Sargassum filipendula*. Bioresour. Technol. 101(14): 5104–5111.

Mahmoud, M. E., Amira, M. F., Seleim, S. M., and Abouelanwar, M. E. 2019. Solvent Free Microwave Synthesis of Nano Polyaniline-Zirconium Silicate Nanocomposite for Removal of Nitro Derivatives. Journal of Industrial and Engineering Chemistry, 77, 371–384.

Mal, N., Srivastava, K., Sharma, Y., Singh, M., Rao, K. M., Enamala, M. K., Chandrasekhar, K., and Chavali, M. 2021. Facets of diatom biology and their potential applications. Biomass Conversion and Biorefinery, https://doi.org/10.1007/s1339-020-01155-5

Malik, A. 2004. Metal bioremediation though growing cells. Environ. Int. 30: 261–278.

Martin-de-Lucia, I., Campos-Mañas, M., Agüera, A., Leganes, F., Fernandez-Piñas, F., and Rosal, R. 2018. Combined toxicity of graphene oxide and wastewater to the green alga Chlamydomonas reinhardtii. Environmental science. Nano. 5. 1729–1744. https://doi.org/10.1039/C8EN00138C.

Mehariya, S., Kumar, P., Marino, T., Casella, P., Lovine, A., Verma, P., Musmarra, D., and Molino, A. 2021a. Aquatic Weeds: A Potential Pollutant Removing Agent from Wastewater and Polluted Soil and Valuable Biofuel Feedstock. In Bioremediation Using Weeds, pp. 59–77. Springer.

Mehariya, S., Goswami, R., Verma, P., Lavecchia, R., and Zuorro, A. 2021b. Integrated approach for wastewater treatment and biofuel production in microalgae biorefineries energies. 14(8): 2282.

Mehariya, S., Goswami, R. K., Karthikeysan, O. P., and Verma, P. 2021c. Microalgae for high-value products: a way towards green nutraceutical and pharmaceutical compounds. Chemosphere. 280: 130553.

Mehta, S. K., and Gaur, J. P. 2001a. Characterization and optimization of Ni and Cu sorption from aqueous solution by *Chlorella vulgaris*. Ecol. Eng. 18(1): 1–13.

Mehta, S. K., and Gaur, J. P. 2001b. Removal of Ni and Cu from single and binary metal solution by free and immobilized *Chlorella vulgaris*. Eur. J. Protistol. 37(3): 261–271.

Mehta, S. K., and Gaur, J. P. 2005. Use of algae for removing heavy metal ions from wastewater: progress and prospects. Crit. Rev. Biotech. 25(3): 113–152.

Mendoza-Cózatl, D. G., and Moreno-Sánchez, R. 2005a. Cd2+ transport and storage in the chloroplast of euglena gracilis. Biochim. Biophys. Acta Bioenerg. 1706(1–2): 88–97.

Mendoza-Cózatl, D. G., Loza-Tavera, H., Hernández-Navarro, A., and Moreno-Sánchez, R. 2005b. Sulfur assimilation and glutathione metabolism under cadmium stress in yeast, protists and plants. FEMS Microbiol. Rev. 29(4): 653–671.

Misbah, M., Samia, S., Hafsa, I., Uzair, H., and Alvina, G. 2014. Production of algal biomass. In: Biomass and Bioenergy: Processing and Properties, pp. 207–227. Springer, Cham.

Monteiro, C. M., Castro, P. M., and Malcata, F. X. 2012. Metal uptake by microalgae: underlying mechanisms and practical applications. Biotechnol. Prog. 28(2): 299–311.

Nagel, K., Adelmeier, U., and Voigt, J. 1996. Subcellular distribution of cadmium in the unicellular green alga *Chlamydomonas reinhardtii*. J. Plant Physiol. 149(1–2): 86–90.

Najafi, M., Yousefi, Y., and Rafati, A. A. 2012. Synthesis, characterization and adsorption studies of several heavy metal ions on amino-functionalized silica nano hollow sphere and silica gel. Separation and Purification Technology, 85, 193–205.

Niu, H. and Volesky, B. 2000, Gold-cyanide biosorption with L-cysteine. J. Chem. Technol. Biotechnol., 75: 436–442. https://doi.org/10.1002/1097-4660(200006)75:6<436::AID-JCTB243>3.0.CO;2-O

Oscar, F. L., Visamaya, S., Arunkumar, M., Thajuddin, N., Dhanasekaran, D. and Nithya, C. 2017. Algal Nanoparticles: Synthesis and Biotechnological Potentials, Algae - Organisms for Imminent Biotechnology, Nooruddin Thajuddin and Dharumadurai Dhanasekaran, IntechOpen, London. DOI: 10.5772/62909. https://www.intechopen.com/chapters/50544.

Pacheco, A.R., Osborne, M.L. and Segrè, D. 2021. Non-additive microbial community responses to environmental complexity. Nat Commun **12**, 2365. https://doi.org/10.1038/s41467-021-22426-3

Pan, X., Liang, X., Yao, L., Wang, X., Jing, Y., Ma, J., Fei, Y., Chen, L., and Mi, L. 2017. Study of the Photodynamic Activity of N-Doped TiO2 Nanoparticles Conjugated with Aluminum Phthalocyanine. Nanomaterials, 7(10), 338. https://doi.org/10.3390/nano7100338

Patel, V., Berthold, D., Puranik, P., and Gantar, M. 2015. Screening of cyanobacteria and microalgae for their ability to synthesize silver nanoparticles with antibacterial activity. Biotechnol. Rep. 5: 112–119.

Pawlik-skowrońska, B. 2001. Phytochelatin production in freshwater algae stigeoclonium in response to heavy metals contained in mining water; effects of some environmental factors. Aquat. Toxicol. 52(3–4): 241–249.

Peña-Castro, J. M., Martínez-Jerónimo, F., Esparza-García, F., and Cañizares-Villanueva, R. O. 2004. Phenotypic plasticity in *Scenedesmus incrassatulus* (chlorophyceae) in response to heavy metals stress. Chemosphere. 57(11): 1629–1636.

Perales-vela, H. V., Pena-Castro, J. M., and Canizares-Villanueva, R. O. 2006. Heavy metal detoxification in eukaryotic microalgae. Chemosphere. 64(1): 1–10.

Pérez-Rama, M., Alonso, J. A., López, C. H., and Vaamonde, E. T. 2002. Cadmium removal by living cells of the marine microalga *Tetraselmis suecica*. Bioresour. Technol. 84(3): 265–270.

Perpetuo, E. A., Souza, C. B., and Nascimento, C. A. O. 2011. Engineering Bacteria for Bioremediation. In Progress in Molecular and Environmental Bioengineering- From Analysis and Modeling to Technology Applications, Angelo Carpi, IntechOpen, London. DOI: 10.5772/19546. Available from: https://www.intechopen.com/chapters/17260

Pogorilyi, R. P., Melnyk, I. A., and Zub, Y. L. 2014. New product from old reaction: Uniform magnetite nanoparticles from iron-mediated synthesis of alkali iodides and their protection from leaching in acidic media. RSC Adv. 4: 22606–22612.

Puranik, P. R., Modak, J. M., and Paknikar, K. M. 1999. A comparative study of the mass transfer kinetics of metal biosorption by microbial biomass. Hydrometallurgy. 52(2): 189–197.

Rajamani, S., Siripornadulsil, S., Falcao, V., Torres, M., Colepicolo, P., and Sayre, R. 2007. Phycoremediation of Heavy Metals Using Transgenic Microalgae. In Transgenic Microalgae as Green Cell Factories, pp. 99–109. Springer, New York, NY.

Rajendran, P., and Gunasekaran, P. 2007. Nanotechnology for Bioremediation of Heavy Metals. In Environmental Bioremediation Technologies, pp. 211–221. Springer, Berlin, Heidelberg.

Ramirez-Merida, L. G., Zepka, L. Q., de Menezes, C. R., and Jacob-Lopes, E. 2015. Microalgae as nanofactory for production of antimicrobial molecules. J. Nanomed. Nanotechnol. 6: 1–3.

Rauwel, P., Küünal, S., Ferdov, S., and Rauwel, E. 2015. A review on the green synthesis of silver nanoparticles and their morphologies studied via TEM. Adv. Mater. Sci. Eng. 682749: 9.

Romero-Gonzalez, M. E., Williams, C. J., and Gardiner, P. H. 2001. Study of mechanisms of cadmium biosorption by delignated seaweed waste. Environ. Sci. Technol. 35: 3025–3030.

Romera, E., González, F., Ballester, A., Blázquez, M. L., and Munoz, J. A. 2007. Comparative study of biosorption of heavy metals using different types of algae. Bioresour. Technol. 98(17): 3344–3353.

Roy, D., Greenlaw, P. N., and Shane, B. S. 1993. Adsorption of heavy metals by green algae and ground rice hulls. J. Environ. Sci. Health. 28: 37–50.

Ruiz, F. A., Marchesini, N., Seufferheld, M., and Docampo, R. 2001. The polyphosphate bodies of *Chlamydomonas reinhardtii* possess a proton-pumping pyrophosphatase and are similar to acidocalcisomes. J. Biol. Chem. 276(49): 46196–46203.

Sandau, E., Sandau, P., Pulz, O. and Zimmermann, M. 1996. Heavy metal sorption by marine algae and algal by-products. Acta Biotechnol., 16: 103–119. https://doi.org/10.1002/abio.370160203

Salahi, A., Mohammadi, T., Behbahani, R.M., and Hemmati, M., 2015. Asymmetric polyethersulfone ultrafiltration membranes for oily wastewater treatment: synthesis, characterization, ANFIS modeling, and performance. J. Environ. Chem. Eng. 3(1): 170–178.

Satoh, M., Karaki, E., Kakehashi, M., Okazaki, E., Gotoh, T., and Oyama, Y. 1999. Heavy-metal induced changes in nonproteinaceous thiol levels and heavy-metal binding peptide in *Tetraselmis tetrathele* (prasinophyceae). J. Phycol. 35(5): 989–994.

Sayre, L. M., Moreira, P. I., Smith, M. A., and Perry, G. 2005. Metal ions and oxidative protein modification in neurological disease. Annali dell'Istituto superiore di sanita, *41*(2), 143–164.

Shanab, S., Essa, A., and Shalaby, E. 2012. Bioremoval capacity of three heavy metals by some microalgae species (Egyptian isolates). Plant Signal Behav. 7: 392–399.

Sharma, A., Sharma, S., Sharma, K., Chetri, S. P. K., Vashishtha, A., Singh, P., Kumar, R., Rathi, B., and Agrawal, V. 2015. Algae as crucial organisms in advancing nanotechnology: a systematic review. J. Appl. Phycol. 2014: 1–16.

Sharma, D., Kanchi, S., and Bisetty, K. 2019. Biogenic synthesis of nanoparticles: A review, Arabian Journal of Chemistry, 12 (8), 3576–3600. https://doi.org/10.1016/j.arabjc.2015.11.002

Sharma, R. M., and Azeez, P.A. 1988. Accumulation of copper and cobalt by blue-green algae at different temperatures. Int. J. Environ. Anal. Chem. 32: 87–95.

Shayesteh, M., Samimi, A., Shafifiee Afarani, M., and Khorram, M. 2016. Synthesis of titania–c-alumina multilayer nanomembranes on performance-improved alumina supports for wastewater treatment. Desalin. Water Treat. 57(20): 9115–9122.

Sheng, P. X., Tan, L. H., and Chen, J. P. 2005. Biosorption performance of two brown marine algae for removal of chromium and cadmium. J. Disper. Sci. Technol. 25(5): 679–686.

Sheng, P. X., Ting, Y.-P., and Chen, J. P. 2007. Biosorption of heavy metal ions (Pb, Cu, and Cd) from aqueous solutions by the marine alga *Sargassum sp.* in single- and multiple-metal systems. Ind. Eng. Chem. Res. 46(8): 2438–2444.

Singh, S., Kumar, V., Romero, R., Sharma, K., and Singh, J. 2019. Applications of Nanoparticles in Wastewater Treatment. R. Prasad et al. (eds.), Nanobiotechnology in Bioformulations, *Nanotechnology in the Life Sciences*, 395–418. https://doi.org/10.1007/978-3-030-17061-5_17

Siripornadulsil, S., Traina, S., Verma, D. P. S., and Sayre, R. T. 2002. Molecular mechanisms of proline-mediated tolerance to toxic heavy metals in transgenic microalgae. Plant Cell. 14(11): 2837–2847.

Skowronski, P., and Ska, B. 2000. Relationship between acid-soluble thiol peptides and accumulated Pb in the green alga *Stichococcus bacillaris*. Aquat. Toxicol. 50: 221–230.

Soldo, D., Hari, R., Sigg, L., and Behra, R. 2005. Tolerance of *Oocystis nephrocytioides* to copper: intracellular distribution and extracellular complexation of copper. Aquat. Toxicol. 71(4): 307–317.

Surendhiran, D., Sirajunnisa, A., and Tamilselvam, K. 2017. Silver–magnetic nanocomposites for water purification. Environ. Chem. Lett. 15(3): 367–386.

Ubando, A.T., Africa, A.D.M., Maniquiz-Redillas, M.C., Culaba, A.B., Chen, W-H., and Chang, J-S. 2021. Microalgal biosorption of heavy metals: A comprehensive bibliometric review, Journal of Hazardous Materials, Volume 402, 123431. https://doi.org/10.1016/j.jhazmat.2020.123431.

Ungureanu, G., Santos, S., Boaventura, R., and Botelho, C. 2015a. Arsenic and antimony in water and wastewater: overview of removal techniques with special reference to latest advances in adsorption. J. Environ. Manag. 151: 326–342.

Ungureanu, G., Santos, S., Boaventura, R., and Botelho, C. 2015b. Biosorption of antimony by brown algae *S. muticum* and *A. nodosum*. Environ. Eng. Manag J. 14: 455–463.

Vatamaniuk, O. K., Mari, S., Lu, Y. P., and Rea, P. A. 2000. Mechanism of heavy metal ion activation of phytochelatin (pc) synthase blocked thiols are sufficient for pc synthase-catalyzed transpeptidation of glutathione and related thiol peptides. J. Biol. Chem. 275(40): 31451–31459.

Vannela, R., and Verma, S. K. 2006. Co2+, Cu2+, and Zn2+ accumulation by cyanobacterium *Spirulina platensis*. Biotechnol. Prog. 22(5): 1282–1293.

Vincy, W., Jasmine, T. M., Sukumaran, S., and Jeeva, S. 2017. Algae as a source for synthesis of nanoparticles- a review. Special Issue - International Conference on Nanotechnology: The Fruition of Science-2017, IJLTET, pp. 005–009.

Volesky, B. 2001. Detoxification of metal-bearing effluents: biosorption for the next century. Hydrometallurgy, 59, 203–216.

Wan Maznah, W. O., Al-Fawwaz, A. T., and Surif, M. 2012. Biosorption of copper and zinc by immobilised and free algal biomass, and the effects of metal biosorption on the growth and cellular structure of *Chlorella sp.* and *Chlamydomonas sp.* isolated from rivers in Penang, Malaysia. J. Environ. Sci. 24(8): 1386–1393.

Wang, J., and Chen, C. 2009. Biosorbents for heavy metals removal and their future. Biotechnol. Adv. 27(2): 195–226.

Xue, H. B., Stumm, W., and Sigg, L. 1988. The binding of heavy metals to algal surfaces. Water Resour. 22: 917–926.

Yadavalli, R., Ratnapuram, H., Peasari, J. R., Reddy, C. N., Ashokkumar, V., and Kuppam, C. 2021. Simultaneous production of astaxanthin and lipids from *Chlorella sorokiniana* in the presence of reactive oxygen species: a biorefinery approach. Biomass Convers. Biorefin. DOI: 10.1007/s13399-021-01276-5

Yadavalli, R., Ratnapuram, H., Motamarry, S., Reddy, C. N., Ashokkumar, V., and Chandrasekhar, K. 2020. Simultaneous production of flavonoids and lipids from *Chlorella vulgaris* and *Chlorella pyrenoidosa*. Biomass Convers. Biorefinery. *DOI: 10.1007/s13399-020-01044-x*

Yalcin, S., Seze, S., and Apak, R. 2012. Characterization and lead (II), cadmium(II), nickel(II) biosorption of dried marine brown macro algae *Cystoseira barbata*. Environ. Sci. Pollut. Res. 19: 3118–3125.

Yang, J., Hou, B., and Wang, J., Tian, B., Bi, J., Wang, N., Li, X., & Huang, X. 2019. Nanomaterials for the removal of heavy metals from wastewater. Nanomaterials. 9(424): 1–39.

Yu, Q., and Kaewsarn, P. 1999. A model for pH dependent equilibrium of heavy metal biosorption. Korean J. Chem. Eng. 16: 753–757.

Zakhama, S., Dhaouadi, H., and M'Henni, F. 2011. Nonlinear modelisation of heavy metal removal from aqueous solution using Ulva lactuca algae. Bioresour. Technol. 102(2): 786–796.

Zhang, Q., Xu, R., Xu, P., Chen, R., He, Q., Zhong, J., and Gu, X. 2014. Performance study of ZrO2 ceramic micro-filtration membranes used in pre-treatment of DMF wastewater. Desalination, 346 (1): 1–8.

Zhao, Y., Hao, Y., and Ramelow, G. J. 1994. Evaluation of treatment techniques for increasing the uptake of metal ions from solution by nonliving seaweed algal biomass. Environ. Monit. Assess. 33: 61–70.

Zhou, J. L., Huang, P. L., and Lin, R. G. 1998. Sorption and desorption of Cu and Cd by macroalgae and microalgae. *Environmental pollution (Barking, Essex: 1987)*, *101*(1), 67–75. https://doi.org/10.1016/s0269-7491(98)00034-7

Zhu, X., Zheng, Y., Chen, Z., Chen, Q., Gao, B., and Yu, S. 2013. Removal of reactive dye from textile effluent through submerged filtration using hollow fiber composite nanofiltlration membrane. Desalin. Water Treat. 51(31–33): 6101–6109.

8 Valued Products from Algae Grown in Wastewater

Durairaj Vijayan, Muthu Arumugam

CONTENTS

8.1 INTRODUCTION

Water, the most abundant component of any life on this earth. "*Neer indri amayathu ulagu* - World does not exist without water" quotes the ancient Tamil literature known as "*Thirukkural*." Freshwater is one of the essential natural commodities required by every living cell in terrestrial habitats. The available freshwater is being utilized and discharged as wastewater, and 80% of this wastewater is inadequately treated and released into ecosystems, leading to emissions of greenhouse gases (GHG) and other

DOI: 10.1201/9781003155713-8

adverse environmental impacts affecting both terrestrial and aquatic habitats. GHG, including carbon dioxide emission, is a primary concern in climate change and global warming, which costs freshwater loss from polar glaciers melting into the ocean. Due to climate change and global warming, freshwater availability is estimated to decline by more than 10% in about 570 cities in the non-polar region, including Cape Town and Melbourne, by 2050 [UN-Water (2020)]. Conservation and efficient management of available freshwater might favor the world in alleviating all these concerns. Adequate wastewater treatments (WWTs) could be one such process to mitigate GHG emission and recycling the used freshwater for appropriate purposes.

Microbial-based WWT could reduce these GHG emissions by utilizing the organic substrate and other surplus nutrients (Kumar, et al., 2019; Agrawal, et al., 2020a, Mehariya, et al., 2021a). Many such conventional microbial-based WWTs include the microalgae in many oxidation ponds such as stabilization ponds, maturation ponds, and facultative ponds. However, the WWT in high-rate algal pond (HRAP), photobioreactors, and any other advanced hybrid methods which consider biomass productivity were called microalgal-based wastewater treatment (MWWT) (Alcántara, et al., 2015). MWWT removes eutrophic nutrients like excess nitrogen, phosphorus, carbon, and other pollutants and heavy metals utilizing the carbon dioxide (GHG) from flue gas as well (Dineshbabu, et al., 2019, Chaturvedi, et al., 2013). It favors the GHG reduction rather than emission and recycling of the treated water. From a long-term perspective, MWWT could favor the eco-friendly treatment system with return on investment, which is possible since the algal biomass generated during the treatment process could yield many commercially valued products (Gouveia, et al., 2016, Kothari, et al., 2017, Mahapatra, et al., 2018, Ángeles, et al., 2019, Li, et al., 2020). This chapter will elaborate on such valued products, including fuel, feed, pigments, nutrients, and other chemicals recovered from algal biomass grown on wastewater and its commercial viability.

8.2 ENVIRONMENTAL IMPACT AND COMMERCIAL VALUE OF ALGAL-BASED WASTEWATER MANAGEMENT

Wastewater from domestic, municipal, and industrial discharges needs to be treated in an environment-friendly process. It may contain toxic chemicals or pollutants and excess nutrients (eutrophic nitrogen, phosphorus, and organic carbon). Industrial wastes are from varied sources such as textiles, dairy, animal farms (swine, aquatic, cattle, poultry), food, leather, paper, pharma, chemicals, and other industries. MWWT of dairy, food, and animal farm wastes are more of eutrophic neutralization, whereas the rest of the MWWTs are more of phycoremediation. MWWT limits the eutrophication and hazardous pollutants released into the aquatic habitats in an eco-friendly manner in both conditions. Also, MWWT with carbon dioxide fixation and/or industrial flue gas conversion to biomass adds value in terms of environment-friendly GHG mitigation and viable commercial biomass production with low operational cost (Arashiro, et al., 2018, Ambat, et al., 2019, Rahman, et al., 2020, Goswami, et al., 2020). Low operational cost, low energy need, less chemical consumption, and enhanced biomass productivity were the significant factors attributing the commercial value of HRAP over other conventional WWTs (Garfí, et al., 2017,

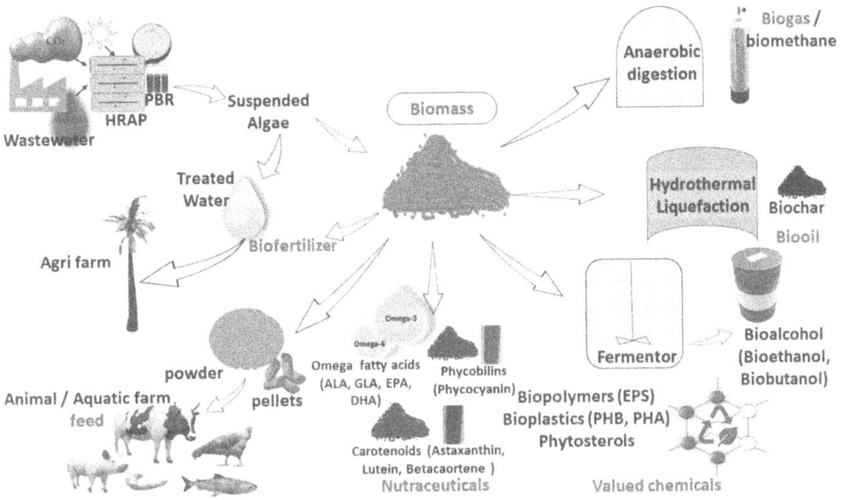

FIGURE 8.1 Schematic representation of valued products from algae grown in wastewater.

Arashiro, et al., 2018). Besides this, the microalgae have an estimated 6.5 billion USD commercial value (Mobin and Alam, 2017). Microalgal biomass comprises both high-value and low-value products (Figure 8.1). The nature, quantity, and worth of the product depend on the microalgal species and the growth/WWT condition.

8.3 BIOENERGY

Bioenergy generates any form of fuel energy such as oil, alcohol, electricity, or gas derived from organic biomass utilization. Algal biomass could be used to produce bioenergy either directly or in a conversion process as a feedstock. Algal biohydrogen production as biogas fuel was established early in the mid-to-late 19th century. In recent decades, concurrent high demand and rapid depletion of fossil fuels have led to innovative alternate fuel sources or modes to meet the essential and commercial fuel demand. Biocrude oil or biodiesel production from algal biomass has comparative advantages: adaptability to non-farmland cultivation, lower area occupancy, higher energy yield, and faster regeneration over alternate oleaginous plant crops (Arumugam, et al., 2011). Furthermore, biomass generated from MWWT could also be converted to biofuel as bio-oil, biogas, bioelectricity, and bioalcohol (Figure 8.1 & Table 8.1). Bioalcohol such as biobutanol or bioethanol fuel production from algal WWT are covered under the sub-section *8.5.2* of this chapter.

8.3.1 Bio-Oil

Bio-oil is liquid lipid obtained from the biomass by either extraction or conversion. Extraction involves chemical solvent extraction with or without pretreatment, and further transesterification in the presence of methanol could get fatty acid methyl ester (FAME) as biodiesel. Conversion is the thermochemical process like

TABLE 8.1
Biofuel Products Derived from Microalgae Grown in Wastewater Treatment

Wastewater Cultivation System	Microalgal Species	Scale (S) Cultivation Time (T)	Growth Condition: Light (L) Aeration (A)	Nutrient/Pollutant Removal Efficiency	Biofuel Product	Productivity: Biomass (B) Product (P)	Reference
Raw urban wastewater in HRAP	*Chlorella sorokiniana* UUIND6	S: 35 L T: 14 d	L: LED lamp with 300 µmole m^{-2} s^{-1} (16h light:8h dark) A: 1800 Lh^{-1}	TN: 97% TP: 95% BOD: 81% COD: 73% TOC: 75%	Biodiesel	B: 1.31 g DCW L^{-1} P: 31% FAME of DCW	Arora, et al. (2020)
Primary settled sewage in the flask	*Mucidosphaerium pulchellum*, *Micractinium pusillum*, *Coleastrum* sp., *Desmodesmus* sp., *Pediastrum boryanum*	S: 1 L T: 5 d & 10 d	L: Simulated summer and winter condition growth chambers A: 1% CO2 in air (0.2 L min^{-1})	NH_3: 71–100% DRP: 57–100%	Biodiesel	B: 14.4±7–188.9±10 mg biomass L^{-1} d^{-1} P: 3.6±1.1–61±2.3 mg lipid L^{-1} d^{-1}; 0.2±0.1–3.9±0.3 KJ energy L^{-1} d^{-1}	Mehrabadi, et al. (2017b)
Effluent from WWT in HRAP	*Scenedesmus obliquus* SAG 276–10	S: 533 L & 266 L T: 10–18 d	L: Sunlight A: with or without CO_2 (N/A)	TN: 0.2–0.79 g m^{-2} d^{-1} TP: 0.03–0.07 g m^{-2} d^{-1}	Biodiesel	B: 10.2–26.2 g TSS m^{-2} d^{-1} P: 3.1–6 g lipids m^{-2} d^{-1}	Arbib, et al. (2017)
Municipal wastewater in photobioreactor	*Chlorella pyrenoidosa*	S: 1 L T: 5 d	L: 40 W fluorescent lamp with 10000 lux (N/A) A: air (50 mL min^{-1})	NH_3: 91%	Biodiesel	B: 70 mg DCW L^{-1} d^{-1} P: 27.3 mg FAME L^{-1} d^{-1}	Zhou, et al. (2020)

Treated municipal sewage wastewater in algal biofilm airlift photobioreactor	*Chlorella vulgaris*	S: 21 L T: 37 days	L: Red/Blue LED lamp in 4:1 ration with 120.8 µmole m^{-2} s^{-1} (continuous) A: 1.5 L min^{-1}	DIN: 1.0±0.16 mg m^{-2} d^{-1} DIP: 0.2±0.03 mg m^{-2} d^{-1}	Biodiesel	B: 15.93±0.46 mg DCW L^{-1} d^{-1} P: 4.09±0.17 mg lipids L^{-1} d^{-1}	Tao, et al. (2017)
Secondary effluent from municipal WWT plant in membrane photobioreactor	*Chlorella vulgaris, Scenedesmus obliquus*	S: 1 L T: 22 d	L: Red/Blue LED lamp in 4:1 ration with 101.5–112.3 µmole m^{-2} s^{-1} (continuous) A: 0.5 L min^{-1}	N/A	Biodiesel	B: 96.3 & 88.8 mg DCW L^{-1} d^{-1} P: 25.76 & 29.57 mg DCW L^{-1} d^{-1}	Gao, et al. (2019)
Raw pharmaceutical wastewater in PBR and PMFC	*Scenedesmus abundans*	S: 900 mL T: 21 d	L: LED with 94.5 µmole m^{-2} s^{-1} A: CO_2 (N/A)	COD: 77–97.2% TN: 82–97.1% TP: 65.9–93.7%	Biodiesel Bioelectricity	B: 63.3–64 mg DCW L-1 d^{-1} P: 24.3–24.8 mg FAME L^{-1} d^{-1}; 838.68 mW power m^{-2}	Nayak and Ghosh (2019)
WWT-HRAPs	*Pediastrum* sp., *Micractinium* sp., *Desmodesmus* sp.	S: 2850000 L T: N/A	N/A	N/A	Biocrude oil	B: N/A P: 3.1–24.9 wt% biocrude oil; 37.5–38.9 KJ energy g^{-1}	Mehrabadi, et al. (2017a)
WWT-HRAPs	*Coleastrum* sp., *Actinustrum* sp., Diatom sp., *Mucidosphaerium pulchellum*	S: 8000 L T: N/A	N/A	N/A	Biocrude oil	B: N/A P: 4.7 wt% biocrude oil; 34.4–37 KJ energy g^{-1}	Mehrabadi, et al. (2017b)
Primary settled municipal wastewater in a photobioreactor	*Galdieria sulphuraria*	S: 700 L T: 90 d	N/A	N/A	Biocrude oil	B: N/A P: 28.1 wt. % biocrude oil; 38–39 KJ energy g^{-1}	Cheng, et al. (2019)

(Continued)

TABLE 8.1 (Continued)
Biofuel Products Derived from Microalgae Grown in Wastewater Treatment

Wastewater Cultivation System	Microalgal Species	Scale (S) Cultivation Time (T)	Growth Condition: Light (L) Aeration (A)	Nutrient / Pollutant Removal Efficiency	Biofuel Product	Productivity: Biomass (B) Product (P)	Reference
Municipal wastewater in attached biofilm PBR	*Phormidium*, *Chlorella pyrenoidosa*	S: 100 L T: 6 d	L: Sunlight A: N/A	COD: 53±2% TN: 87±5% NH_3: 81±3% TP: 75±2%	Biocrude oil Biomethane	B: 3.48±0.44 g m^{-2} d^{-1} P: 43±2.0 wt% biocrude oil (24.8 KJ energy g^{-1}); 346.59 ± 5 mL biomethane g^{-1} (9.9 KJ energy g^{-1})	Naaz, et al. (2019)
Paper, textile, leather and municipal WWT in flask and PBR	*Chlorella sorokiniana* strain DBWC2, *Chlorella* sp. strain DBWC7	S: 500 mL & 4 L T: 18 d	L: Fluorescent lamp with 250 µE m–2 s–1 (16 h light: 8 h dark) A: 1 vvm	TN: 55.66–89.3% TP: 100% COD: 52–94.23%	Biocrude oil	B: 4.1 g L^{-1} P: 15 wt% biocrude oil	Goswami, et al. (2019)
Domestic sewage WWT in flat-panel PBR	*Monoraphidium* sp. KMC4	S: 25 L T: N/A	N/A	N/A	Biocrude oil	B: N/A P: 39.38 wt% biocrude oil; 39.47 KJ energy g^{-1}	Mishra and Mohanty (2020)
Primary settled WWT in PBR	*Galdieria sulphuraria*	S: 700 L T: N/A	N/A	N/A	Biocrude oil	B: N/A P: 56.6–65.5 wt% biocrude oil; 43.7 KJ energy g^{-1}	Cui, et al. (2020)
Municipal WWT in HRAP	*Chlorella* sp., *Stigeoclonium* sp., *Nitzschia* sp., *Navicula* sp.,	S: 470 L T: 260	L: Sunlight A: N/A	TN: 49±17–48±16% TP: 37±52–25±52% COD: 62±22–65±23% NH3: 91±8–93±6%	Biomethane	B: 15±6–20±7 g VSS $m^{-2}d^{-1}$ P: 188.7–258.3 mL CH_4 g–1	Arashiro, et al. (2019)
Dairy industry WWT in PBR	*Arthrospira platensis*	S: 150 mL T: 5 d	L: lamp with 140 µmole m^{-2} s^{-1} (12 h light: 12 h dark) A: CO_2 (N/A)	TN: 88.41% TP: 97.01%	Biomethane	B: 1.26±0.31 g DCW L^{-1} d^{-1} P: 482.54±8.27 mL CH_4 g^{-1}	Álvarez, et al. (2020)

Secondary effluent from HRAP & anaerobic digestion centrate in PBR	*Nostoc* sp., *Phormidium* sp., *Geitlerinema* sp., *Chroococcus* sp., *Aphanocapsa sp.*, *Gloeocapsa* sp., *Calothrix* sp.	S: 2 L T: 45 d	L: fluorescent lamp with 350 µmole m^{-2} s^{-1} (12 h light: 12 h dark) A: air (2 L min^{-1})	NH_3: 86% TP: 100% COD: 52%	Biomethane	B: N/A P: 159–199 mL CH_4 g^{-1}	Arashiro, et al. (2020)
Secondary WWT effluent in PBR	*Tribonema* sp.	S: lab-scale (N/A) T: 5 d	L: fluorescent lamp with 150 µmole m^{-2} s^{-1} (10 h light: 14 h dark) A: N/A	N/A	Biomethane	B: N/A P: 293–580 mL CH_4 g^{-1}	Hu, et al. (2021)
Anaerobic digestor centrate in HRAP	*Chlorella* spp., *Pseudanabaena* sp.	S: 180 L T: 92 d	L: sunlight A: N/A	TN: 80±8–87±4% TP: 84±5–92±2% NH_3: 99±1–100±0% TIC: 72±8–95±1%	Biomethane	B: 665±79–1078±84 mg TSS L^{-1} P: 90–94% CH_4 concentrated from 70%	Posadas, et al. (2017)
Anodic effluent in PMFC	*Chlorella vulgaris*	S: 100 mL T: 11 d	L: fluorescent lamp with 200 µmole m^{-2} s^{-1} (12 h light: 12 h dark) A: air (31 mL min^{-1})	COD: 0.19 g L^{-1} d^{-1} NH_3: 5 mg L^{-1} d^{-1}	Bioelectricity	B: N/A P: 34.2±10.0 mW power m^{-2}	Commault, et al. (2017)
MFC-treated dairy wastewater – cathode connected with PBR	*Chlorella* spp.	S: N/A T: 60 d	L: LED lamp A: air / CO_2	N/A	Bioelectricity	B: N/A P: 1.9±0.5–2.8±0.9 W power m^{-3}	Bolognesi, et al. (2021)

N/A, not available/applicable; HRAP, high-rate algal pond; TSS, total suspended solids; VSS, volatile suspended solids; TN, total nitrate; TP, total phosphorus; DRP, dissolved reactive phosphorus; DIN, dissolved inorganic nitrogen; DIP, dissolved inorganic phosphorus; TOC, total organic carbon; COD, chemical oxygen demand; BOD, biological oxygen demand; NH_3, total ammonia; DCW, dry cell weight; FAME, fatty acid methyl ester; PBR, photobioreactor; TIC, total inorganic carbon; PMFC, photosynthetic microbial fuel cell.

hydrothermal liquefaction (HTL) or pyrolysis, which yields biocrude oil along with biochar, gas, and aqueous phase. Biocrude oil could further be subjected to a biorefinery approach to get many valuable petroleum products. Abundant nutrient and water availability in wastewater makes it a cheap and renewable culturing medium for obtaining biodiesel from algal biomass. It was seen that the abundant nutrients, if available under favorable growth conditions, result in enhanced biomass and lipid productivity. *Chlorella sorokiniana* grown in raw urban wastewater for 14 days in custom-built HRAP (35 L capacity) showed a five-fold increase in neutral lipid content than that produced in bold basal medium (BBM) (Arora, et al., 2020). The selection of strain is always vital in any commercial-scale production. Mehrabadi, et al. (2017b) experimented with five dominant algal species: *Desmodesmus* sp., *Coleastrum* sp., *Mucidosphaerium pulchellum*, *Micractinium pusillum*, and *Pediastrum boryanum* isolated from pilot-scale WWT-HRAP to compare their efficiency in primary settled sewage treatment. They observed that *Micractinium pusillum* had the most significant potential for low-cost energy production in WWT with better settleability, high biomass, lipid and energy yield during simulated summer and winter conditions in the growth chamber.

WWT-HRAP operational parameters such as depth (0.15 m and 0.3 m) and CO_2 supplementation were compared by Arbib, et al. (2017) in both batch and continuous (chemostat) mode using dominant *Scenedesmus obliquus*. Besides nutrient removal efficiency, WWT-HRAP with 0.3 m depth in continuous cultivation under CO_2 supplementation showed better biomass and lipid productivity. Algae always attract bacteria during WWT in a synergetic or symbiotic relationship on a beneficial aspect, resulting in enhanced nutrient removal. *Chlorella pyrenoidosa* co-cultivated with ammonia-oxidizing bacteria *Kluyvera* sp. FN5 showed improved nutrient removal efficiency with high biomass and lipid yield suitable for biodiesel production (Zhou, et al., 2020). An algal biofilm airlift photobioreactor (ABA-PBR) with the solid carrier (fiber) for *Chlorella vulgaris*-treated municipal sewage wastewater had shown enhanced biomass and lipid yield than that of conventional airlift PBR (Tao, et al., 2017). It was seen that even a small operational amendment in airlift PBR with solid carrier incorporation resulted in a significant difference in the biomass and lipid yield. An integrated process always helps in the overall economic value of any WWT. Raw pharmaceutical wastewater was treated with *Scenedesmus abundans* simultaneously in PBR and photosynthetic microbial fuel cell (PMFC), which produced lipid-rich biomass (24.3–24.8 mg FAME L^{-1} d^{-1}) besides electricity generation and nutrient removal efficiency (Nayak and Ghosh, 2019).

Biocrude oil of up to 24.9 wt% was obtained from the HTL of WWT-HRAP biomass slurry, and its yield and high heating value (HHV) were also analyzed at varied temperatures of 150–300°C (Mehrabadi, et al., 2017a). Similarly, 28.1 wt% biocrude oil with energy value 38–89 KJ g^{-1} was obtained by Cheng, et al. (2019) from HTL of wastewater-grown *Galdieria sulphuraria* biomass. However, pyrolysis (300–500°C) of the WWT-HRAP grown dried biomass yielded lower 4.7 wt% biocrude oil (Mehrabadi, et al., 2017b). HTL operational parameters significantly influenced the oil yield. Parameters such as reaction time (15–30 minutes), temperature (275–350°C), and biomass:water ratio (1:20, 1:10, and 1:5) were compared for biocrude oil production from WWT-HRAP biomass. It was seen that reaction time of 15 minutes, >300°C temperature, and 1:10 biomass:water ratio yields good biocrude oil up to 44.4 wt% (Couto, et al., 2018).

Microalgal consortium (*Phormidium* sp. and *Chlorella pyrenoidosa*) biomass grown in attached biofilm PBR municipal WWT system were subjected to biocrude oil yield by HTL and biomethane yield by anaerobic co-digestion for comparing their net energy ratio (Naaz, et al., 2019). The HTL process with 43±2 wt% biocrude oil yield was more energy-efficient than anaerobic digestion. Algal-bacterial consortium consists of two *Chlorella* spp. and two bacteria used in four different WWTs such as paper, textile, leather industries, and municipality to get biocrude oil by HTL (Goswami, et al., 2019, Makut, et al., 2020). Mishra and Mohanty (2020) processed the *Monoraphidium* sp. WWT biomass and domestic sewage sludge in co-HTL, which yielded 39.38 wt% biocrude oil with 39.47 KJ bioenergy g^{-1}. This study also showed an enhanced crude oil distillation profile of various co-products. Similarly, the solvent-free co-HTL process of *Galdieria sulphuraria* WWT biomass and crude glycerol produced high biocrude oil of 56.2–65.5 wt% and HHV of upgraded liquid biofuel as 43.7 KJ energy g^{-1} (Cui, et al., 2020). This refined liquid biofuel competes in HHV and low minimal fuel selling price (MFSP) as \$0.38–1.24 L^{-1} gasoline-equivalent on commercial perspective. Apart from MWWT meeting discharge regulations, optimized algal harvesting time and component-specific conversion to biocrude and other byproducts could lead to efficient resource recovery and considerably improved system economics (Li, et al., 2019, Goswami, et al., 2021a). A plethora of studies emphasize bio-oil recovery from MWWT. Still, minimal information is available regarding commercial-scale production and bio-oil recovery for assessing the feasibility of being a part of the alternate fuel system. Finding an economically feasible and net energy balanced process favoring the bio-oil productivity of required volume to meet at least a partial alternate fuel demand is an immense challenge that needs to be appraised.

8.3.2 Biogas

Through biogas production, energy generation is generally anaerobic conversion of organic substrate to a gaseous form of carbon, hydrogen, or hydrocarbon. However, microalgae can generate hydrogen and oxygen directly in oxygenic photosynthesis by photolysis of water. Biohydrogen could also be generated indirectly from the anaerobic digestion of algal biomass (Goswami et al. 2021b). In direct photolytic conversion or indirect anaerobic fermentation, several challenges need to be addressed with technological improvements, making it a sustainable and cost-effective biohydrogen production from microalgae (Singh and Das, 2020). Moreover, the organic substrate-rich WWT favors the generation of gaseous carbon or hydrocarbon.

Recent interest in hydrocarbon or methane extraction as an alternate fuel source made the biomethane or biohydrocarbon gas production from WWT attract more commercial value for its renewable and cheap or cost-effective substrate availability. Arashiro, et al. (2019) compared WWT removal efficiency, biomass productivity, and biomethane productivity between primary treated WWT and non-treated WWT under HRAP. This study showed no significant difference in the removal efficiency in both conditions, but the non-primary treated WWT had higher biomass productivity up to 20 g VSS m^{-2} d^{-1}. Also, biomethane yield was higher up to 258 mL CH_4 g^{-1} in anaerobic digestor when biomass co-digested with primary sludge than that of algal biomass mono-digestion.

Arthrospira platensis biomass grown in dairy industry WWT co-digested in anaerobic digestor with cattle manure generated 482.54±8.27 mL of CH_4 g^{-1} (Álvarez, et al., 2020). Similarly, *Tribonema* sp. biomass from secondary WWT effluent co-digested with pig manure produced 350–580 mL CH_4 g^{-1} (Hu, et al., 2021). Furthermore, after anaerobic digestion, the liquid digest was also tested for the *Tribonema* sp. growth, which yielded 441 mg biomass L^{-1} d^{-1}, enabling digestate valorization. The consortium of cyanobacterial species was grown with HRAP-WWT in secondary effluent and anaerobic digestor centrate used together as a medium in PBR. The biomass generated was used for natural phycobiliprotein pigment extraction. It was then subjected to anaerobic digestion of residual biomass conversion into biomethane (159–199 mL CH_4 g^{-1}) (Arashiro, et al., 2020). This study shows the sequential integrated treatment process has higher biogas yield with the additional valued product.

Besides anaerobic fermentation, biogas upgradation was also getting recent interest through hybrid MWWT with absorption column (AC) in which generated biogas was getting upgraded or concentrated to enhanced biomethane. Secondary settled effluent (culture-free liquid) from the anaerobic digestor centrate WWT-HRAP was passed through AC along with synthetic biogas (0.5% H_2S, 29.5% CO_2, and 70% CH_4) (Posadas, et al., 2017); 50–95% of CO_2 and 100% H_2S were removed, upgraded, and then concentrated in AC as CH_4 up to 94%. Alkalinity, liquid-biogas ratio, seasonal environment variation, and diffuser material type may influence the upgraded biogas' quality irrespective of the methane concentration (Marín, et al., 2018, Marín, et al., 2019, Marín, et al., 2020). During biogas upgrading, oxygen and nitrogen from algal WWT need to be removed, which contaminates the biomethane and affects its application. Removal of oxygen and nitrogen gases in biogas scrubbing and polydimethylsiloxane (PDMS) gas-liquid membrane allows the upgraded biomethane to comply with international standards (Ángeles, et al., 2020). The commercial viability of MWWT-based biogas (biohydrogen or biomethane) generation relies on several factors. It includes appropriate treatment or operational conditions, an integrated upgrading system with minimal contaminants, and value addition by the complete valorization of the residual biomass and anaerobic digestate (Kumar and Verma, 2021).

8.3.3 Bioelectricity

Bioelectricity generation through a PMFC involves two chambers. It includes anode and cathode chambers connected through a cation exchange membrane. Both electrolytes are linked with a resistor for power generation. The anodic section mostly contains wastewater for anaerobic fermentation and exchange the cation to the cathode during the process. In the cathode, the oxygen, generated either by aeration or algal photosynthesis, acts as an acceptor to form water and electricity (Kumar, et al. 2018). Nayak and Ghosh (2019) generated up to 838.68 mW power m^{-2} from PMFC with *Scenedesmus abundans* using treated pharmaceutical wastewater.

Similarly, *Chlorella* sp.–bacteria MFC yielded about 34.2±10.0 mW power m^{-2} (Commault, et al., 2017). Bolognesi, et al. (2021) combined two-chambered MFC with PBR where only culture-free broth gets exchanged between MFC and PBR, which was aerated with air or CO_2. MFC was fed with dairy wastewater, which yielded 1.9±0.5–2.8±0.9 W power m^{-3}. Various operational parameters and conditions were also analyzed, and algal-connected MFC influenced more on net energy recovery and power generation. Bioelectricity generation could support value addition to any bioresource recovery integrated process of MWWT.

8.4 NUTRIENTS

Nutrients are any substance that supplies energy to support the growth and proliferation of any creature. Microalgae are known to have almost all nutrients like carbohydrates, protein, lipid, vitamins, minerals, and water. Most of the algal nutritive products are generally recognized as safe (GRAS) substances complying with Food and Drug Administration (FDA) regulatory guidelines. Single-cell protein (SCP) from *Spirulina* and *Chlorella* are well-known for their protein-rich food or feed supplementation. Wastewater-grown algal biomass was also found to have such nutritive valued biomolecules (Figure 8.1 & Table 8.2). It meets the FDA regulations such as toxic metal limitation and pathogen level in dietary nutrient supplements to be given as feed or food supplement (Moheimani, et al., 2018, Sun, et al., 2019, Cardoso, et al., 2020, Cheng, et al., 2020, Rodrigues de Assis, et al., 2020).

8.4.1 Fatty Acids

Microalgae are rich in lipids composed of fatty acids at the stationary phase, especially under stress conditions (Sulochana and Arumugam, 2016, Anusree, et al., 2017, Sulochana and Arumugam, 2020). Some microalgae possess a high amount of essential fatty acids such as polyunsaturated omega fatty acids (Udayan and Arumugam, 2017, Udayan, et al., 2018, Udayan, et al., 2020). Wastewater-grown algal biomass also comprises these omega fatty acids: alpha or gamma-linolenic acid (C18:3 - ALA or GLA), eicosapentaenoic acid (C20:5 - EPA), and docosahexaenoic acid (C22:6 - DHA). Omega fatty acids have nutraceutical values with antioxidant properties and other therapeutic values. *Schizochytrium* sp. S31 was grown in tofu whey wastewater supplemented with glucose and yeast extract in heterotrophic fermentation to yield DHA of 0.24 g L^{-1} d^{-1} (S-K Wang, et al., 2020). Similarly, Humaidah, et al. (2020) experimented with *Aurantiochytrium* sp. for food processing wastewater, which generated biomass with up to 97.8 mg DHA g^{-1} and 1.79 mg EPA g^{-1} for aquaculture feed. Anaerobic digested swine wastewater effluent after the weak post-electric field was adjusted with different N:P ratios (10:1–50:1) in intermittent-vacuum stripping filamentous microalgae *Tribonema* sp. cultivation. Oleaginous biomass rich in omega 3 fatty acids, including EPA and DHA, was obtained (Huo, et al., 2020). AWW was treated with or without sludge amendments for co-cultivation of *Euglena gracilis* and *Selenastrum* sp. In this, sludge-amended AWW removed nutrients efficiently and produced biomass rich in tocopherol (vitamin E) and omega fatty acids such as arachidonic acid (ARA), EPA, and DHA (Tossavainen, et al., 2019). Textile

TABLE 8.2
Nutritive Products Derived from Microalgae Grown in Wastewater Treatment

Wastewater Cultivation System	Microalgal Species	Scale (S) Cultivation Time (T)	Growth Condition: Light (L) Aeration (A)	Nutrient / Pollutant Removal Efficiency (%)	Nutritive Product	Productivity: Biomass (B) Product (P)	Reference
Anaerobically digested piggery wastewater effluent in open tank and PBR	*Chlorella vulgaris*	S: 25 L (indoor tank) & 75 L (outdoor PBR) T: 25 d	L: lamp with 150 μmole m^{-2} s^{-1} (12 h light: 12 h dark) & sunlight (PBR) A: 0.2 vvm (2% CO_2 in air) & 0.2 vvm (air)	TN: 32.44±7.59% & 72.48±10.50% TP: 100.00±0.00% & 86.93±2.49 % COD: N/A & 85.94±1.70%	Feed	B: 0.63±0.05 g DCW L^{-1} & 0.61±0.04 g DCW L^{-1} P: 22.81±3.83 wt% FAME & 26.32±4.41 wt% FAME; 36.87±1.69 wt% protein & 29.22±2.38 wt% protein	Sun, et al. (2019)
Anaerobically digested piggery wastewater effluent in HRAP	*Chlorella* sp., *Scenedesmus* sp.	S: 1650 L T: 30 d	L: sunlight A: N/A	TN: 0.88±1.64 g m^{-2} d^{-1} NH_3: 1.97±0.32 g m^{-2} d^{-1} COD: 5.83±1.37 g m^{-2} d^{-1}	Feed	B: 2.20±0.49 g DCW m^{-2} d^{-1} P: 39.2 wt% protein; 35.3 wt% carbohydrates; 6.48 wt% hydrolyzed fat; Vitamins 0.95–61.36 mg Kg^{-1}	Moheimani, et al. (2018)
Domestic sewage in HRAP & hybrid biofilm reactor connected HRAP	*Chlorella vulgaris*, *Tetradesmus obliquus*, *Eutetramorus fottie*, *Desmodesmus* sp., *Monoraphidium contortum*, *Navicula* sp.	S: 3300 L T: 212 d	L: sunlight A: N/A	TP: 13.4–16.2 % NH_3: 75.3–77.3 % COD: 57.9–58.8%	Feed	B: 2.88-6.13 g m^{-2} d^{-1} P: 35.21–35.46 wt% protein; 10.19–10.43 wt% carbohydrates; 15.01 –16.95 wt% lipids	Rodrigues de Assis, et al. (2020)

Aquaculture wastewater (AWW) in the tank reactor	*Spirulina* sp. LEB18	S: 5 L T: 10 d	L: sunlight A: N/A	TN: 81.1% TP: 99.97% COD: 89.34%	Feed	B: 0.22±0.02–0.26±0.01 g biomass L^{-1} d^{-1} P: 10–15 wt% lipid; 19.41 wt% protein; 65–70 wt% carbohydrates	Cardoso, et al. (2020)
Anaerobically digested swine wastewater effluent in biofilm attached reactor	*Chlorella pyrenoidosa*	S: N/A T: 8 d	L: lamp with 75 - 85 µmole m^{-2} s^{-1} (12 h light: 12 h dark) A: N/A	COD: 73.7–86.8% NH_3: 82.2–94.1% TN: 83.6–85.2% TP: 62.1–84.3%	Feed	B: 0.33–4.21 g m^{-2} d^{-1} P: 57.3 wt% protein; 14.87 wt% polysaccharides; 16.2 wt% lipid	Cheng, et al. (2020)
Tofu whey wastewater in flask fermentation	*Schizochytrium* sp. S31 (ATCC 20888)	S: 100 mL T: 8 d	L: N/A A: N/A	TN: 66% TP: 59.3% COD: 64.7%	DHA	B: 1.89 g biomass L^{-1} d^{-1} P: 0.24 g DHA L^{-1} d^{-1}	S-K Wang, et al. (2020)
Bean boiling miso-processing wastewater in flask fermentation	*Aurantiochytrium* sp. strain L3W	S: 100 mL T: 3 d	L: N/A A: N/A	N/A	DHA EPA	B: 0.7–0.8 g L^{-1} P: 96.2–97.8 mg DHA g^{-1}; 1.4–1.79 mg EPA g^{-1}	Humaidah, et al. (2020)
Anaerobically digested and electric field treated swine wastewater effluent in a flask	*Tribonema* sp.	S: 1 L T: 14 d	L: lamp with 45–50 µmole m^{-2} s^{-1} A: 0.5 vvm air	TN: 96.5% TP: 74.2% NH_3: 100% COD: 52.5%	Omega (ω-3) fatty acids EPA DHA	B: 75.5±0.3–141.2–0.3 mg biomass L^{-1} d^{-1} P: 28.5±1.2–64.4±3.9 mg oil L^{-1} d^{-1}; 11.7–21.8 % ω-3 fatty acids in oil; 1.9–6.3 % DHA & EPA in oil	Huo, et al. (2020)

(Continued)

TABLE 8.2 (Continued)
Nutritive Products Derived from Microalgae Grown in Wastewater Treatment

Wastewater Cultivation System	Microalgal Species	Scale (S) Cultivation Time (T)	Growth Condition: Light (L) Aeration (A)	Nutrient / Pollutant Removal Efficiency (%)	Nutritive Product	Productivity: Biomass (B) Product (P)	Reference
Aquaculture wastewater in Reactor	*Euglena gracilis* sp. (CCAP 1224/5Z) and *Selenastrum* sp. (SCCAP K–1877)	S: 2 L reactor T: 14 d	L: lamp with 250 µmole m^{-2} s^{-1} A: 0.5 L min^{-1} (2% CO_2 in air)	TN: 75.4–89.2% TP: 84.3–95.7% COD: 43.2–67.5% NH3: 98.9–99.5%	ARA DHA EPA Tocopherol (vitamin E)	B: 1.5 g DCW L^{-1} P: 4.6±1.4 mg ARA L^{-1}; 5.0±1.3 mg DHA L^{-1}; 2.3±0.6 mg EPA L^{-1}; 877 µg tocopherol L^{-1}	Tossavainen, et al. (2019)
Textile industry wastewater (diluted in tap water) in the open tank	*Chlorella variabilis* (ATCC PTA 12198)	S: 100 L T: 6 d	L: sunlight A: N/A	TP: 78.17% TIP: 25.22% TC: 86.43%	Lipid GLA (ω-6 fatty acid) ε-polylysine	B: 74.96±2.62 g m^{-2} d^{-1} P: 20.1±2.2 wt% lipid; 32.61 % GLA in lipid; 35.44 mg ε-polylysine g^{-1} of hydrolyzed sugars	Bhattacharya, et al. (2017)
Anaerobically digested swine wastewater effluent in the reactor	*Chlorella* spp.	S: 500 L T: 12 d	L: sunlight A: N/A	TP: 100% NH_3: 100%	Protein Carbohydrates	B: 247. 34 mg DCW L^{-1} P: 50.1±0.7% protein; 34.4±0.4% carbohydrates; 22% amino acid in protein	Michelon, et al. (2021)
Industrial WWT from the potato processing plant in a glass reactor	*Chlorella sorokiniana*	S: 600 mL T: 20 h	L: LED lamp with 2700 µmol m^{-2} s^{-1} A: CH_4 (60%) with CO_2 (40%)	TN: 40% TP: 62% COD: 96%	Protein	B: 683.25±0.17 mg biomass L^{-1} P: 45 wt% protein; 30 wt% lipid; 23 wt% carbohydrates	Rasouli, et al. (2018)

Swine farm WWT in PBR	*Chlorella sorokiniana* AK–1	S: 1 L T: 15 d	L: LED lamp with 150 µmol m^{-2} s^{-1} A: 0.1 vvm (2% CO_2 in air)	COD: 84.3±4.9% TN: 90.1±1.4% TP: >99.6%	Protein Lutein	B: 0.36±0.04 g L^{-1} d^{-1} P: 0.27±0.01 g protein L^{-1} d^{-1}; 2.20±0.39 mg lutein L^{-1} d^{-1}	Chen, et al. (2020)
Anaerobically digested dairy manure wastewater in PBR	*Chlorella vulgaris*	S: 19 L T: 14 & 28d	L: lamp (16h light: 8h dark) A: air	TN: 17.38% TP: 45.95% NH_3: 78.24%	Protein Chlorophyll	B: N/A P: 35 wt% protein; 34.62 µg chlorophyll mL^{-1}	Taufikurahman, et al. (2020)
Anaerobically digested swine wastewater effluent in glass reactor	*Chlorella*	S: 1 L T: 20 d	L: Lamp with 60 µmol m^{-2} s^{-1} A: N/A	TP: 33.7±4.2–93.4±1.2% NH_3: 0.4±1.1–27.5±1.8 %	Protein Carbohydrates Lipid	B: 0.015±0.021–0.115±0.021 g L^{-1} d^{-1} P: 3.6±0.5–62.2±0.4 mg protein L^{-1} d^{-1}; 1.2±1.7–24.8±4.5 mg carbohydrates L^{-1} d^{-1}; 4.7±0.5–29.0±1.8 mg oil L^{-1} d^{-1}	Huo, et al. (2021)
Wastewater in photofermentor	*Chlorella pyrenoidosa* 15–2070	S: 35 L T: 2 d	L: LED lamp with 2000 µmol m^{-2} s^{-1} A: 10 L min^{-1}	NH_3: 95.5±0.2%	Protein	B: 11.8±0.6 g biomass L^{-1} d^{-1} P: 4.81±0.25 g protein L^{-1} d^{-1}	Q Wang, et al. (2020)
Urban municipal wastewater in PBR	*Nostoc sp.*, *Tolypothrix sp.*, *Calothrix sp.*,	S: 7 L T: 25 & 30 d	L: lamp with 220 µmol m^{-2} s^{-1} (12h light: 12h dark) A: N/A	TN: >95% TP: 35–78% TOC: >93% TIC: >82%	Carbohydrates	B: 0.05–0.07 mg biomass L^{-1} d^{-1} P: 48 wt% carbohydrates	Arias, et al. (2020)

N/A, not available/applicable; HRAP, high-rate algal pond; TN, total nitrate; TP, total phosphorus; DIN, dissolved inorganic nitrogen; TIP, total inorganic phosphorus; TOC, total organic carbon; TIC, total inorganic carbon; COD, chemical oxygen demand; TC, total Carbon; NH_3, total ammonia; DCW, dry cell weight; FAME, fatty acid methyl ester; PBR, photobioreactor; EPA, eicosapentaenoic acid; DHA, docosahexaenoic acid; ARA, arachidonic acid; GLA – Gamma-linolenic acid wastewater was treated with *Chlorella variabilis* in an open tank reactor under natural sunlight for six days in summer. Besides remediating boron, aluminium, nickel, iron, cobalt, sodium, potassium, calcium, and magnesium, it also produced lipid-rich biomass with high omega 6 fatty acid gamma-linolenic acid (32.61% GLA of total lipids) (Bhattacharya, et al., 2017). Fatty acid composition of either saturated or unsaturated fatty acid richness favors bioenergy or nutritional application of the MWWT-grown algal biomass. As nutrition, MWWT fatty acid's commercial feasibility is required to meet the daily dietary limits and their demanding functional nutraceutical effects.

8.4.2 Protein

Proteins are composed of free amino acids, peptides, and/or enzymes. They are very much essential for the proliferation and enzyme-mediated biochemical metabolism in living beings. Microalgal SCP-based products comprising additional nutritious biomolecules are very well-acknowledged in the global market. The protein content of wastewater-grown algal biomass is usually high due to enormous ammonia or nitrogenous supplementation. Michelon et al. (2021) produced *Chlorella* spp. in anaerobically digested swine wastewater for 12 days in a 500 L reactor under sunlight, which yielded high protein (50 wt%) biomass comprising 22% of essential amino acids in total protein and 25 bioactive peptides. Industrial wastewater from potato processing plant was treated with *Chlorella sorokiniana* by Rasouli, et al. (2018) in a lab-scale reactor aerated with the recirculated gas composition of 40% CO_2 and 60% CH_4 for 20 hours. Biomass comprises high protein content of about 45 wt% of total biomass, which contains essential amino acids suitable for fish and chicken feed. Textile wastewater treated with *Chlorella variabilis* biomass was valorized after oil extraction to ε-polylysine by utilizing the hydrolyzed reduced sugar fermentation (Bhattacharya, et al., 2017). ε-polylysine is a cationic peptide of industrial importance used as a food preservative as it has a natural antimicrobial effect. Chen, et al. (2020) treated 50% swine wastewater with *Chlorella sorokiniana* AK-1, which yielded protein and lutein-rich biomass. *Chlorella vulgaris* was grown in anaerobically digested dairy wastewater effluent and produced up to 35% protein of dry biomass (Taufikurahman, et al., 2020). High salinity wastewater with ammonia supplementation generated high protein (57.6 wt%) of mixotrophic *Chlorella pyrenoidosa* (Q Wang, et al., 2020). Anaerobic digested swine manure was clarified with Fenton's reaction using H_2O_2 and then treated with adequate N:P ratio adjustment to cultivate *Chlorella* sp. to get nutritional biomass rich in protein (Huo, et al., 2021). Protein is a building block of any living thing, and MWWT biomass consists of essential free amino acids and active peptides. Protein from algal biomass is generally recognized as safe and used in various animal and aquatic farms.

8.4.3 Carbohydrates, Vitamins, and Other Minerals

Carbohydrates are the primary energy source for almost every living being. Microalgae store their reserve food as starch and generate a lot of polysaccharides as well. Nutrient repleted condition favors the carbohydrate richness in algal biomass.

Wastewater-grown biomass was favored by plenty of available organic carbon in addition to inorganic CO_2 to produce carbohydrate-rich biomass. Microalgae grown in various wastewater had high nutritional carbohydrates, enabling them as food or feed supplementation (Moheimani, et al., 2018, Cardoso, et al., 2020, Cheng, et al., 2020, Rodrigues de Assis, et al., 2020). Arias, et al. (2020) treated urban wastewater in 3.5 and 35 L PBR in batch and semi-continuous cyanobacteria cultivation. It produced carbohydrate (up to 48 wt%) rich biomass from dominant *Nostoc* sp., *Tolypothrix* sp. and *Calothrix* sp. Carbohydrate content in the biomass hydrolyzed into reduced sugars and further fermented to biogas generation (refer to section 8.3.2), bioalcohol (see section 8.5.2), bioactive peptide, or other valued products.

Vitamins are essential micronutrients required in minimal quantity for the growth of any organism. They are very much necessary in proper health functional roles such as anti-oxidizing agents and co-factors. Wastewater-grown microalgae contain these essential vitamins at required dietary daily limits (Moheimani, et al., 2018, Tossavainen, et al., 2019). Similarly, several micro and trace elements like calcium, potassium, sodium, selenium, zinc, and magnesium were also incorporated in the algal biomass, depending on their growth medium salts. Overall, the nutritional value of MWWT biomass, irrespective of its protein or lipid, or carbohydrate richness, also possesses other biomolecules: enough vitamins, micro, and trace elements required to fulfill the daily supplementary limit at least in a fraction for feed or food. The commercial value of biomass rich in nutraceutical value makes them a high-value product with economic viability (Mehariya, et al., 2021b).

8.5 VALUED CHEMICALS

Valued chemicals derived from microalgae were mostly intracellular products. They were obtained from fermentation and as byproducts in processes like biorefinery using algae as a feedstock. Wastewater-grown algae could generate many value-added chemicals as well, including pigments, biopolymer, bioplastics, biochar, phytosterols, and bioalcohol (Figure 8.1 & Table 8.3).

8.5.1 Pigments

Pigments are colored chemicals acting as photoreceptors as well as a photodamage protector in microalgae. In microalgae, photosynthesis occurs due to chlorophyll, which receives the light energy from the blue and red spectrum in the visible region of the natural or artificial light source. Phycobilins are also known as accessory pigments in cyanobacteria. Phycobilins include both phycoerythrin and phycocyanin. They are known for their light-absorbance from the green and orange regions of visible light. Carotenoids are also present in many microalgae as an accessory pigment.

8.5.1.1 Carotenoids

Carotenoids are tetraterpenoids present in plants, bacteria, and microalgae. There are more than 600 types of carotenoids, which are yellow, orange, or reddish. They have an accessory role in photosynthetic light absorbance. In microalgae and some photosynthetic anaerobic bacteria, they prevent photooxidative damage during excess light

TABLE 8.3
Valued Chemicals Derived from Microalgae Grown in Wastewater Treatment

Wastewater Cultivation System	Microalgal Species	Scale (S) Cultivation Time (T)	Growth Condition: Light (L) Aeration (A)	Nutrient/Pollutant Removal Efficiency (%)	Valued Chemical Product	Productivity: Biomass (B) Product (P)	Reference
Anaerobic digested poultry wastewater effluent in the flask	*Dunaliella* sp. FACHB-558	S: 500 mL T: 12 d	L: lamp with 50–250 μmol m^{-2} s^{-1} (12h light: 12h dark) A: N/A	TN: 63.8% TP: 87.2% TOC: 64.1%	Beta-carotene	B: 678 mg biomass L^{-1} P: 4.02 mg beta-carotene L^{-1}	Han, et al. (2019)
Primary settled swine wastewater in the flask	*Haematococcus pluvialis* SAG 34-1d	S: 300 mL T: 25 d	L: lamp with 150 μE m^{-2} s^{-1} (continuous) A: 0.2 vvm air	TN: 96.5–97.9% TP: 97–98.1% NH_3: 97.5–99.3% COD: 23.4–26.4%	Astaxanthin	B: 45–63 mg biomass L^{-1} d^{-1} P: 0.92±0.03–1.27±0.02 wt% astaxanthin	Ledda, et al. (2016)
Primary treated minkery wastewater in PBR	*Haematococcus pluvialis* CPCC 93	S: 2.5 L T: 24 d	L: lamp with 50–200 μmole m^{-2} s^{-1} (continuous) A: Air	TN: 99.7±0.3–99.8±0.2% TP: 20.7±2.1–24.8±1.8	Astaxanthin	B: 113.5±3.6–151.1±5.7 mg biomass L^{-1} d^{-1} P: 16.64±0.57–39.72±1.69 mg astaxanthin L^{-1}	Liu and Yildiz (2019)
Anaerobic treated potato juice wastewater in the flask	*Haematococcus pluvialis* SAG 192.80	S: 500 mL T: 15 d	L: lamp with 1000 & 6000 lux (12h light: 12h dark & continuous) A: N/A	NH_3: 69.4–83.4% TP: 86.5–98.3% TN: 42.5–53.5%	Astaxanthin	B: 0.3–0.41 g biomass L^{-1} P: 11.5–27.9 mg astaxanthin g^{-1}	Pan, et al. (2021)

Industrial wastewater in reactors	*Phaeodactylum tricornuotom*, *Nannochloropsis salina* SAG 4.85, *Nannochloropsis limetica* SAG 18.99, *Chlorella sorokiniana*, *Dunaniella salina*, *Desmodesmus* sp.	S: 5–10 L T: N/A	L: 200–2000 µmole m^{-2} s^{-1} A: 2% CO_2 in the air	N/A	Carotenoid Lutein Phenols Tocopherol Flavonoids	B: N/A P: 2.56±0.02–6.70±0.01 mg carotenoid g^{-1}; 2069±34–5111±61 µg lutein g^{-1}; 3.16±0.04–7.72±0.08 mg phenolics g^{-1}; 13.12±0.01–361.9±23 µg tocopherol g^{-1}; 0.84±0.12–4.03±1.10 mg flavonoids g^{-1}	Safafar, et al. (2015)
Municipal wastewater in PBR & HRAP	*Muriellopsis* sp., *Scenedesmus almeriensis*, *Chlamydomonas segnis*, *Chlorella pyrenoidosa*, *Chlorella vulgaris*	S: 20 & 800 L T: 4 d	L: sunlight A: N/A	TN: 84% TP: 93%	Carotenoids Lutein Protein	B: 104.25 mg biomass L^{-1} d^{-1} P: 51 wt% protein; 0.6 wt% carotenoids; 0.3 wt% lutein.	Cavieres, et al. (2020)
Synthetic dairy wastewater in the flask	*Chlorella variabilis* CCAP 211/84, *Scenedesmus obliquus* CCAP 276/3C	S: 200 mL T: 6–14 d	L: 10–40 µmole m^{-2} s^{-1} A: N/A	TP: 70.19% NH_3: 86.22% COD: 54.72%	Lutein	B: 29.13 mg biomass L^{-1} d^{-1} P: 12.59 mg lutein g^{-1}	Gatamaneni Loganathan, et al. (2020)
Wastewater medium with piggery litter waste in the reactor	*Chlorella minutissima*	S: 500 mL T: 7 d	L: 15000 lux A: 5% CO_2 in air	TN: 93.52% NH_3: 92.65% TP: 90.69%	Lutein Lipid	B: 292.21±0.016 mg biomass L^{-1} d^{-1} P: 169.29 mg lipid L^{-1} d^{-1}; 7.21 mg lutein L^{-1} d^{-1}	De Bhowmick, et al. (2019)
Shrimp cultivation wastewater in the reactor	*Chlorella sorokiniana* MB-1 M12	S: 1 L T: 65 d	L: 150 µmole m^{-2} s^{-1} A: 2% CO_2 in air	N/A	Lutein	B: 1.085±0.149–1.295±0.200 g biomass L^{-1} d^{-1} P: 4.12±0.42–5.02±0.76 mg lutein L^{-1} d^{-1}	Chen, et al. (2019)

(Continued)

TABLE 8.3
Valued Chemicals Derived from Microalgae Grown in Wastewater Treatment

Wastewater Cultivation System	Microalgal Species	Scale (S) Cultivation Time (T)	Growth Condition: Light (L) Aeration (A)	Nutrient/Pollutant Removal Efficiency (%)	Valued Chemical Product	Productivity: Biomass (B) Product (P)	Reference
Fisheries farm water in flask & PBR	*Spirulina maxima*	S: 250 mL & 10 L T: 30 d	L: 200 μmole m^{-2} s^{-1} (12h light: 12h dark) A: N/A	N/A	Phycocyanin	B: 1.18 & 1.05 g biomass L^{-1} P: 0.23 & 0.21 g phycocyanin L^{-1}	Gámez-Ortiz, et al. (2019)
Fish farm wastewater in the reactor	*Spirulina maxima*, *Oscillatoria sp.*	S: 1.5 L T: 30 d	L: N/A (12h light: 12h dark) A: 0.6 vvm (1% CO_2 in air)	N/A	Phycobilins	B: 1-3 g biomass L^{-1} P: 11 wt% phycobiliproteins	García-Martíneza, et al. (2019)
Synthetic ash dam wastewater in biofilm-algal turf scrubber and bag PBR	*Tolypothrix sp.* NQAIF319	S: 2.2.m^2 (turf) & 500 L (bag) T: 16 d	L: sunlight A: 0.05 L air L^{-1} min^{-1} (with or without 15 % CO_2)	Al: 99.8% As: 67.4–75.7% Cu: 72.5–83.2% Fe: 65.5–81.0% Mo: 27.5–98.9% Ni: 54.6–80.1% Se: 12.3–84.2% Sr: 98.4–99.3% Zn: 87–100%	Phycocyanin Phycoerythrin	B: 34–42 g DCW m^{-2} & 870–1310 g DCW m^{-2} P: 0.15–0.3 g phycocyanin m^{-2} d^{-1} & 2–4 g phycocyanin m^{-2} d^{-1}; 0.1–0.25 g phycoerythrin m^{-2} d^{-1} & 1.5–3.0 g phycoerythrin m^{-2} d^{-1}	Velu, et al. (2020)
Primary treated swine wastewater in PBR	*Thermosynechococcus* sp. CL-1 (TCL-1)	S: 1 L T: 12 h	L: 500–2000 μE m^{-2} s^{-1} A: 0.5 vvm air	N/A	Phycocyanin Beta-carotene Zeaxanthin	B: 0.828±0.020–1.001±0.104 g biomass L^{-1} P: 13 wt% C-Phycocyanin	Narindri Rara Winayu, et al. (2021)

Municipal wastewater in the flask	*Nannochloropsis oculata*, *Tetraselmis suecica*	S: 25 mL T: 14 d	L: 1300 lux A: N/A	N/A	Bioethanol	B: 0.975–1.285 g DCW L^{-1} & 0.48–1.055 g DCW L^{-1} P: 0.41% to 7.26% ethanol	Reyimu and Özçimen (2017)
Municipal wastewater in PBR	*Nannochloropsis gaditana* DEE003	S: 1 L T: 20 d	L: 80 µmole m^{-2} s^{-1} A: 0.5 L air min^{-1}	N/A	Bioethanol	B: 0.092±0.001–0.167±0.008 g biomass L^{-1} d^{-1} P: 70.3±2.4–94.3±5.5 mg ethanol g^{-1}	Onay (2018a)
Municipal wastewater in the flask	*Chlorella sorokiniana* NITTS3	S: N/A T: 10 d	L: 33 µE m^{-2} s^{-1} (12h light: 12h dark) A: N/A	N/A	Bioethanol	B: N/A P: 86.70±0.52 mg ethanol g^{-1}	Dhandayuthapani, et al. (2021)
Swine wastewater in PBR	*Chlamydomonas sp.* QWY37	S: 500 mL T: N/A	L: 250–750 µmole m^{-2} s^{-1} A: 0.05 vvm (2.5 % CO_2 in air)	TN: 96% TP: 100% COD: 81%	Bioethanol	B: 1445±35 mg biomass L^{-1} d^{-1} P: 61 g ethanol L^{-1} (62 %)	Qu, et al. (2020)
Anaerobic digested vinasse in the flask	*Chlamydomonas reinhardtii* CC-1093	S: 300 mL T: 4 d	L: 8140 Lux A: 0.75 L min^{-1} (2.5 % CO_2 in air)	TC: 26.09–62.02% TN: 63.56–83.48%	Bioethanol	B: 583–1129 mg biomass L^{-1} d^{-1} P: 166.4–172.1 mg ethanol g^{-1} (61.9% & 68.3 %)	Tasic, et al. (2020)
Swine wastewater in glass PBR	*Neochloris aquatica* CL-M1	S: 200 mL T: N/A	L: 50–200 µmole m^{-2} s^{-1} A: 0.2 vvm (2.5 % CO_2 in air)	NH_3: 92.5% COD: 73.5% TP: 86.6 %	Biobutanol Biohydrogen	B: 6.1 g biomass L^{-1} P: 12 g butanol L^{-1} (0.89 g butanol L^{-1} h^{-1}); 68 mL hydrogen L^{-1} h^{-1}	Wang, et al. (2017)
Municipal wastewater in PBR	*Chlorella sp.* DEE006	S: 1 L T: N/A	L: 180 µmole m^{-2} s^{-1} A: 0.3 L min^{-1} air	N/A	Biobutanol Bioethanol	B: 0.19±0.003–0.28±0.001 g biomass L^{-1} d^{-1} P: 6.23 g biobutanol L^{-1}	Onay (2018b)

(Continued)

TABLE 8.3
Valued Chemicals Derived from Microalgae Grown in Wastewater Treatment

Wastewater Cultivation System	Microalgal Species	Scale (S) Cultivation Time (T)	Growth Condition: Light (L) Aeration (A)	Nutrient/Pollutant Removal Efficiency (%)	Valued Chemical Product	Productivity: Biomass (B) Product (P)	Reference
Cheese whey water in the flask	*Oscillatoria* sp.	S: N/A T: 22 d	L: 1600 lux A: N/A	N/A	Biobutanol	B: 0.0742 g biomass L^{-1} d^{-1} P: 4.2 g biobutanol L^{-1}	Kallarakkal, et al. (2021)
Brewery wastewater in PBR	*Leptolyngbya* sp.	S: 1 L T: 30 d	L: 27 μmole m^{-2} s^{-1} (continuous) A: N/A	TN: 80.11% TP: 72.62% COD: 95.25% NH_3: 92.23%	EPS	B: N/A P: 99% auto flocculation	Papadopoulos, et al. (2020)
Synthetic industrial wastewater and municipal wastewater in revolving algal biofilm reactor and PBR	Mixed microalgae population from raceway pond	S: 1.2 L T: N/A	L: 110–120 μmole m^{-2} s^{-1} (continuous) A: 1 L air min^{-1}	TDS: 21±4–2,783±192 mg L^{-1} d^{-1}	EPS	B: 79±13–364±43 g DCW L^{-1} d^{-1} P: 20–160 mg EPS g^{-1}	Peng, et al. (2020)
Shrimp wastewater in PBR	*Synechocystis* sp. PCC 6803 ΔSphU	S: 10 L T: 14 d	L: 40 μE m^{-2} s^{-1} A: 1 L air min^{-1}	TN: 80.10±1.8% TP: 96.99±0.5%	PHB	B: 35.49±1.1 mg biomass L^{-1} d^{-1} P: 12.73±1.2 mg PHB L^{-1} d^{-1} (32.48±1.7 wt%)	Krasaesueb, et al. (2019)

Agricultural runoff in hybrid HRAP-PBR	Mixed microalgae dominant cyanobacteria	S: 11700 L T: ~215 d	L: sunlight A: CO_2	TN: >95% TP: >95%	PHB Carbohydrates	B: 419±323.4 mg TSS L^{-1}; 275.6±190.8 mg VSS L^{-1} P: 19.2 g PHB d^{-1} (4.5 wt% PHB in VSS); 69 wt% carbohydrates	Rueda, et al. (2020b)
Municipal wastewater in HRAP	*Chlorella* sp., *Scenedesmus* sp.	S: 200 L T: 10 d	L: sunlight A: 5 L min^{-1} (CO_2 or Simulated flue gas air)	TN: 94.3 & 94.9%	Biofertilizer	B: 467 & 479 mg biomass L^{-1} P: N/A	Das, et al. (2019)
Domestic wastewater in HRAP	*Scenedesmus* sp.	S: 60 L T: 7 days	L: sunlight A: flue gas (2.5% CO_2)	N/A	Biofertilizer	B: 97.1 mg biomass L^{-1} d^{-1} P: N/A	Nayak, et al. (2019)
Paddy-soaked rice mill wastewater in HRAP	*Chlorella pyrenoidosa*	S: 20 L T: 7 d	L: ~50 33 µmole m^{-2} s^{-1} (12h light: 12h dark) A: N/A	TP: 64.76±2.59% NH_3: 69.39±3.34 %	Biofertilizer	B: 0.11 g biomass L^{-1} d^{-1} P: N/A	Umamaheswari and Shanthakumar (2021)
Synthetic dairy wastewater in the flask	*Chlorella variabilis* CCAP 211/84, *Scenedesmus obliquus* CCAP 276/3 C	S: 2 L T: N/A	L: 25 ± 3 µmol $m^{-2}s^{-1}$ A: N/A	N/A	Biofertilizer	B: N/A P: N/A	Loganathan, et al. (2020)

N/A, not available/applicable; HRAP, high-rate algal pond; TSS, total suspended solids; VSS, volatile suspended solids; TN, total nitrate; TP, total phosphorus; TOC, total organic carbon; COD, chemical oxygen demand; BNH_3, total ammonia; DCW, dry cell weight; PBR, photobioreactor; EPS, extracellular polymeric substances; PHB, polyhydroxybutrates; PHA, polyhydroxyalkanoates; TDS, total dissolved solids exposure. They are highly valued products due to their high antioxidative property and other therapeutic applications as nutraceuticals. Most carotenoids originate from lycopene. The primary carotenoids observed in microalgae include beta-carotene, astaxanthin, and lutein (Goswami, et al., 2021c). Beta-carotene, also known as pro-vitamin A, is a precursor for vitamin A, which is essential for good eyesight in humans. Microalgae, specifically *Dunaliella* spp., produce abundant beta-carotene, including from WWT. Anaerobic digested poultry wastewater digestate treatment using *Dunaliella* sp. made beta-carotene up to 4.02 mg L^{-1} under optimized salinity and light intensity conditions. Enhanced beta-carotene of 7.26 mg L^{-1} was obtained with wastewater-grown algae further grown in algal medium (Han, et al., 2019).

8.5.1.1.1 Astaxanthin

Astaxanthin is a high-value carotenoid with robust antioxidant activity. Natural astaxanthin has significant uniqueness over other synthetic forms of astaxanthin. It is due to its stable stereoisomerism and better bioavailability. Thus, microalgae, specifically raw astaxanthin-rich *Haematococcus pluvialis* biomass, has been well established in terms of commercial production. Primary settled and filtered swine wastewater effluent was cultivated in 12.5–100% dilution in the algal medium, which accumulated up to 1.27 wt% astaxanthin (Ledda, et al., 2016). Similarly, primary settled, filtered, and diluted minkery wastewater treated in 2 L controlled PBR for six days of growth phase followed by 18 days of induction phase produced 6.64±0.57–39.72±1.69 mg astaxanthin L^{-1} (Liu and Yildiz, 2019). Pan, et al. (2021) generated 11.5–27.9 mg astaxanthin g^{-1}, treating two kinds of effluents using *H. pluvialis* cultivation. It included anaerobically processed potato juice wastewater in methanogenesis and acidification. Although these WWTs contained additional nutrients, astaxanthin content in the biomass was still minimal compared to that grown in other algal mediums. However, a further attempt to optimize the conditions and parameters in WWT as experimented by Pan, et al. (2021) and Ledda, et al. (2016) could favor cost-effective algal astaxanthin recovery. Also, astaxanthin productivity needs to be improved as lower content was observed in most wastewater-grown microalgae.

8.5.1.1.2 Lutein

Lutein is another carotenoid with antioxidant activity and very much essential for eye health. Several microalgae also contain lutein derived from lycopene as a precursor. Recently, lutein was produced in many WWT studies using harvested algal biomass. *Chlorella variabilis* and *Scenedesmus obliquus* consortia grown in synthetic dairy wastewater showed higher lutein productivity (12.59 mg lutein g^{-1}) (Gatamaneni Loganathan, et al., 2020). Wastewater medium supplemented with acetate and piggery litter and operational conditions such as optimized light intensity and CO_2 aeration were considered parameters in experimental model design for optimization. Under optimized conditions, it generated 169.29 mg lipid L^{-1} d^{-1} and 7.21 mg lutein L^{-1} d^{-1} (De Bhowmick, et al., 2019). Shrimp cultivation wastewater was treated with BG11 grown *Chlorella sorokiniana* under semi-continuous cultivation mode. It yielded 1.085±0.149–1.295±0.200 g biomass L^{-1} d^{-1} and 4.12±0.42–5.02±0.76 mg lutein L^{-1} d^{-1} productivity (Chen, et al., 2019). Similarly, *Monascus* fermentation broth treated with *Chlorella protothecoides* under heterotrophic mode produced 7.34 mg lutein L^{-1} d^{-1} (Wang, et al., 2019).

Based on lab-scale municipal WWT efficiency of microalgae, namely *Muriellopsis* sp., *Chlamydomonas segnis, Scenedesmus almeriensis, Chlorella pyrenoidosa* and *Chlorella vulgaris*, pilot-scale 800 L HRAP was treated with *Muriellopsis* sp. and *Scenedesmus almeriensis* (Cavieres, et al., 2020). *Muriellopsis* sp. outperformed *Scenedesmus almeriensis* with high biomass (104.25 mg L^{-1} d^{-1}), protein (51 wt%), carotenoid (0.6 wt%) and lutein (0.3 wt%) contents. Likewise, Safafar, et al. (2015) also treated industrial wastewater with six different microalgal species: *Phaeodactylum tricornutum, Nannochloropsis salina* SAG 4.85, *Nannochloropsis limnetica* SAG 18.99, *Chlorella sorokiniana, Dunaliella salina, and Desmodesmus* sp. The total contents obtained for carotenoids, lutein, phenolics, tocopherol, and flavonoids were 2.56±0.02–6.70±0.01 mg g^{-1}, 2069±34–5111±61 μg g^{-1}, 3.16±0.04–7.72±0.08 mg g^{-1}, 13.12±0.01–361.9±23 μg g^{-1}, and 0.84±0.12–4.03±1.10 mg g^{-1}, respectively, to their biomass content. Besides, they also assayed for the antioxidant activity in which *Desmodesmus* sp. showed higher antioxidative activity due to high content of all these chemicals, in comparison to other microalgal species. Unlike astaxanthin, there were many alternative plant crop sources rich in lutein. However, lutein recovery with high productivity from wastewater-grown algae might give competency in terms of marketable value.

8.5.1.2 Chlorophyll and Phycobiliproteins

Chlorophyll is the primary photosynthetic green-colored pigment present in algae and plants. It was recently recognized as a value-added product because of its applications such as a colorant and therapeutic agent with antioxidant and anti-inflammatory activities. Anaerobically digested wastewater-grown *Chlorella vulgaris* produced 34.62 μg chlorophyll mL^{-1} (Taufikurahman, et al., 2020). Diluted slaughterhouse wastewater treated with *Chlorella pyrenoidosa* had 15.31 μg chlorophyll mL^{-1} (Azam, et al., 2020). Similarly, domestic wastewater treated with consortia of *Chlorella* sp. and *Scenedesmus* sp. made up to 27.03 μg chlorophyll mL^{-1} (Silambarasan, et al., 2021).

Phycobiliproteins are colored pigmented proteins known for their light-harvesting support to chlorophyll. Blue-colored phycocyanin and red-colored phycoerythrin are two accessory pigments present in cyanobacteria. The auto-florescent ability and attractive natural colorant applications made it a commercially valued chemical. Cyanobacteria *Spirulina maxima* grown in fisheries farm wastewater produced 0.21–0.23 g phycocyanin L^{-1} (Gámez-Ortiz, et al., 2019). Similarly, the fisheries wastewater treated with *S. maxima* and *Oscillatoria* sp. generated 10.9 and 11 wt% phycobiliproteins (García-Martíneza, et al., 2019). Swine wastewater was treated in an anoxic chamber with or without aeration using cyanobacteria *Thermosynechococcus* sp. They both generated 13 wt% phycocyanin and carotenoid such as beta-carotene and zeaxanthin as well (Narindri Rara Winayu, et al., 2021). Velu, et al. (2020) treated synthetic ash dam wastewater using *Tolypothrix* sp. in biofilm-forming algal turf scrubber and 500 L bag photobioreactor. It removed most of the heavy metals and generated phycobiliproteins. Phycocyanin and phycoerythrin yield in biofilm-turf scrubber PBR was 0.15–0.3 g m^{-2} d^{-1} and 2–4 g m^{-2} d^{-1}, respectively. In bag PBR, the yield was much higher: 0.1–0.25 g phycocyanin m^{-2} d^{-1} and 1.5–3.0 g phycoerythrin m^{-2} d^{-1}.

8.5.2 Bioalcohol

Alcohol such as ethanol and butanol was recovered from the fermentation of hydrolyzed reducing sugars of algal biomass (Agrawal, et al., 2020b, Bhardwaj, et al., 2020). Both ethanol and butanol are currently being used as a blend or, rarely, 100% transport vehicle fuel. Apart from this, they have several other uses as solvents, disinfectants, cosmetics ingredients, swelling agents in the textile industry, and many other applications. Wastewater-grown algal biomass could be valorized to bioalcohol by acetone-butanol-ethanol (ABE) fermentation by bacteria or yeast. *Tetraselmis suecica* and *Nannochloropsis oculata* grown in municipal wastewater and their hydrolyzed biomass yielded 0.41–7.26% bioethanol in yeast fermentation (Reyimu and Özçimen, 2017). Similarly, municipal wastewater-grown *Nannochloropsis gaditana* biomass produced 70.3±2.4–94.3±5.5 mg ethanol g^{-1} (Onay, 2018a). Dhandayuthapani, et al. (2021) also treated municipal wastewater with *Chlorella sorokiniana*, and after lipid extraction, defatted biomass obtained was fermented after optimized ultrasonic pretreatment to yield 86.70±0.52 mg bioethanol g^{-1}.

Swine wastewater at different ratios was treated with *Chlamydomonas* sp. in semi-continuous cultivation mode, which produced up to 61 g L^{-1} ethanol upon yeast fermentation (Qu, et al., 2020). Anaerobically digested vinasse (ADV) was purified and treated to make chlorine-stress ADV (Cl-ADV) and ammonium (N) cum sulphate (S)-stress ADV (NS-ADV). These Cl-ADV and NS-ADV were treated under aerobic cultivation of *Chlamydomonas reinhardtii for four days* and subsequent anaerobic self-fermentation for 24 h to generate ethanol (166.4–172.1 mg ethanol g^{-1}) (Tasic, et al., 2020). The self-fermentation process in ethanol production makes this unique among other yeast-mediated hydrolyzed algal biomass fermentation. This process also has several advantages in bypassing the pretreatment and saccharification process before fermentation.

Biobutanol was produced by fermentation of the hydrolyzed biomass of the microalgae grown in WWTs. Microalgae rich in carbohydrates were also utilized for biobutanol production. *Neochloris aquatica* biomass grown in swine wastewater was rich in carbohydrates. It was pretreated and ABE-fermented to produce acetone:butanol:ethanol in 2:7:1 ratio with high butanol generation (12 g L^{-1} butanol concentration with 0.89 g butanol L^{-1} h^{-1} productivity (Wang, et al., 2017). Also, it produced biohydrogen with a yield of 68.8 mL hydrogen L^{-1} h^{-1}. *Oscillatoria* sp. was grown in cheese whey water diluted with BG 11 (60% NP deprived) medium, which produced about 4.2 g biobutanol L^{-1} (Kallarakkal, et al., 2021). Onay, (2018b) treated municipal wastewater at different dilutions using *Chlorella* sp., which generated maximum bioalcohol up to 6.23±0.19 g biobutanol concentration L^{-1} and bioethanol yield 0.16±0.005 g g^{-1} (of sugar). Bioalcohol production by self-fermentation or genetically modified microalgae strain improvement could favor fuel recovery from wastewater in a renewable, sustainable process for alternate fuel production at a commercial scale (Chaturvedi, et al., 2021).

8.5.3 Biopolymers and Bioplastics

Algae produce different kinds of extracellular polymeric substances (EPS). Biopolymers, including polyhydroxybutrates (PHB) and polyhydroxyalkanoates (PHA), were considered bioplastics. EPS is used widely as a flocculant, thickener, preservative, bio-lubricant, gelling agent, and for many other applications. Brewery wastewater was treated with *Leptolyngbya* sp. dominant cyanobacteria-bacteria consortia, which removed over 70% of the wastewater nutrients. It also produced an EPS showing 99% auto-flocculation efficiency of the treated culture in 30 minutes (Papadopoulos, et al., 2020). The treated wastewater had nutrients at the permitted limit as per regulation and showed better food shrimp growth and crop growth with no toxic effect upon further application. Synthetic industrial wastewater and municipal wastewater were treated with mixed HRAP microalgal population in revolving algal biofilm (RAB) reactor and bubble column PBR (Peng, et al., 2020). High TDS removal efficiency and EPS content up to 20–160 mg EPS g^{-1} were obtained.

Bioplastics are polymers of polyesters such as PHB and PHA produced by microbes, including cyanobacteria. PHB (32.44 wt%) was generated by the cyanobacteria *Synechocystis* sp. grown in shrimp wastewater (Krasaesueb, et al., 2019). Wastewater-borne cyanobacteria *Synechococcus* sp. and *Synechocystis* sp. were grown in BG11 medium for glycogen and PHB accumulation (Rueda, et al., 2020a). Maximum glycogen and PHB production observed in *Synechocystis* sp. was 69 wt% and 65.04 wt%, respectively. Besides, it appeared that the surplus dissolved inorganic carbon availability increased both glycogen and PHB content. In contrast, at low availability, glycogen gets converted to PHB, which gives insight into possible strategic mechanisms for improving PHB productivity. The same researchers (Rueda, et al., 2020b) treated agricultural runoff in a sequential hybrid pond connected with horizontal tube PBR (semi-closed) with 11700 L capacity for approximately 215 days. An experiment was done in three such hybrid PBRs sequentially connected to enrich cyanobacteria in the first PBR and enrich carbohydrates and PHB in the other two PBRs, producing up to 4.5 wt% PHB in VSS and 69 wt% carbohydrates in VSS.

Apart from these, valued chemicals such as phytosterols and biochar can be produced from wastewater-grown algae. Cyanobacteria *Phormidium autumnale* generated phytosterols, including squalene, stigmasterol, cholesterol, and β-sitosterol, in WWT (Fagundes, et al., 2021). Sterols have immense commercial value and therapeutic applications, and were currently being produced from yeast fermentation. Selected phytosterol rich only in plant or algae could be focused for commercial competency using wastewater-grown algae. Similarly, biochar could be obtained from thermochemical process (HTL)-treated algal biomass after bio-oil or biofuel separation as an added byproduct. Biochar could be used as carbon material similar to charcoal in many applications such as battery fill.

8.6 ORGANIC BIOFERTILIZER

Biofertilizers are the nutritious manure comprising microbial viable cells or debris. Cyanobacterial species found usually associated with the rhizosphere which helps in increasing the yield of wetland crops. For example, *Azolla-Anabaena* symbiosis was known for decades for its nitrogen fixation and supplementation to crops like *Oryza sativa*. Apart from this, microalgal biomass itself or its residue after specific product separation could be a nutritious source for crops (Table 8.3). Municipal wastewater was treated in outdoor HRAP with either CO_2 or simulated flue gas mixer and grown *Chlorella* sp. and *Scenedesmus* sp. (Das, et al., 2019). Flue gas-supplemented wastewater-grown algal biomass after acid treatment was used further as manure for wheat, which showed better crop growth than that of standard algal medium-grown biomass and NPK synthetic manure. Similarly, *Chlorella pyrenoidosa* biomass grown in paddy-soaked rice mill wastewater was used as fertilizer for Indian okra (*Abelmoschus angulosus*). It showed better crop growth than chemical fertilizer amendment (Umamaheswari and Shanthakumar, 2021).

Application of wastewater-grown *Scenedesmus* sp. biomass (MA) valorized after oil separation. Then, it was applied for *Oryza sativa* (rice) cultivation as a biofertilizer and compared with vermicompost (VC) and chemical fertilizer (CF) (Nayak and Ghosh, 2019). Overall production cost of the rice crop grown only with algal biomass manure (MA100) was lower than that of vermicompost (VC100) and found competetive with that of chemical fertilizer (CF100). However, when the combination of algal biomass (MA50) and chemical fertilizer (CF50) in equal proportion was used, it further reduced the overall crop production cost than that of CF100. This cost-effectiveness confers its potential as an alternative option for better crop productivity with sustainability. Algal consortia of *Scenedesmus obliquus* and *Chlorella variabilis* was grown in synthetic dairy WWT and used as biofertilizer. Either as suspended cells or as cell-free supernatant in 40% and 60% concentration, it was applied with plant growth medium to corn (*Zea mays*) and soybean (*Glycine max*) crops (Loganathan, et al., 2020). Both concentrations of algal consortia-suspended cells improved the crops' growth rate with higher flavonoid content, antioxidant activity, phenolic content, and minerals compared to control and cell-free supernatant treatment. Algal-bacterial consortia in a biofilm reactor was also treated with selenium-rich AWW by Han, et al. (2020). The study also explained the N, P, and selenium-rich biomass potential obtained from the treated water as plant biofertilizers. Valorized biomass after value-added product separation has a supplementary commercial worth as a biofertilizer. It would improve the techno-economical feasibility of algal WWT.

8.7 FUTURE PERSPECTIVE

Most of the studies covered in this chapter were carried out either on a bench scale or pilot scale, which showed their potential to scale-up to the commercial level. Effective utilization of wastewater to valued products recovered from algae at high productivity and sustainability with cost-effective and energy-efficient processes would ensure its commercial viability. Many factors need to be considered to make sustainable process development successful. For instance, to point out in HRAP operation, the scale of operation, water depth, and CO_2 supplementation significantly affected WWT-based algal valued product recovery (Kim, et al., 2018, Uggetti, et al., 2018, Sutherland, et al., 2020). Further research should consider multiple product recovery-focused sustainable processes with maximum biomass utilization and zero or minimal waste generation that complies with dischargeable regulatory limits of treated wastewater. This would also make it a much needed renewable, eco-friendly approach. Apart from this, the valued products must be handled or processed to meet their required quality for appropriate application.

8.8 CONCLUSION

Potential valued commercial product recoveries from microalgae WWTs were portrayed extensively in this chapter. It also discussed the commercial viability of such valued products, including biodiesel, biocrude oil, biohydrogen, biomethane, bioelectricity, omega fatty acids, amino acids, bioactive peptides, carbohydrates, vitamins, minerals, carotenoids, phycobiliproteins, bioethanol, biobutanol, biopolymers, bioplastics, phytosterols, biochar, and organic biofertilizers. Most of the WWT studies experimented with *Chlorella* spp. (refer Tables 8.1–8.3), specifically in municipal wastewater, showing its high potential and limited investigation using other algal species. Similarly, nutrients and nutraceuticals were obtained mostly from aquatic or animal farm WWT. Thus, it ensures sustainable renewable processes in an eco-friendly manner. Microalgal species selection for appropriate WWT and developing sustainable processes with high biomass productivity will ensure augmented product(s) recovery with commercial viability.

ABBREVIATIONS

ALA, alpha-linolenic acid; ARA, arachidonic acid; AWW, aquaculture wastewater; MWWT, Microalgal-based biological wastewater treatment; COD, chemical oxygen demand; DHA, docosahexaenoic acid; EPA, eicosapentaenoic acid; EPS, extracellular polymeric substances; FAME, fatty acid methyl ester; FDA, Food and Drug Administration; GHG, greenhouse gases; GLA, gamma-linolenic acid; HRAP, high-rate algal pond; HTL, hydrothermal liquefaction; MFC, microbial fuel cell; PBR, photobioreactor; PHA, polyhydroxyalkanoates; PHB, polyhydroxybutrates; PMFC, photosynthetic microbial fuel cell; SCP, single cell protein; TDS, total dissolved solid; VSS, volatile suspended solid; WWT, wastewater treatment

ACKNOWLEDGEMENT

Author Durairaj Vijayan gratefully acknowledges the financial support from the Science & Engineering Research Board (SERB), Department of Science and Technology, Government of India under the National post-doctoral fellowship scheme project (PDF/2018/000889).

CONFLICT OF INTEREST STATEMENT

Authors have no conflict of interest in any matter relevant to this book chapter contribution.

REFERENCES

Agrawal, Komal, Ankita Bhatt, Nisha Bhardwaj, Bikash Kumar, and Pradeep Verma. "Integrated Approach for the Treatment of Industrial Effluent by Physico-chemical and Microbiological Process for Sustainable Environment." In *Combined Application of Physico-Chemical & Microbiological Processes for Industrial Effluent Treatment Plant*, 119–143: Springer, Singapore, 2020a.

Agrawal, Komal, Ankita Bhatt, Nisha Bhardwaj, Bikash Kumar, and Pradeep Verma. "Algal Biomass: Potential Renewable Feedstock for Biofuels Production–Part I." In *Biofuel Production Technologies: Critical Analysis for Sustainability*, 203–237: Springer, Singapore, 2020b.

Alcántara, Cynthia, Esther Posadas, Benoit Guieysse, and Raúl Muñoz. "Microalgae-Based Wastewater Treatment A2 - Kim, Se-Kwon." In *Handbook of Marine Microalgae*, 439–55: Academic Press, Boston, 2015.

Álvarez, Xavier, Olga Arévalo, Miriam Salvador, Ingrid Mercado, and Borja Velázquez-Martí. "Cyanobacterial Biomass Produced in the Wastewater of the Dairy Industry and Its Evaluation in Anaerobic Co-Digestion with Cattle Manure for Enhanced Methane Production." *Processes* 8, no. 10 (2020): 1290.

Ambat, Indu, Walter Z. Tang, and Mika Sillanpää. "Statistical Analysis of Sustainable Production of Algal Biomass from Wastewater Treatment Process." *Biomass and Bioenergy* 120 (2019): 471–78.

Ángeles, Roxana, Rosario Rodero, Andrea Carvajal, Raúl Muñoz, and Raquel Lebrero. "Potential of Microalgae for Wastewater Treatment and Its Valorization into Added Value Products." In *Application of Microalgae in Wastewater Treatment*, 281–315: Springer, 2019.

Ángeles, Roxana, Ángel Rodríguez, Christian Domínguez, Juan García, Pedro Prádanos, Raúl Muñoz, and Raquel Lebrero. "Strategies for N2 and O2 Removal During Biogas Upgrading in a Pilot Algal-Bacterial Photobioreactor." *Algal Research* 48 (2020): 101920.

Anusree, V.M., B.S. Sujitha, J. Anand, and M. Arumugam. "Dissolved Inorganic Carbonate Sustain the Growth, Lipid and Biomass Yield of Scenedesmus quadricauda under Nitrogen Starved Condition." *Indian Journal if Experimental Biology* 55 (2017): 702–10.

Arashiro, Larissa T., Ivet Ferrer, Catalina C. Pániker, Juan Luis Gómez-Pinchetti, Diederik P.L. Rousseau, Stijn W.H. Van Hulle, and Marianna Garfí. "Natural Pigments and Biogas Recovery from Microalgae Grown in Wastewater." *ACS Sustainable Chemistry & Engineering* 8, no. 29 (2020): 10691–701.

Arashiro, Larissa T., Ivet Ferrer, Diederik P.L. Rousseau, Stijn W.H. Van Hulle, and Marianna Garfí. "The Effect of Primary Treatment of Wastewater in High Rate Algal Pond Systems: Biomass and Bioenergy Recovery." *Bioresource Technology* 280 (2019): 27–36.

Arashiro, Larissa Terumi, Neus Montero, Ivet Ferrer, Francisco Gabriel Acién, Cintia Gómez, and Marianna Garfí. "Life Cycle Assessment of High Rate Algal Ponds for Wastewater Treatment and Resource Recovery." *Science of the Total Environment* 622 (2018): 1118–30.

Arbib, Zouhayr, Ignacio de Godos, Jesús Ruiz, and José A. Perales. "Optimization of Pilot High Rate Algal Ponds for Simultaneous Nutrient Removal and Lipids Production." *Science of the Total Environment* 589 (2017): 66–72.

Arias, Dulce María, Enrica Uggetti, and Joan García. "Assessing the Potential of Soil Cyanobacteria for Simultaneous Wastewater Treatment and Carbohydrate-Enriched Biomass Production." *Algal Research* 51 (2020): 102042.

Arora, Neha, Krishna Kumar Jaiswal, Vinod Kumar, M.S. Vlaskin, Manisha Nanda, Vikas Pruthi, and P.K. Chauhan. "Small-Scale Phyco-Mitigation of Raw Urban Wastewater Integrated with Biodiesel Production and Its Utilization for Aquaculture." *Bioresource Technology* 297 (2020): 122489.

Arumugam, M., A. Agarwal, M.C. Arya, and Z. Ahmed. "Microalgae: A Renewable Source for Second Generation Biofuels." *Current Science* 100, no. 8 (2011): 1141–42.

Azam, Rifat, Richa Kothari, Har Mohan Singh, Shamshad Ahmad, Veeramuthu Ashokkumar, and V.V. Tyagi. "Production of Algal Biomass for Its Biochemical Profile Using Slaughterhouse Wastewater for Treatment under Axenic Conditions." *Bioresource Technology* 306 (2020): 123116.

Bhardwaj, Nisha, Komal Agrawal and Pradeep Verma. "Algal Biofuels: An Economic and Effective Alternative of Fossil Fuels." In *Microbial Strategies for Techno-economic Biofuel Production*, 59–83: Springer, 2020.

Bhattacharya, Sourish, Sumit Kumar Pramanik, Praveen Singh Gehlot, Himanshu Patel, Tejal Gajaria, Sandhya Mishra, and Arvind Kumar. "Process for Preparing Value-Added Products from Microalgae Using Textile Effluent through a Biorefinery Approach." *ACS Sustainable Chemistry & Engineering* 5, no. 11 (2017): 10019–28.

Bolognesi, Silvia, Daniele Cecconet, Arianna Callegari, and Andrea G. Capodaglio. "Combined Microalgal Photobioreactor/Microbial Fuel Cell System: Performance Analysis under Different Process Conditions." *Environmental Research* 192 (2021): 110263.

Cardoso, Lucas Guimarães, Jessica Hartwig Duarte, Bianca Bomfim Andrade, Paulo Vitor França Lemos, Jorge Alberto Vieira Costa, Janice Izabel Druzian, and Fabio Alexandre Chinalia. "*Spirulina* Sp. Leb 18 Cultivation in Outdoor Pilot Scale Using Aquaculture Wastewater: High Biomass, Carotenoid, Lipid and Carbohydrate Production." *Aquaculture* 525 (2020): 735272.

Cavieres, L., J. Bazaes, P. Marticorena, K. Riveros, P. Medina, C. Sepúlveda, and C. Riquelme. "Pilot-Scale Phycoremediation Using *Muriellopsis* Sp. for Wastewater Reclamation in the Atacama Desert: Microalgae Biomass Production and Pigment Recovery." *Water Science and Technology* 83 (2021): 331–43.

Chaturvedi, Venkatesh, Monika Chandravanshi, Manoj Rahangdale and Pradeep Verma. "An Integrated Approach of Using Polystyrene Foam as an Attachment System for Growth of Mixed Culture of Cyanobacteria with Concomitant Treatment of Copper Mine Waste Water." *Journal of Waste Management* 2013 (2013): 1–7.

Chaturvedi, Venkatesh, Rahul Kumar Goswami, and PradeepVerma. "Genetic Engineering for Enhancement of Biofuel Production in Microalgae." In *Biorefineries: A Step Towards Renewable and Clean Energy*, 539–559. Springer, 2021.

Chen, Chun-Yen, En-Wei Kuo, Dillirani Nagarajan, Shih-Hsin Ho, Cheng-Di Dong, Duu-Jong Lee, and Jo-Shu Chang. "Cultivating *Chlorella Sorokiniana* Ak-1 with Swine Wastewater for Simultaneous Wastewater Treatment and Algal Biomass Production." *Bioresource Technology* 302 (2020): 122814.

Chen, Jih-Heng, Yuichi Kato, Mami Matsuda, Chun-Yen Chen, Dillirani Nagarajan, Tomohisa Hasunuma, Akihiko Kondo, Cheng-Di Dong, Duu-Jong Lee, and Jo-Shu Chang. "A Novel Process for the Mixotrophic Production of Lutein with *Chlorella Sorokiniana* Mb-1-M12 Using Aquaculture Wastewater." *Bioresource Technology* 290 (2019): 121786.

Cheng, Feng, Jacqueline M. Jarvis, Jiuling Yu, Umakanta Jena, Nagamany Nirmalakhandan, Tanner M. Schaub, and Catherine E. Brewer. "Bio-Crude Oil from Hydrothermal Liquefaction of Wastewater Microalgae in a Pilot-Scale Continuous Flow Reactor." *Bioresource Technology* 294 (2019): 122184.

Cheng, Pengfei, Ruirui Chu, Xuezhi Zhang, Lirong Song, Dongjie Chen, Chengxu Zhou, Xiaojun Yan, Jay J. Cheng, and Roger Ruan. "Screening of the Dominant *Chlorella pyrenoidosa* for Biofilm Attached Culture and Feed Production While Treating Swine Wastewater." *Bioresource Technology* 318 (2020): 124054.

Commault, Audrey S., Olivier Laczka, Nachshon Siboni, Bojan Tamburic, Joseph R. Crosswell, Justin R. Seymour, and Peter J. Ralph. "Electricity and Biomass Production in a Bacteria-Chlorella Based Microbial Fuel Cell Treating Wastewater." *Journal of Power Sources* 356 (2017): 299–309.

Couto, Eduardo Aguiar, Filomena Pinto, Francisco Varela, Alberto Reis, Paula Costa, and Maria Lúcia Calijuri. "Hydrothermal Liquefaction of Biomass Produced from Domestic Sewage Treatment in High-Rate Ponds." *Renewable Energy* 118 (2018): 644–53.

Cui, Zheng, Jonah M. Greene, Feng Cheng, Jason C. Quinn, Umakanta Jena, and Catherine E. Brewer. "Co-Hydrothermal Liquefaction of Wastewater-Grown Algae and Crude Glycerol: A Novel Strategy of Bio-Crude Oil-Aqueous Separation and Techno-Economic Analysis for Bio-Crude Oil Recovery and Upgrading." *Algal Research* 51 (2020): 102077.

Das, P., M.A. Quadir, M.I. Thaher, G.S.H.S. Alghasal, and H.M.S.J. Aljabri. "Microalgal Nutrients Recycling from the Primary Effluent of Municipal Wastewater and Use of the Produced Biomass as Bio-Fertilizer." *International Journal of Environmental Science and Technology* 16, no. 7 (2019): 3355–64.

De Bhowmick, Goldy, Ajit K. Sarmah, and Ramkrishna Sen. "Performance Evaluation of an Outdoor Algal Biorefinery for Sustainable Production of Biomass, Lipid and Lutein Valorizing Flue-Gas Carbon Dioxide and Wastewater Cocktail." *Bioresource Technology* 283 (2019): 198–206.

Dhandayuthapani, K., V. Sarumathi, P. Selvakumar, Tatek Temesgen, P. Asaithambi, and P. Sivashanmugam. "Study on the Ethanol Production from Hydrolysate Derived by Ultrasonic Pretreated Defatted Biomass of *Chlorella sorokiniana* Nitts3." *Chemical Data Collections* 31 (2021): 100641.

Dineshbabu, Gnanasekaran, Durairaj Vijayan, Vaithiyalingam Shanmugasundaram Uma, Bidhu Bhusan Makut, and Debasish Das. "Microalgal Systems for Integrated Carbon Sequestration from Flue Gas and Wastewater Treatment." In *Application of Microalgae in Wastewater Treatment*, 339–70: Springer, 2019.

Fagundes, Mariane Bittencourt, Gerardo Alvarez-Rivera, Raquel Guidetti Vendruscolo, Mônica Voss, Patricia Arrojo da Silva, Juliano Smanioto Barin, Eduardo Jacob-Lopes, Leila Queiroz Zepka, and Roger Wagner. "Green Microsaponification-Based Method for Gas Chromatography Determination of Sterol and Squalene in Cyanobacterial Biomass." *Talanta* 224 (2021): 121793.

Gámez-Ortiz, L.P., M.J. Gónzález-Soto, M.E. Perez-Roa, J.B. García-Martinez, N.A. Urbina-Suarez, and C.E. Diaz-Castañeda. "Bioconversion of Post-Culture Wastewater from Farm Fisheries for the Production of High-Value Algal Biomass." *Journal of Physics: Conference Series* 1388 (2019): 012036.

Gao, Feng, Wei Cui, Jing-Ping Xu, Chen Li, Wei-Hong Jin, and Hong-Li Yang. "Lipid Accumulation Properties of *Chlorella vulgaris* and *Scenedesmus obliquus* in Membrane Photobioreactor (Mpbr) Fed with Secondary Effluent from Municipal Wastewater Treatment Plant." *Renewable Energy* 136 (2019): 671–76.

García-Martíneza, Janet B., Nestor A. Urbina-Suarezb, Antonio Zuorroc, Andres F. Barajas-Solanob, and Viatcheslav Kafarova. "Fisheries Wastewater as a Sustainable Media for the Production of Algae-Based Products." *Chemical Engineering* 76 (2019).

Garfí, Marianna, Laura Flores, and Ivet Ferrer. "Life Cycle Assessment of Wastewater Treatment Systems for Small Communities: Activated Sludge, Constructed Wetlands and High Rate Algal Ponds." *Journal of Cleaner Production* 161 (2017): 211–19.

Gatamaneni Loganathan, Bhalamurugan, Valerie Orsat, and Mark Lefsrud. "Phycoremediation and Valorization of Synthetic Dairy Wastewater Using Microalgal Consortia of *Chlorella variabilis* and *Scenedesmus obliquus*." *Environmental Technology* (2020): 1–14.

Goswami, Gargi, Bidhu Bhusan Makut, and Debasish Das. "Sustainable Production of Bio-Crude Oil Via Hydrothermal Liquefaction of Symbiotically Grown Biomass of Microalgae-Bacteria Coupled with Effective Wastewater Treatment." *Scientific Reports* 9, no. 1 (2019): 15016.

Goswami, Rahul Kumar, Komal Agrawal, Sanjeet Mehariya, Antonio Molino, Dino Musmarra, and Pradeep Verma. "Microalgae-Based Biorefinery for Utilization of Carbon Dioxide for Production of Valuable Bioproducts." In *Chemo-Biological Systems for CO_2 Utilization*, 199–224: Taylor & Francis, 2020.

Goswami, Rahul Kumar, Sanjeet Mehariya, Pradeep Verma, Roberto Lavecchia, and Antonio Zuorro. "Microalgae-Based Biorefineries for Sustainable Resource Recovery from Wastewater." *Journal of Water Process Engineering* 40 (2021a): 101747.

Goswami, Rahul Kumar, Sanjeet Mehariya, Parthiba Karthikeyan Obulisamy, and Pradeep Verma. "Advanced Microalgae-Based Renewable Biohydrogen Production Systems: A Review." *Bioresource Technology* 320, A (2021b): 124301.

Goswami, Rahul Kumar, Agrawal, Komal. and Pradeep Verma, An Overview of Microalgal Carotenoids: Advances in the Production and Its Impact on Sustainable Development. In *Bioenergy Research: Evaluating Strategies for Commercialization and Sustainability*, (2021c) pp. 105–128.

Gouveia, Luísa, Sofia Graça, Catarina Sousa, Lucas Ambrosano, Belina Ribeiro, Elberis P. Botrel, Pedro Castro Neto, Ana F. Ferreira, and Carla M. Silva. "Microalgae Biomass Production Using Wastewater: Treatment and Costs: Scale-up Considerations." *Algal Research* 16 (2016): 167–76.

Han, Ting, Haifeng Lu, Yu Zhao, Hong Xu, Yuanhui Zhang, and Baoming Li. "Two-Step Strategy for Obtaining *Dunaliella* Sp. Biomass and B-Carotene from Anaerobically Digested Poultry Litter Wastewater." *International Biodeterioration & Biodegradation* 143 (2019): 104714.

Han, Wei, Yufeng Mao, Yunpeng Wei, Peng Shang, and Xu Zhou. "Bioremediation of Aquaculture Wastewater with Algal-Bacterial Biofilm Combined with the Production of Selenium Rich Biofertilizer." *Water* 12, no. 7 (2020): 2071.

Hu, Yuansheng, Manoj Kumar, Zhongzhong Wang, Xinmin Zhan, and Dagmar B. Stengel. "Filamentous Microalgae as an Advantageous Co-Substrate for Enhanced Methane Production and Digestate Dewaterability in Anaerobic Co-Digestion of Pig Manure." *Waste Management* 119 (2021): 399–407.

Humaidah, Nurlaili, Satoshi Nakai, Wataru Nishijima, Takehiko Gotoh, and Megumi Furuta. "Application of *Aurantiochytrium* Sp. L3w for Food-Processing Wastewater Treatment in Combination with Polyunsaturated Fatty Acids Production for Fish Aquaculture." *Science of the Total Environment* 743 (2020): 140735.

Huo, Shuhao, Junzhi Liu, Feifei Zhu, Sajid Basheer, David Necas, Renchuan Zhang, Kun Li, Dongjie Chen, Pengfei Cheng, Krik Cobb, Paul Chen, Bailey Brandel, and Roger Ruan. "Post Treatment of Swine Anaerobic Effluent by Weak Electric Field Following Intermittent Vacuum Assisted Adjustment of N:P Ratio for Oil-Rich Filamentous Microalgae Production." *Bioresource Technology* 314 (2020): 123718.

Huo, Shuhao, David Necas, Feifei Zhu, Dongjie Chen, Jun An, Nan Zhou, Wei Liu, Lu Wang, Yanling Cheng, Yuhuan Liu, and Roger Ruan. "Anaerobic Digestion Wastewater Decolorization by H2o2-Enhanced Electro-Fenton Coagulation Following Nutrients Recovery Via Acid Tolerant and Protein-Rich *Chlorella* Production." *Chemical Engineering Journal* 406 (2021): 127160.

Kallarakkal, Khadeeja Parveen, Karuppan Muthukumar, Arun Alagarsamy, Arivalagan Pugazhendhi, and Samsudeen Naina Mohamed. "Enhancement of Biobutanol Production Using Mixotrophic Culture of *Oscillatoria* Sp. in Cheese Whey Water." *Fuel* 284 (2021): 119008.

Kim, Byung-Hyuk, Jong-Eun Choi, Kichul Cho, Zion Kang, Rishiram Ramanan, Doo-Gyung Moon, and Hee-Sik Kim. "Influence of Water Depth on Microalgal Production, Biomass Harvest, and Energy Consumption in High Rate Algal Pond Using Municipal Wastewater." *Journal of Microbiology and Biotechnology* 28, no. 4 (2018): 630–37.

Kothari, Richa, Arya Pandey, Shamshad Ahmad, Ashwani Kumar, Vinayak V. Pathak, and V.V. Tyagi. "Microalgal Cultivation for Value-Added Products: A Critical Enviro-Economical Assessment." *3 Biotech* 7, no. 4 (2017): 243.

Krasaesueb, Nattawut, Aran Incharoensakdi, and Wanthanee Khetkorn. "Utilization of Shrimp Wastewater for Poly-Β-Hydroxybutyrate Production by *Synechocystis* Sp. Pcc 6803 Strain Δsphu Cultivated in Photobioreactor." *Biotechnology Reports* 23 (2019): e00345.

Kumar, Bikash, Komal Agrawal, Nisha Bhardwaj, Venkatesh Chaturvedi, and Pradeep Verma. "Advances in Concurrent Bioelectricity Generation and Bioremediation Through Microbial Fuel Cells." In *Microbial Fuel Cell Technology for Bioelectricity*, 211–239: Springer, Cham, 2018.

Kumar, Bikash, Komal Agrawal, Nisha Bhardwaj, Venkatesh Chaturvedi, and Pradeep Verma "Techno-Economic Assessment of Microbe-Assisted Wastewater Treatment Strategies for Energy and Value-Added Product Recovery." In *Microbial Technology for the Welfare of Society*, 147–181: Springer, 2019.

Kumar, Bikash, and Pradeep Verma. "Techno-Economic Assessment of Biomass-Based Integrated Biorefinery for Energy and Value-Added Product." In *Biorefineries: A Step Towards Renewable and Clean Energy*, 581–616: Springer, 2021.

Ledda, Claudio, Jessica Tamiazzo, Maurizio Borin, and Fabrizio Adani. "A Simplified Process of Swine Slurry Treatment by Primary Filtration and *Haematococcus pluvialis* Culture to Produce Low Cost Astaxanthin." *Ecological Engineering* 90 (2016): 244–50.

Li, Hugang, Jamison Watson, Yuanhui Zhang, Haifeng Lu, and Zhidan Liu. "Environment-Enhancing Process for Algal Wastewater Treatment, Heavy Metal Control and Hydrothermal Biofuel Production: A Critical Review." *Bioresource Technology* 298 (2020): 122421.

Li, Yalin, Sydney A. Slouka, Shanka M. Henkanatte-Gedera, Nagamany Nirmalakhandan, and Timothy J. Strathmann. "Seasonal Treatment and Economic Evaluation of an Algal Wastewater System for Energy and Nutrient Recovery." *Environmental Science: Water Research & Technology* 5, no. 9 (2019): 1545–57.

Liu, Yu, and Ilhami Yildiz. "Bioremediation of Minkery Wastewater and Astaxanthin Production by *Haematococcus pluvialis*." *International Journal of Global Warming* 19, no. 1–2 (2019): 145–57.

Loganathan, Bhalamurugan Gatamaneni, Valerie Orsat, and Mark Lefsrud. "Evaluation and Interpretation of Growth, Biomass Productivity and Lutein Content of *Chlorella variabilis* on Various Media." *Journal of Environmental Chemical Engineering* 8, no. 3 (2020): 103750.

Mahapatra, Durga Madhab, V. Sudharsan Varma, Shanmugaprakash Muthusamy, and Karthik Rajendran. "Wastewater Algae to Value-Added Products." In *Waste to Wealth*, 365–393: Springer, 2018.

Makut, Bidhu Bhusan, Gargi Goswami, and Debasish Das. "Evaluation of Bio-Crude Oil through Hydrothermal Liquefaction of Microalgae-Bacteria Consortium Grown in Open Pond Using Wastewater." *Biomass Conversion and Biorefinery* (2020): 1–15.

Marín, David, Alessandro A. Carmona-Martínez, Raquel Lebrero, and Raúl Muñoz. "Influence of the Diffuser Type and Liquid-to-Biogas Ratio on Biogas Upgrading Performance in an Outdoor Pilot Scale High Rate Algal Pond." *Fuel* 275 (2020): 117999.

Marín, David, Antonio Ortíz, Rubén Díez-Montero, Enrica Uggetti, Joan García, Raquel Lebrero, and Raúl Muñoz. "Influence of Liquid-to-Biogas Ratio and Alkalinity on the Biogas Upgrading Performance in a Demo Scale Algal-Bacterial Photobioreactor." *Bioresource Technology* 280 (2019): 112–17.

Marín, David, Esther Posadas, Patricia Cano, Víctor Pérez, Raquel Lebrero, and Raúl Muñoz. "Influence of the Seasonal Variation of Environmental Conditions on Biogas Upgrading in an Outdoors Pilot Scale High Rate Algal Pond." *Bioresource Technology* 255 (2018): 354–58.

Mehariya, Sanjeet, Rahul Kumar Goswami, Pradeep Verma, Roberto Lavecchia, and Antonio Zuorro. "Integrated Approach for Wastewater Treatment and Biofuel Production in Microalgae Biorefineries" *Energies 14*, (8) (2021a): 2282.

Mehariya, Sanjeet, Rahul Kumar Goswami, Obulisamy Parthiba Karthikeysan, and Pradeep Verma. "Microalgae for high-value products: A way towards green nutraceutical and pharmaceutical compounds." *Chemosphere 280* (2021b): 130553.

Mehrabadi, Abbas, Rupert Craggs, and Mohammed M. Farid. "Wastewater Treatment High Rate Algal Pond Biomass for Bio-Crude Oil Production." *Bioresource Technology* 224 (2017a): 255–64.

Mehrabadi, Abbas, Mohammed M. Farid, and Rupert Craggs. "Potential of Five Different Isolated Colonial Algal Species for Wastewater Treatment and Biomass Energy Production." *Algal Research* 21 (2017b): 1–8.

Michelon, William, Marcio Luis Busi da Silva, Alexandre Matthiensen, Cristiano José de Andrade, Lidiane Maria de Andrade, and Hugo Moreira Soares. "Amino Acids, Fatty Acids, and Peptides in Microalgae Biomass Harvested from Phycoremediation of Swine Wastewaters." *Biomass Conversion and Biorefinery* (2021).

Mishra, Sanjeev, and Kaustubha Mohanty. "Co-Htl of Domestic Sewage Sludge and Wastewater Treatment Derived Microalgal Biomass – an Integrated Biorefinery Approach for Sustainable Biocrude Production." *Energy Conversion and Management* 204 (2020): 112312.

Mobin, Saleh, and Firoz Alam. "Some Promising Microalgal Species for Commercial Applications: A Review." *Energy Procedia* 110 (2017): 510–17.

Moheimani, Navid Reza, Ashiwin Vadiveloo, Jeremy Miles Ayre, and John R. Pluske. "Nutritional Profile and in Vitro Digestibility of Microalgae Grown in Anaerobically Digested Piggery Effluent." *Algal Research* 35 (2018): 362–69.

Naaz, Farah, Arghya Bhattacharya, Kamal K. Pant, and Anushree Malik. "Investigations on Energy Efficiency of Biomethane/Biocrude Production from Pilot Scale Wastewater Grown Algal Biomass." *Applied Energy* 254 (2019): 113656.

Narindri Rara Winayu, Birgitta, Ko Tung Lai, Hsin Ta Hsueh, and Hsin Chu. "Production of Phycobiliprotein and Carotenoid by Efficient Extraction from *Thermosynechococcus* Sp. Cl-1 Cultivation in Swine Wastewater." *Bioresource Technology* 319 (2021): 124125.

Nayak, Jagdeep K., and Uttam K. Ghosh. "Post Treatment of Microalgae Treated Pharmaceutical Wastewater in Photosynthetic Microbial Fuel Cell (Pmfc) and Biodiesel Production." *Biomass and Bioenergy* 131 (2019): 105415.

Nayak, Manoranjan, Dillip Kumar Swain, and Ramkrishna Sen. "Strategic Valorization of De-Oiled Microalgal Biomass Waste as Biofertilizer for Sustainable and Improved Agriculture of Rice (*Oryza Sativa* L.) Crop." *Science of the Total Environment* 682 (2019): 475–84.

Onay, Melih. "Bioethanol Production from *Nannochloropsis gaditana* in Municipal Wastewater." *Energy Procedia* 153 (2018a): 253–57.

Onay, Melih. "Investigation of Biobutanol Efficiency of *Chlorella* Sp. Cultivated in Municipal Wastewater." *Journal of Geoscience and Environment Protection* 6, no. 10 (2018b): 40–50.

Pan, Minmin, Xinyu Zhu, Gang Pan, and Irini Angelidak. "Integrated Valorization System for Simultaneous High Strength Organic Wastewater Treatment and Astaxanthin Production from *Haematococcus Pluvialis*." *Bioresource Technology* 326 (2021): 124761.

Papadopoulos, Konstantinos P., Christina N. Economou, Stefanos Dailianis, Nikolina Charalampous, Natassa Stefanidou, Maria Moustaka-Gouni, Athanasia G. Tekerlekopoulou, and Dimitris V. Vayenas. "Brewery Wastewater Treatment Using Cyanobacterial-Bacterial Settleable Aggregates." *Algal Research* 49 (2020): 101957.

Peng, Juan, Kuldip Kumar, Martin Gross, Thomas Kunetz, and Zhiyou Wen. "Removal of Total Dissolved Solids from Wastewater Using a Revolving Algal Biofilm Reactor." *Water Environment Research* 92, no. 5 (2020): 766–78.

Posadas, Esther, David Marín, Saúl Blanco, Raquel Lebrero, and Raúl Muñoz. "Simultaneous Biogas Upgrading and Centrate Treatment in an Outdoors Pilot Scale High Rate Algal Pond." *Bioresource Technology* 232 (2017): 133–41.

Qu, Wenying, Pau Loke Show, Tomohisa Hasunuma, and Shih-Hsin Ho. "Optimizing Real Swine Wastewater Treatment Efficiency and Carbohydrate Productivity of Newly Microalga *Chlamydomonas* Sp. Qwy37 Used for Cell-Displayed Bioethanol Production." *Bioresource Technology* 305 (2020): 123072.

Rahman, Ashiqur, Saumya Agrawal, Tabish Nawaz, Shanglei Pan, and Thinesh Selvaratnam. "A Review of Algae-Based Produced Water Treatment for Biomass and Biofuel Production." *Water* 12, no. 9 (2020): 2351.

Rasouli, Zahra, Borja Valverde-Pérez, Martina D'Este, Davide De Francisci, and Irini Angelidaki. "Nutrient Recovery from Industrial Wastewater as Single Cell Protein by a Co-Culture of Green Microalgae and Methanotrophs." *Biochemical Engineering Journal* 134 (2018): 129–35.

Reyimu, Zubaidai, and Didem Özçimen. "Batch Cultivation of Marine Microalgae *Nannochloropsis Oculata* and *Tetraselmis Suecica* in Treated Municipal Wastewater toward Bioethanol Production." *Journal of Cleaner Production* 150 (2017): 40–46.

Rodrigues de Assis, Letícia, Maria Lúcia Calijuri, Paula Peixoto Assemany, Thiago Abrantes Silva, and Jamily Santos Teixeira. "Innovative Hybrid System for Wastewater Treatment: High-Rate Algal Ponds for Effluent Treatment and Biofilm Reactor for Biomass Production and Harvesting." *Journal of Environmental Management* 274 (2020): 111183.

Rueda, Estel, María Jesús García-Galán, Rubén Díez-Montero, Joaquim Vila, Magdalena Grifoll, and Joan García. "Polyhydroxybutyrate and Glycogen Production in Photobioreactors Inoculated with Wastewater Borne Cyanobacteria Monocultures." *Bioresource Technology* 295 (2020a): 122233.

Rueda, Estel, María Jesús García-Galán, Antonio Ortiz, Enrica Uggetti, Javier Carretero, Joan García, and Rubén Díez-Montero. "Bioremediation of Agricultural Runoff and Biopolymers Production from Cyanobacteria Cultured in Demonstrative Full-Scale Photobioreactors." *Process Safety and Environmental Protection* 139 (2020b): 241–50.

Safafar, Hamed, Jonathan Van Wagenen, Per Møller, and Charlotte Jacobsen. "Carotenoids, Phenolic Compounds and Tocopherols Contribute to the Antioxidative Properties of Some Microalgae Species Grown on Industrial Wastewater." *Marine Drugs* 13, no. 12 (2015): 7339–56.

Silambarasan, Sivagnanam, Peter Logeswari, Ramachandran Sivaramakrishnan, Aran Incharoensakdi, Pablo Cornejo, Balu Kamaraj, and Nguyen Thuy Lan Chi. "Removal of Nutrients from Domestic Wastewater by Microalgae Coupled to Lipid Augmentation for Biodiesel Production and Influence of Deoiled Algal Biomass as Biofertilizer for *Solanum Lycopersicum* Cultivation." *Chemosphere* 268 (2021): 129323.

Singh, Harshita, and Debabrata Das. "Chapter 15 - Biohydrogen from Microalgae." In *Handbook of Microalgae-Based Processes and Products*, edited by Eduardo Jacob-Lopes, Mariana Manzoni Maroneze, Maria Isabel Queiroz and Leila Queiroz Zepka, 391–418: Academic Press, 2020.

Sulochana, Sujitha Balakrishnan, and Muthu Arumugam. "Influence of Abscisic Acid on Growth, Biomass and Lipid Yield of Scenedesmus Quadricauda under Nitrogen Starved Condition." *Bioresource Technology* 213 (2016): 198–203.

Sulochana, Sujitha Balakrishnan, and Muthu Arumugam. "Targeted Metabolomic and Biochemical Changes During Nitrogen Stress Mediated Lipid Accumulation in Scenedesmus Quadricauda Casa Cc202." *Frontiers in Bioengineering and Biotechnology* 8 (2020): 1223.

Sun, Zhong-liang, Li-qin Sun, and Guo-zhong Chen. "Microalgal Cultivation and Nutrient Removal from Digested Piggery Wastewater in a Thin-Film Flat Plate Photobioreactor." *Applied Biochemistry and Biotechnology* 187, no. 4 (2019): 1488–501.

Sutherland, Donna L., Jason Park, Stephan Heubeck, Peter J. Ralph, and Rupert J. Craggs. "Size Matters – Microalgae Production and Nutrient Removal in Wastewater Treatment High Rate Algal Ponds of Three Different Sizes." *Algal Research* 45 (2020): 101734.

Tao, Qiong, Feng Gao, Chun-Yuan Qian, Xiao-Ze Guo, Zhou Zheng, and Zhao-Hui Yang. "Enhanced Biomass/Biofuel Production and Nutrient Removal in an Algal Biofilm Airlift Photobioreactor." *Algal Research* 21 (2017): 9–15.

Tasic, Marija B., Anderson de Jesus Bonon, Maria Ingrid Rocha Barbosa Schiavon, Bruno Colling Klein, Vlada B. Veljković, and Rubens Maciel Filho. "Cultivation of *Chlamydomonas Reinhardtii* in Anaerobically Digested Vinasse for Bioethanol Production." *Waste and Biomass Valorization* 12, no. 2 (2021): 857–865.

Taufikurahman, Taufik, Muhammad Arief Ardiansyah, Novi Tri Astutiningsih, and Eko Agus Suyono. "Cultivation of *Chlorella Vulgaris* in Anaerobically Digested Dairy Manure Wastewater (Addmw) for Protein and Chlorophyll Production." *Sains Malaysiana* 49, no. 9 (2020): 2035–42.

Tossavainen, Marika, Katariina Lahti, Minnamari Edelmann, Reetta Eskola, Anna-Maija Lampi, Vieno Piironen, Pasi Korvonen, Anne Ojala, and Martin Romantschuk. "Integrated Utilization of Microalgae Cultured in Aquaculture Wastewater: Wastewater Treatment and Production of Valuable Fatty Acids and Tocopherols." *Journal of Applied Phycology* 31, no. 3 (2019): 1753–63.

Udayan, Aswathy, and Muthu Arumugam. "Selective Enrichment of Eicosapentaenoic Acid (20: 5n-3) in N. Oceanica Casa Cc201 by Natural Auxin Supplementation." *Bioresource Technology* 242 (2017): 329–33.

Udayan, Aswathy, S Kathiresan, and Muthu Arumugam. "Kinetin and Gibberellic Acid (Ga3) Act Synergistically to Produce High Value Polyunsaturated Fatty Acids in Nannochloropsis Oceanica Casa Cc201." *Algal Research* 32 (2018): 182–92.

Udayan, Aswathy, Hariharan Sabapathy, and Muthu Arumugam. "Stress Hormones Mediated Lipid Accumulation and Modulation of Specific Fatty Acids in Nannochloropsis Oceanica Casa Cc201." *Bioresource Technology* 310 (2020): 123437.

Uggetti, Enrica, Bruno Sialve, Jérôme Hamelin, Anaïs Bonnafous, and Jean-Philippe Steyer. "Co2 Addition to Increase Biomass Production and Control Microalgae Species in High Rate Algal Ponds Treating Wastewater." *Journal of CO2 Utilization* 28 (2018): 292–98.

Umamaheswari, Jagannathan, and Subramainam Shanthakumar. "Paddy-Soaked Rice Mill Wastewater Treatment by Phycoremediation and Feasibility Study on Use of Algal Biomass as Biofertilizer." *Journal of Chemical Technology & Biotechnology* 96, no. 2 (2021): 394–403.

UN-Water, UNESCO. *United Nations World Water Development Report 2020: Water and Climate Change*: UNESCO: Paris, France, 2020.

Velu, Chinnathambi, Samuel Cirés, Diane L. Brinkman, and Kirsten Heimann. "Bioproduct Potential of Outdoor Cultures of *Tolypothrix* Sp.: Effect of Carbon Dioxide and Metal-Rich Wastewater." *Frontiers in Bioengineering and Biotechnology* 8, no. 51 (2020).

Wang, Qingke, Zongyi Yu, and Dong Wei. "High-Yield Production of Biomass, Protein and Pigments by Mixotrophic *Chlorella Pyrenoidosa* through the Bioconversion of High Ammonium in Wastewater." *Bioresource Technology* 313 (2020): 123499.

Wang, Shi-Kai, Xu Wang, Yong-Ting Tian, and Yue-Hua Cui. "Nutrient Recovery from Tofu Whey Wastewater for the Economical Production of Docosahexaenoic Acid by Schizochytrium Sp. S31." *Science of the Total Environment* 710 (2020): 136448.

Wang, Yue, Shih-Hsin Ho, Chieh-Lun Cheng, Dillirani Nagarajan, Wan-Qian Guo, Chiayi Lin, Shuangfei Li, Nanqi Ren, and Jo-Shu Chang. "Nutrients and Cod Removal of Swine Wastewater with an Isolated Microalgal Strain *Neochloris Aquatica* Cl-M1 Accumulating High Carbohydrate Content Used for Biobutanol Production." *Bioresource Technology* 242 (2017): 7–14.

Wang, Zhenyao, Rong Zhou, Yufang Tang, Ziting Wang, Bo Feng, and Yuqin Li. "The Growth and Lutein Accumulation in Heterotrophic *Chlorella Protothecoides* Provoked by Waste *Monascus* Fermentation Broth Feeding." *Applied Microbiology and Biotechnology* 103, no. 21 (2019): 8863–74.

Zhou, Xu, Wenbiao Jin, Qing Wang, Shida Guo, Renjie Tu, Song-fang Han, Chuan Chen, Guojun Xie, Fanqi Qu, and Qilin Wang. "Enhancement of Productivity of *Chlorella Pyrenoidosa* Lipids for Biodiesel Using Co-Culture with Ammonia-Oxidizing Bacteria in Municipal Wastewater." *Renewable Energy* 151 (2020): 598–603.

9 Seaweeds Used in Wastewater Treatment: Steps to Industrial Commercialization

Sara Pardilhó, João Cotas, Ana M. M. Gonçalves, Joana Maia Dias, Leonel Pereira

CONTENTS

9.1 INTRODUCTION

Seaweeds are a diverse group of multicellular photosynthetic organisms that live in aquatic habitats, freshwater, and marine environments (Bharathiraja et al. 2015). These organisms are commonly known to be fast-growing, which is comparable to terrestrial plants, with the difference of having a more simplistic morphological structure, besides the ecosystem they inhabit (McHugh 2003).

Seaweeds may reflect the ecological status of coastal environments. Algae are related to water pollution in several ways, being considered good bio-monitors of water quality in coastal seawaters. The enrichment of water nutrients due to organic effluents leads to the growth of algal species, resulting in massive surface growths (or "blooms") reducing water quality and affecting its use (Luo et al. 2020, Sen et al. 2013). Due to their inherent characteristics, seaweeds are exploited in a wide range of industrial applications: as a product, ingredient, or as part of a management process/solution to reduce industry impact in the surrounding environment (Arumugam et al. 2018, Kılınç et al. 2013, Leandro et al. 2020b). Accordingly, seaweeds can play an important role in the purification of water bodies once they

DOI: 10.1201/9781003155713-9

are able to accumulate organic and inorganic toxic compounds (Deniz et al. 2018; Goswami et al. 2021a, Luo et al. 2020, Sen et al. 2013). Such solutions might pass by seaweed-based processes to mitigate eutrophication in aquatic systems or as a bioremediation tool to reduce heavy metals, toxic dyes, and other harmful compounds in wastewater or industrial effluents (Holdt et al. 2014, Kılınç et al. 2013, Kim et al. 2017; Kumar et al. 2019).

As a result of their bioaccumulation abilities (mostly related to their cell wall composition), algae have become an important organism for wastewater treatment (WWT) (Ortiz-Calderon et al. 2017). Seaweeds are largely exploited as a substitute for functional activated carbon, as an absorption element of toxic compounds and excessive nutrients from wastewater (Arumugam et al. 2018). Nowadays widely accepted, algal WWT systems are as effective as the conventional ones, being considered a low-cost alternative to complex and expensive treatment systems (Sen et al. 2013).

Agriculture and aquaculture wastewater is used to cultivate seaweed for feed and food, since ancient times, due to their richness in nutrients such as nitrates, potassium, and phosphorus (García-Poza et al. 2020, Le Gouvello et al. 2017, Reid et al. 2020, Torres et al. 2019). However, seaweed biomass from WWT (SWWT) can be more difficult to be commercially exploited, mostly when it comes to wastewater from industries, mainly due to the higher concentration of harmful compounds (Arumugam et al. 2018, Sá Monteiro et al. 2019).

In this chapter, the potential of exploiting SWWT is approached, including the quality controls which could be implemented, giving a general overview about this complex theme, contributing, in the future, to a blue-green and circular economy (Figure 9.1).

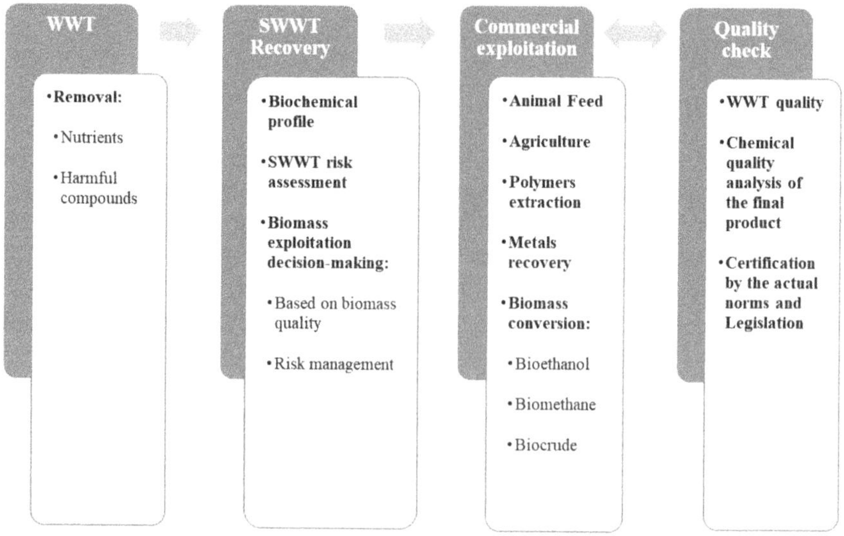

FIGURE 9.1 Purpose for the seaweed biomass from wastewater treatment (SWWT) until a safe commercial exploration.

9.2 SEAWEED AS A WASTEWATER TREATMENT TOOL

Nowadays, there are a high number of documented seaweed uses that attract the attention of many researchers. Among these applications is the use of seaweeds in WWT.

There are several treatment technologies to remove organic or inorganic pollutants from wastewater. Besides that, adsorption is considered an excellent method to treat wastewater from multiple industries, having advantages such as availability, easy handling, and higher efficiency at low cost, when compared with other treatments (Arumugam et al. 2018).

Biosorption using seaweeds has shown promising results in the removal of pollutants like dyes (from paper, textile, or printing industries) and phosphorous, nitrogen, phenolic compounds, and heavy metals from several industries. This is considered an eco-friendly process that conjugates the availability and low cost of the raw material, becoming an alternative technology for effectively removing pollutants from wastewater. The use of seaweed allows the replacement of activated carbon, mostly used, which presents a high price and limited reuse (Arumugam et al. 2018, Deniz et al. 2018).

Most of the dyes used in textile industries are not biodegradable, being resistant to aerobic digestion processes. Nowadays, different types of adsorbents are researched to be employed for dye removal in WWT, to replace the most commonly used: activated carbon. Considering the adsorption of dyes, seaweeds are capable of performing this biosorption process due to the presence of active functional groups in the biomass (hydroxyl, carboxyl, carbonyl, amine, and sulfate) (Arumugam et al. 2018).

9.2.1 Removal of Excess Nitrogen and Phosphorus: Treatment of Eutrophic Water

Eutrophication is a primary problem in WWT, being extremely harmful to aquatic ecosystems. Seaweeds use excess nutrients to grow rapidly. Therefore, these WWTs can be used as a growth medium and seaweeds can be normally exploited considering the common pathways used for commercial seaweeds (Arumugam et al. 2018). The increase in seaweed production for commercial exploration has made the combination with WWT management appealing. Due to their nutrient concentration, wastewater is considered a good nutrient source to enhance seaweed production, with low costs. Moreover, the control of wastewater chemical and biochemical profile might allow cultivation of seaweeds, with adequate quality (Cole et al. 2016).

Table 9.1 presents some studies in which seaweeds demonstrated a great potential to assimilate nutrients from eutrophicated wastewaters. The referenced studies focused on the normal seaweed pathways, namely towards the production of animal feed and fertilizers.

The results demonstrate the capacity of seaweeds to remove the characteristic excess nutrients from wastewater, which might be combined with the seaweed's growth in WWT plants, representing an opportunity for the future of seaweed cultivation (Figure 9.2.a).

TABLE 9.1
Review of Seaweeds Used to Remove Nutrients in Wastewater Treatment

Phylum	Species	Industry/ Origin	Nutrients	Removal Performance	Reference
Rhodophyta	*Gracilariopsis lemaneiformis* (formerly *Gracilaria lemaneiformis*)	Aquaculture water (bay water)	Nitrogen, Phosphate	Nitrogen = 21.0% Phosphate = 28.6%	Wei et al. 2017
	Gracilariopsis longissima (formerly *Gracilaria verrucosa*)	Aquaculture wastewater	Nitrite, Nitrate, Ammonia, Phosphate	Nitrite = 94.5% Nitrate = 91.4% Ammonia = 99.3% Phosphate = 100%	Devi et al. 2007
	Gracilariopsis longissima	Smoked-fish processing industry	Nitrate, Ammonia, Phosphate	Nitrate = 37.1% Ammonia = 67.6% Phosphate = 96%	Sari et al. 2020
	Gracilariopsis longissima	Shrimp pond wastewater	Nitrate, Ammonia, Phosphate	Nitrate = 22.2% (20 days) Ammonia = 90% (10 days) Phosphate = 20.1% (30 days)	Trianti et al. 2020
Chlorophyta	*Ulva rigida*	Rejected water from a sludge-fed biogas plant	Ammonia, Carbon, Phosphate	Ammonia = 5.31% dry weight Carbon = 36.89% dry weight Phosphate = 0.24% dry weight	Sode et al. 2013
	Ulva flexuosa (formerly *Enteromorpha flexuosa*)	Aquaculture wastewater	Nitrite, Nitrate, Ammonia, Phosphate	Nitrite = 87.2% Nitrate = 94.5% Ammonia = 82.5% Phosphate = 82.5%	Devi et al. 2007
	Oedogonium sp.	Municipal wastewater	Nitrogen, Phosphorus	Nitrogen = 62.0% Phosphorus = 75.0%	Neveux et al. 2016
	Ulva prolifera	Coking wastewater	Nitrogen, Phosphorus	Nitrogen = 26.1% Phosphorus = 68.5%	Gu et al. 2020

9.2.2 Removal of Harmful Compounds and Pollutants

Besides nutrient-removal, seaweeds are used in WWT to reduce or remove toxic heavy metals, other harmful compounds, and toxins (He et al. 2014), which have adverse effects to the ecosystems and also to animal and human health (Arumugam et al. 2018).

Anthropogenic sources are mainly responsible for the generation of such harmful substances, transposing them into unconfined water bodies and thus greatly impacting aquatic ecosystems (Alloway 2013). Accordingly, there is a need to develop

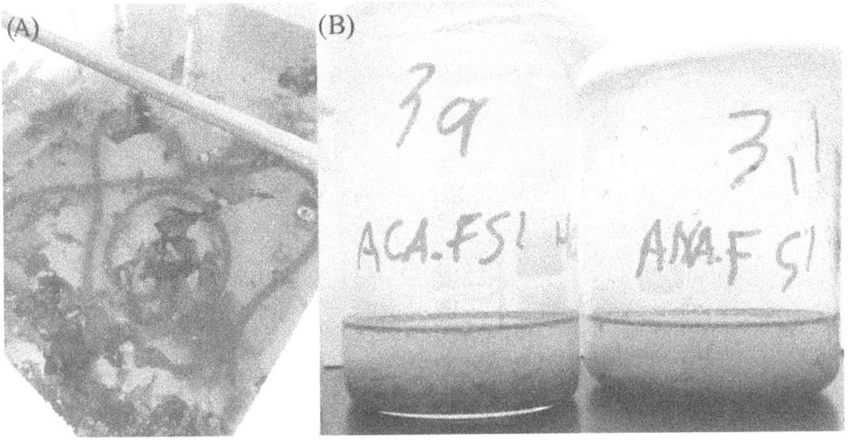

FIGURE 9.2 Example of a seaweed nutrient removal system in WWT and desorption treatment method to recover heavy metals from SWWT: a) Nutrient removal from eutrophicated area by a prototype cultivation system designed by Lusalgae, Lda in Mondego River estuary (Figueira da Foz, Coimbra, Portugal); b) Brown SWWT desorption to recover metals, with a hydrochloric acid solution.

ecofriendly techniques to remove such compounds from wastewater in WWT plants. As previously stated, seaweeds appear a promising low-cost alternative to remove harmful compounds from wastewater, reducing its negative impacts (Table 9.2).

Seaweeds can also be exploited to recover metals of high value from wastewater (Figure 9.2.b), as a low-cost technology, enabling their recovery and making the use of metals more efficient (Birungi et al. 2020, Ortiz-Calderon et al. 2017).

Studies show that the three groups of macroalgae (brown, red, and green) allow the biosorption of different metals and harmful compounds like synthetic dyes. Macroalgae have sulphated polysaccharides in their cell walls, which have a high capacity to bind to pollutants such as metals (by mainly ionic charges). The presence of heavy metals in wastewater becomes harmful for environment and living organisms, especially when exceeding the standard limits, once they are non-biodegradable (Arumugam et al. 2018). Brown algae seem to have higher metal sorption capacity when compared to other algae groups, due to the presence of alginic acid (Neveux et al. 2018). However, studies found that red macroalgae are used only for pigment removal (El-Naggar et al. 2020, Omar et al. 2018); thus, establishing a relationship between the different groups of algae and the removal of such harmful compounds is inviable. Green seaweeds are the most exploited due to the easy control, rapid growth, and cultivation method (Arumugam et al. 2018).

The study by Senthilkumar et al. (2019) demonstrated that *Ulva lactuca* can be regenerated and able to remove zinc from wastewater with little differences (considering three cycles of reuse of the first inoculum regeneration) from the first biomass used, demonstrating that the same biomass can be used more than once with negligible loss of efficiency (3.2%).

TABLE 9.2
Review of Seaweeds Used to Remove Harmful Compounds in Wastewater Treatment

Phylum	Species	Industry/ Origin	Harmful Compound	Removal Performance	Reference
Rhodophyta	*Gracilaria corticata*	Textile wastewater	Malachite green dye	93.3% (T = 25°C, t = 60 min, pH of solution) 92.5% (T = room temperature, t = 60 min, pH = 8) 96% (T = room temperature, t = 150 min, pH of solution)	Omar et al. 2018
	Gracilaria sp. dried biomass	Simulated wastewater	Nickel, Methylene bluc dyc	Nickel = 97.53% Methylene blue dye = 94.86%	El-Naggar et al. 2020
Chlorophyta	*Ulva lactuca*	Phosphating industry wastewater	Zinc	1st cycle = 65.1% 2nd cycle = 63.3% 3rd cycle = 61.9%	Senthilkumar et al. 2019
	Ulva prolifera	Coking wastewater	Arsenic, Cadmium, Chromium, Copper, Lead, Zinc	Arsenic = 56% Cadmium = 42.4% Chromium = 3.7% Copper = 83.4% Lead = 65.8% Zinc = 11.7%	Gu et al. 2020
Phaeophyceae	*Treptacantha baccata* (formely *Cystoseira baccata*) (beach cast)	Synthetic wastewater	Aluminum	Aluminum >80%	Lodeiro et al. 2010
	Sargassum sp. with clay (40:60)	Electroplating industry wastewater	Chromium	Chromium = 99%	Aprianti et al. 2017

Lodeiro et al. (2010) did a circular economy approach using a beach cast seaweed (mainly composed by *Treptacantha baccata*) as adsorption material to remove and recover aluminum in synthetic and real wastewater, comparing the results with activated carbon. Using seaweeds, aluminum absorption surpassed 80% (reached with less than 30 min of contact), around 10 times more effective when compared with commercial activated carbon. Aluminum was easily further recovered from the seaweed with an acid stripping solution; however, this step caused a decrease in the adsorption capacity of the algae, which is not considered problematic. Besides the WWT, the use of seaweeds contributed in mitigating the problem in Galician Coast (Spain) related to the occurrence of beach cast seaweeds, which has caused difficulties for fishing and tourism industries.

Aprianti et al. (2017) studied the combined use of clay and *Sargassum* sp. seaweed as biosorbent for Chromium (Cr (VI)) removal from electroplating industry wastewater. The results showed an improvement in the adsorption capacity by combining both materials, reaching 99.39% removal when a ratio of 40:60 (*Sargassum*:clay) was employed for 10 h.

9.3 SEAWEEDS USED IN WASTEWATER TREATMENT: INDUSTRIAL POTENTIAL

The landfilling of seaweed biomass used in WWT processes can cause pollution problems by leaching of toxic metals or other organic pollutants into the soil and surface or groundwater systems. For this reason, it is urgent to find other applications for such biomass that do not compromise the environment quality.

The recovery of SWWT after bioremediation to obtain products with commercial value is of high importance, making the biosorption process more economically attractive (Mehariya et al. 2021a). Traditionally, seaweed biomass is burned or landfilled, which besides being inappropriate for the environment, represents a loss of a resource for the economic sector. The chemical composition of seaweed biomass makes it suitable for the production of biofuels or other value-added products. Thus, after use, SWWT can be further recovered for biotechnological applications.

The enrichment of biomass with microelements, from the biosorption process, makes it suitable for a new niche application as feed supplement and for agriculture purposes as nitrogen and phosphorous supplements (Ortiz-Calderon et al. 2017, Sen et al. 2013); such benefits must be balanced with possible negative impacts depending upon the type of biomass. The production of dietary supplements from dried seaweed biomass obtained this way has already demonstrated good results in animals, mainly of *Ulva* sp. and *Gracilaria* sp. (Michalak et al. 2013, Saeid et al. 2013). Another approach is the use seaweed biomass for energy production (methane), by anaerobic digestion, resulting in a digestate by-product which can be used as fertilizer, usually after a stabilization treatment (Salerno et al. 2009, Sen et al. 2013).

Seaweeds can be used as feed supplement if the heavy metals and other harmful compounds do not surpass the regulated limits. In fact, seaweeds are known to have various benefits to the animal's welfare (Morais et al. 2020). Seaweed feed supplementation can reduce the use of medical and antibiotics costs during animal production, thus improving the quality of animal welfare and also the quality of the final product (Makkar et al. 2016).

In a study performed by Roberts et al. (2015), a biochar was produced from algae used in WWT (through thermal treatment under oxygen-limited conditions) and evaluated for the suitability of applying the dried biomass directly and the biochar for soil amelioration. The study concluded that the use of biomass impaired the germination and growth of radish due to leaching of metals to the soil. On the other hand, biochar produced from the same biomass leached negligible amounts of metals to the soil, favoring the germination and growth of radishes. The study also showed that biochar improved the fertility of the soil and the availability of nutrients (P, K, Ca, and Mg).

Another study, by Cole et al. (2017), used biomass from municipal WWT (combined with sugarcane bagasse) to produce compost and biochar (through slow pyrolysis at 450°C) for agriculture use. The application of the compost in a low fertility soil significantly increased the production of sweet corn, and when biochar was applied in conjunction with the compost, there was an additional increase of 15% in corn productivity. The study concluded that algal compost alone or with biochar could be used to replace most of the nutrients provided by synthetic fertilizers, enhancing the productivity of agricultural soils.

The establishment of integrated biorefineries is a hot topic nowadays to explore SWWT more efficiently (Mehariya et al. 2021b). Integrated seaweed biorefinery theory is a macro-cascade of sequential extractions to monetize all the biomass compounds at the maximum. Consequently, a production platform to exploit the seaweed into secure and value-added products, for example, polysaccharides and pigments, might be explored. The final seaweed residue might be directed to biofermentation (to obtain biomethane or bioethanol) (Agrawal et al. 2020; Bhardwaj et al. 2020), with the remaining fractions of the processes used to obtain a soil biofertilizer (Milledge et al. 2019).

Seaweed polysaccharides are one of the most commercial seaweed compounds, exploited by diverse industries. The main polysaccharides (agar, carrageenan, and alginate) in seaweeds can vary up to 40–50% of the seaweed dry weight (Torres et al. 2019). The polysaccharide usage varies according to the purity state; even so, the low-grade polysaccharides have demand in diverse industrial areas (Venkatesan et al. 2017). Seaweed polysaccharide extraction industry is well developed and has extraction protocols which are cost-effective to guarantee the quality to different industries (Kraan 2012, Mortensen et al. 2016, Pereira et al. 2020, Rychen et al. 2017, Sá Monteiro et al. 2019, Younes et al. 2018).

Biotransformation is the process of biomass conversion with the objective to generate energy by converting organic compounds into condensed energy forms, for example, biogas, bioethanol, biohydrogen or crude bio-oil (Goswami et al. 2020; Goswami et al. 2021b). Due to the low lipid content of seaweeds, biofermentation presents the best opportunity to produce energy from the SWWT (Milledge et al. 2019, Salerno et al. 2009, Torres et al. 2019). The pre-extraction of seaweed polysaccharides will enhance biomass fermentation yields and product quality, reducing co-extraction and impurities in the process (Wang et al. 2020).

Neveux et al. (2016) used the *Oedogonium* sp. cultivated in municipal WWT to produce seaweed biocrude. This seaweed cultivation resulted in biocrude production representing around 26% of the seaweed dry weight. Additionally, seaweed treatment by anaerobic digestion before hydrothermal liquefaction demonstrated to be a remarkable pathway to convert the algal biomass into biocrude, without the needed to dry the SWWT. Also, in this study, simulation of biocrude production was made, demonstrating biocrude yield (of high quality) equivalent to 52–75 barrels/ha/year, considering a continuous *Oedogonium* sp. biomass production from municipal WWT of 25.5–36.5 t/ha/year.

9.4 HOW CAN THE SWWT QUALITY BE CHECKED?

Wastewater discharge must consider standard environmental regulations. A stable supply and high quality of materials are essential for unlocking the value-added potential of seaweeds. European policies point towards a circular economy, namely by nutrient recycling. However, the current wastewater legal framework is outside the scope of waste legislation, where the definition of "waste" excludes sewage and wastewater. The different regulations might thus make it more difficult to incorporate circular economy aspects resulting from the combination of those activities.

Wastewater origin of the biomass can limit further applications in some sectors. However, the advent of circular economy policies may help in supporting that. Cooperation of the wastewater sector and algae industries might thus be of high importance to study bottlenecks and propose technological applications to value such unused biomass.

The traditional methods to check biomass quality is through legislation, varying with the intended final application. In some cases, a clear biomass quality check exists, through government regulations, which can be different between countries. When seaweed is used as resource, there is a need to determine the biochemical profile, heavy metals and other pollutant contents, such as arsenic content for animal feed and agriculture (Adamse et al. 2017, Holdt et al. 2011, Leandro et al. 2020a, Sá Monteiro et al. 2019). In the case of seaweed compounds, there is a need to ensure the purity (to avoid co-extractions and contaminations), chemical norms of the extraction or fermentation process and also, the presence of heavy metals (Adamse et al. 2017, Holdt et al. 2011, Leandro et al. 2020a, Sá Monteiro et al. 2019).

Biomass quality can be analyzed with spectroscopic and chromatographic methods to guarantee the quality. More importantly, this analysis can be a key tool to ensure the best SWWT exploitation.

Seaweed biomass cultivated to remove the nutrient concentration in aquaculture systems is expected to be used in animal feed, without problem. However, there is the need to analyze the biomass and check it as a normal seaweed source market.

Regarding seaweed polymers and pigments, there are existing legislation and regulation mechanisms to be applied to guarantee the quality for some usages, for example, food, feed, agriculture, textile, and printing (Imeson 2011, Jönsson et al. 2020, Morais et al. 2020, Sá Monteiro et al. 2019, Younes et al. 2018).

The production of animal feed from SWWT needs to be carefully analyzed taking into account bacteriological, biochemical, and heavy metal profiles so that the use in feed supplements is approved, considering the seaweed origin and the water quality. There is a general list of contaminants which are prohibited in animal feed (Verstraete 2013). Risk assessment of the SWWT needs to be based on scientific evidences and undertaken in an independent, objective, and transparent manner (Verstraete 2013). Animal welfare is privileged and protected by regulatory laws (Verstraete 2013). Evidence given by revised studies shows that seaweed biomass can potentially be used for animal feed after biochemical and chemical analysis.

The use of seaweeds as fertilizer must take into account the evaluation of the quality of the final product according to applicable regulations. At the European Union (EU) level, Regulation 2019/1009 of the European Parliament and the Council of 5 June 2019 lay down rules for the marketing of fertilizing products in the EU. In this document, for the first time, is referenced the use of algae [excluding blue-green algae (cyanobacteria)], which are included as plants and not as a new organism group (European Parliament, 2019).

TABLE 9.3
Information Concerning the Spanish Regulation for Seaweed Extracts and Fertilizers (Adapted from European Biostimulants Industry Council)

Type of Fertilizer	Required Information to be Declared	Restricted Compounds	Technical Information in the Product Sheets and Technical Labelling
Solid seaweed extract	Detailed information about the extraction method and species used	Alginic acid <9% Mannitol <3% Arsenic <50 mg/Kg	Concentration of alginic acid, mannitol, nitrogen (if >1%), free amino acids (if >1 %) Identification of the species
Liquid seaweed extract	Detailed information about the extraction method and species used	Alginic acid <1.5% Mannitol <0.5% Arsenic <50 mg/Kg	Concentration of alginic acid, mannitol, nitrogen (if >1%), free amino acids (if >1%) Identification of the species
Fertilizers with seaweeds extracts	Detailed information about fertilizer composition and seaweed species used	Alginic acid <0.3% Mannitol <0.1 Arsenic <50 mg/Kg Nitrogen + Phosphorus pentoxide + Potassium oxide <7%, if in mineral form	pH Electrical conductivity Concentration of alginic acid, mannitol, amino acids (if >1%) Denomination of the type of fertilizer, such as: • Foliar application • Nutrient solution • Fertigation

In Spain, a specific regulation (No 1774/2002 from European Commission) for seaweed fertilizers exists, imposing analytical methods to quantify the elements in seaweed fertilizers. This regulation is very specific for brown seaweeds, and the main specifications are presented in Table 9.3 (European Biostimulants Industry Council).

The Spanish regulation is considered a role model in Europe regarding legislation aimed at seaweed extract use in agriculture. Nevertheless, the chemical and security sheets of agricultural products are vital to ensure the safe application. In this case, SWWT needs to be certificated by analytical methods to be further exploited. The biorefinery concept previously presented can be a key contributor to extract harmful compounds, after which the remaining SWWT can be used as a fertilizer.

SWWT can be used to extract polysaccharides, where the quality control is already strong taking into account the food safety regulations. Thus, the biorefinery cascade can be used for polysaccharide extraction for food/feed purposes. Consequently, the biorefinery approach can have a significant role in the use of SWWT. The main restrictions are the heavy metal concentrations (such as arsenic, lead, cadmium, and mercury) and the microbiological requirements (CBI Ministry of Foreign Affairs 2020) (namely the presence of bacteria in the extract). If the polysaccharides extracted from SWWT fulfill the requirements of the Food Safety Agencies [such as Food and Drug Administration (USA) or European Food Safety

Authority (EU)], they are approved for food and feed industry application. If not, there are more options with lower restrictions where seaweeds polysaccharides can be used (Draget et al. 2005, Hentati et al. 2020, Leandro et al. 2020a). However due to several constraints in the other uses, fermentation is considered the best option for SWWT exploitation, namely to produce biogas or bioethanol. From the fermentation process, solids can be further recovered, for example, as fertilizer.

Nowadays, many of the vehicles currently fueled by biomethane use retrofitted kits to allow petrol engines to be fueled by this renewable fuel. At the EU level, there is a lack of standards for transport biomethane; however, some national standards are beginning to emerge. In Austria, there is a standard OVGW1 G31 for gas upgrading and injection into the natural gas network, used by many as a quality assurance standard (Burton et al. 2009). It is reasonable to assume that as seaweeds begin to be harnessed for biogas production, parallel developments will stimulate the creation of a European standard to incorporate upgraded biogas in the market, guaranteeing security and quality of the product.

The use of bioethanol is much more widespread, due to the fact that it can be blended with petrol fuel. According to the EU standard EN 228, this can be done up to a 5% blend and still maintain petrol fuel quality standards. There is also a standard related to the requirements and test methods for ethanol as a blending component for petrol in automotive fuels (EN 15376:2007+A1:2009). Contrary to biogas, there is a large worldwide market for ethanol, once it is a standardized industrial chemical with many applications (Burton et al. 2009).

Considering the incorporation of biogas and bioethanol in the transport sector, the main bottleneck is the economic viability. With the current low oil price, it is difficult to make algal fuels competitive to fossil fuels, and their exploitation might depend on incentives. On the other hand, the establishment of biorefinery approaches might significantly increase the economic viability of exploring such relevant biomass resource.

9.5 CONCLUSION

Seaweeds demonstrate a high potential as bioremediation tools in WWT systems, and thus, if possible, SWWT must be fully exploited, allowing the implementation of circular economy practices. However, there is a need to guarantee the quality and safety of the seaweed biomass used in WWT. Nonetheless, seaweed biomass from eutrophicated waters is already exploited commercially, cultivated in aquaculture to remove excess nutrients resulting from cultivation of other species (such as fish and shrimp). Bioremediation of harmful compounds using seaweeds is still in its beginning; however, revised studies have shown good results. Nonetheless, the subsequent use of such biomass must ensure that negative impacts are prevented (e.g. transference of harmful compounds).

Nowadays, there is a clear gap worldwide regarding legislation for the use of seaweed biomass from WWT plants. In the future, this gap needs to be shortened, to make possible and clear how seaweeds from WWT can be explored, reducing waste generation and impacts of less environmentally friendly end-of-life solutions.

ACKNOWLEDGMENTS

This work was financially supported by Base Funding - UIDB/00511/2020 of the Laboratory for Process Engineering, Environment, Biotechnology and Energy (LEPABE) - funded by national funds through the FCT/MCTES (PIDDAC). It was also financed by national funds through FCT (Foundation for Science and Technology), I.P., within the scope of the projects UIDB/04292/2020 (MARE, Marine and Environmental Sciences Centre) and UIDP/50017/2020+UIDB/50017/2020 (CESAM, Centre for Environmental and Marine Studies). The authors also acknowledge FCT for funding Sara Pardilhó' (SFRH/BD/139513/2018) Ph.D. fellowship. João Cotas thanks the European Regional Development Fund through the Interreg Atlantic Area Program, under the project NASPA 523 (EAPA_451/2016). Ana M. M. Gonçalves acknowledges University of Coimbra for the contract IT057-18-7253.

CONFLICT OF INTEREST

The authors declare no conflict of interest.

REFERENCES

Adamse, Paulien, et al. 2017. "Cadmium, lead, mercury and arsenic in animal feed and feed materials – trend analysis of monitoring results." *Food Additives & Contaminants: Part A* 34 (8):1298–1311. doi: https://doi.org/10.1080/19440049.2017.1300686.

Agrawal, Komal. et al. 2020. Algal Biomass: Potential Renewable Feedstock for Biofuels Production–Part I. In *Biofuel Production Technologies: Critical Analysis for Sustainability*, 203–237. Singapore: Springer.

Alloway, Brian J. 2013. "Sources of Heavy Metals and Metalloids in Soils." In *Heavy Metals in Soils: Trace Metals and Metalloids in Soils and their Bioavailability*, edited by Brian J. Alloway, 11–50. Dordrecht: Springer Netherlands.

Aprianti, Tine, et al. 2017. "Studies on adsorption capacity of clay–Sargassum sp biosorbent for Cr (VI) removal in wastewater from electroplating industry." 3rd International Conference on Construction and Building Engineering (ICONBUILD), Indonesia, 14–17 August 2017.

Arumugam, Nithiya, et al. 2018. "Treatment of wastewater using seaweed: a review." *International Journal of Environmental Research and Public Health* 15 (12):2851. doi: https://doi.org/10.3390/ijerph15122851.

Bharathiraja, B., et al. 2015. "Aquatic biomass (algae) as a future feed stock for bio-refineries: A review on cultivation, processing and products." *Renewable and Sustainable Energy Reviews* 47:634–653. doi: https://doi.org/10.1016/j.rser.2015.03.047.

Bhardwaj, Nisha. et al 2020. Algal Biofuels: An Economic and Effective Alternative of Fossil Fuels. In *Microbial Strategies for Techno-economic Biofuel Production*, 59–83. Springer.

Birungi, Zainab S., et al. 2020. "Recovery of Rare Earths, Precious Metals and Bioreduction of Toxic Metals from Wastewater Using Algae." In *Emerging Eco-friendly Green Technologies for Wastewater Treatment*, edited by Ram Naresh Bharagava, 267–297. Singapore: Springer.

Burton, Tom, et al. 2009. "A review of the potential of marine algae as a source of biofuel in Ireland." *Sustainable Energy Ireland, Dublin, Ireland.* https://pure.au.dk/portal/en/publications/a-review-of-the-potential-of-marine-algae-as-a-source-of-biofuel-in-ireland(6d034080-2b72-11df-9806-000ea68e967b).html

CBI Ministry of Foreign Affairs. 2020. Entering the European market for seaweed extracts. https://www.cbi.eu/market-information/natural-food-additives/seaweed-extracts-food-0/market-entry.

Cole, Andrew J., et al. 2016. "Adding value to the treatment of municipal wastewater through the intensive production of freshwater macroalgae." *Algal Research* 20:100–109. doi: https://doi.org/10.1016/j.algal.2016.09.026.

Cole, Andrew J., et al. 2017. "Good for sewage treatment and good for agriculture: Algal based compost and biochar." *Journal of Environmental Management* 200:105–113. doi: https://doi.org/10.1016/j.jenvman.2017.05.082.

Deniz, Fatih, and Elif Tezel Ersanli. 2018. "An ecofriendly approach for bioremediation of contaminated water environment: Potential contribution of a coastal seaweed community to environmental improvement." *International Journal of Phytoremediation* 20 (3):256–263. doi: https://doi.org/10.1080/15226514.2017.1374335.

Devi, I Rajarajasri Pramila, and V.S. Gowri. 2007. "Biological treatment of aquaculture discharge waters by seaweeds." *I Control Pollution* 23 (1):135–140.

Draget, Kurt Ingar, et al. 2005. "Alginates from Algae." *Biopolymers Online*. doi: https://doi.org/10.1002/3527600035.bpol6008.

El-Naggar, Noura El-Ahmady, and Nashwa H. Rabei. 2020. "Bioprocessing optimization for efficient simultaneous removal of methylene blue and nickel by Gracilaria seaweed biomass." *Scientific Reports* 10 (1):17439. doi: https://doi.org/10.1038/s41598-020-74389-y.

European Biostimulants Industry Council. Spain - Marketing Requirements. https://biostimulants.weebly.com/spain.html#.

European Parliament. 2019. "Regulation (EU) 2019/1009 of the European Parliament and of the Council of 5 June 2019 laying down rules on the making available on the market of EU fertilising products and amending Regulations (EC) No 1069/2009 and (EC) No 1107/2009 and repealing Regulation (EC) No 2003/2003." *Official Journal of the European Union* L170:1–114.

García-Poza, Sara, et al. 2020. "The evolution road of seaweed aquaculture: Cultivation technologies and the industry 4.0." *International Journal of Environmental Research and Public Health* 17 (18):6528. doi: https://doi.org/10.3390/ijerph17186528.

Goswami, Rahul K., et al. 2020. "Microalgae-Based Biorefinery for Utilization of Carbon Dioxide for Production of Valuable Bioproducts." In *Chemo-Biological Systems for CO2 Utilization*, 203–228. Taylor & Francis.

Goswami, Rahul K., et al. 2021b. "Advanced microalgae-based renewable biohydrogen production systems: A review." *Bioresource Technology* 320 (A):124301. doi: https://doi.org/10.1016/j.biortech.2020.124301.

Goswami, Rahul K., et al. 2021a. "Microalgae-based biorefineries for sustainable resource recovery from wastewater." *Journal of Water Process Engineering* 40:101747. https://doi.org/10.1016/j.jwpe.2020.101747.

Gu, Wenhui, and Guangce Wang. 2020. "Absorptive process and biological activity of Ulva prolifera and algal bioremediation of coking effluent." *BioResources* 15 (2):2605–2620.

He, Jinsong, and J. Paul Chen. 2014. "A comprehensive review on biosorption of heavy metals by algal biomass: materials, performances, chemistry, and modeling simulation tools." *Bioresource Technology* 160:67–78. doi: https://doi.org/10.1016/j.biortech.2014.01.068.

Hentati, Faiez, et al. 2020. "Rheological investigations of water-soluble polysaccharides from the Tunisian brown seaweed Cystoseira compressa." *Food Hydrocolloids* 103:105631. doi: https://doi.org/10.1016/j.foodhyd.2019.105631.

Holdt, Susan L., and Maeve D. Edwards. 2014. "Cost-effective IMTA: a comparison of the production efficiencies of mussels and seaweed." *Journal of Applied Phycology* 26 (2):933–945. doi: https://doi.org/10.1007/s10811-014-0273-y.

Holdt, Susan Løvstad, and Stefan Kraan. 2011. "Bioactive compounds in seaweed: functional food applications and legislation." *Journal of Applied Phycology* 23 (3):543–597. doi: https://doi.org/10.1007/s10811-010-9632-5.

Imeson, Alan. 2011. *Food Stabilisers, Thickeners and Gelling Agents*. John Wiley & Sons.

Jönsson, Madeleine, et al. 2020. "Extraction and modification of macroalgal polysaccharides for current and next-generation applications." *Molecules* 25 (4):930. doi: https://doi.org/10.3390/molecules25040930.

Kılınç, Berna, et al. 2013. "Seaweeds for food and industrial applications." In: *Food Industry*. 735–748. Intech.

Kim, Jang K., et al. 2017. "Seaweed aquaculture: cultivation technologies, challenges and its ecosystem services." *ALGAE* 32 (1):1–13. doi: https://doi.org/10.4490/algae.2017.32.3.3.

Kraan, Stefan. 2012. "Algal polysaccharides, novel applications and outlook." In *Carbohydrates-Comprehensive Studies on Glycobiology and Glycotechnology*, 489–532. InTech.

Kumar, Bikash. et al 2019. "Techno-Economic Assessment of Microbe-Assisted Wastewater Treatment Strategies for Energy and Value-Added Product Recovery." In *Microbial Technology for the Welfare of Society*, 147–181. Singapore: Springer.

Le Gouvello, Raphaëla, et al. 2017. "Aquaculture and marine protected areas: Potential opportunities and synergies." *Aquatic Conservation: Marine and Freshwater Ecosystems* 27:138–150. doi: https://doi.org/10.1002/aqc.2821.

Leandro, Adriana, et al. 2020a. "Seaweed's bioactive candidate compounds to food industry and global food security." *Life* 10 (8):140. doi: https://doi.org/10.3390/life10080140.

Leandro, Adriana, et al. 2020b. "Diverse applications of marine macroalgae." *Marine Drugs* 18 (1):17. doi: https://doi.org/10.3390/md18010017.

Lodeiro, Pablo, et al. 2010. "Aluminium removal from wastewater by refused beach cast seaweed. Equilibrium and dynamic studies." *Journal of Hazardous Materials* 178 (1):861–866. doi: https://doi.org/10.1016/j.jhazmat.2010.02.017.

Luo, Hongtian, et al. 2020. "Potential bioremediation effects of seaweed Gracilaria lemaneiformis on heavy metals in coastal sediment from a typical mariculture zone." *Chemosphere* 245:125636. doi: https://doi.org/10.1016/j.chemosphere.2019.125636.

Makkar, Harinder P.S., et al. 2016. "Seaweeds for livestock diets: A review." *Animal Feed Science and Technology* 212:1–17. doi: https://doi.org/10.1016/j.anifeedsci.2015.09.018.

McHugh, Dennis J. 2003. "A guide to the seaweed industry." FAO Fisheries Technical Paper.

Mehariya, Sanjeet et al. 2021a. "Microalgae for high-value products: A way towards green nutraceutical and pharmaceutical compounds." *Chemosphere* 280:130553. doi: https://doi.org/10.1016/j.chemosphere.2021.130553.

Mehariya, Sanjeet., et al. 2021b. "Aquatic Weeds: A Potential Pollutant Removing Agent from Wastewater and Polluted Soil and Valuable Biofuel Feedstock." In *Bioremediation Using Weeds*, 59–77. Springer.

Michalak, Izabela, et al. 2013. "State of the art for the biosorption process—a review." *Applied Biochemistry and Biotechnology* 170 (6):1389–1416. doi: https://doi.org/10.1007/s12010-013-0269-0.

Milledge, John J., et al. 2019. "A brief review of anaerobic digestion of algae for bioenergy." *Energies* 12 (6):1166. doi: https://doi.org/10.3390/en12061166.

Morais, Tiago, et al. 2020. "Seaweed potential in the animal feed: A review." *Journal of Marine Science and Engineering* 8 (8):559. doi: https://doi.org/10.3390/jmse8080559.

Mortensen, Alicja, et al. 2016. "Re-evaluation of agar (E 406) as a food additive." *EFSA Journal* 14 (12):e04645. doi: https://doi.org/10.2903/j.efsa.2016.4645.

Neveux, N., et al. 2016. "The treatment of municipal wastewater by the macroalga Oedogonium sp. and its potential for the production of biocrude." *Algal Research* 13:284–292. doi: https://doi.org/10.1016/j.algal.2015.12.010.

Neveux, Nicolas, et al. 2018. "The Bioremediation Potential of Seaweeds: Recycling Nitrogen, Phosphorus, and Other Waste Products." In *Blue Biotechnology: Production and Use of Marine Molecules*, 217–241. Wiley.

Omar, Hanan, et al. 2018. "Bioremoval of toxic dye by using different marine macroalgae." *Turkish Journal of Botany* 42 (1):15–27. doi: https://doi.org/10.3906/bot-1703-4.

Ortiz-Calderon, Claudia, et al. 2017. "Metal Removal by Seaweed Biomass." In *Biomass Volume Estimation and Valorization for Energy*. InTech.

Pereira, Leonel, and João Cotas. 2020. "Introductory Chapter: Alginates-A General Overview." In *Alginates-Recent Uses of This Natural Polymer*. InTech.

Reid, Gregor K., et al. 2020. "Performance measures and models for open-water integrated multi-trophic aquaculture." *Reviews in Aquaculture* 12 (1):47–75. doi: https://doi.org/10.1111/raq.12304.

Roberts, David A., et al. 2015. "From waste water treatment to land management: Conversion of aquatic biomass to biochar for soil amelioration and the fortification of crops with essential trace elements." *Journal of Environmental Management* 157:60–68. doi: https://doi.org/10.1016/j.jenvman.2015.04.016.

Rychen, Guido, et al. 2017. "Safety and efficacy of sodium and potassium alginate for pets, other non food-producing animals and fish." *EFSA Journal* 15 (7):e04945. doi: https://doi.org/10.2903/j.efsa.2017.4945.

Sá Monteiro, M., et al. 2019. "Analysis and risk assessment of seaweed." *EFSA Journal* 17 (S2):e170915. doi: https://doi.org/10.2903/j.efsa.2019.e170915.

Saeid, A., et al. 2013. "Biomass of Spirulina maxima enriched by biosorption process as a new feed supplement for swine." *Journal of Applied Phycology* 25 (2):667–675. doi: https://doi.org/10.1007/s10811-012-9901-6.

Salerno, Michael, et al. 2009. "Biogas production from algae biomass harvested at wastewater treatment ponds." *Civil and Environmental Engineering*. doi: https://doi.org/10.13031/2013.28877.

Sari, Dhia Rahma, and Ratih Ida Adharini. 2020. "The utilization of Gracilaria verrucosa as fish processing wastewater biofilter." *E3S Web of Conferences* 147 (3):02022. doi: https://doi.org/10.1051/e3sconf/202014702022.

Sen, Bulent, et al. 2013. "Relationship of Algae to Water Pollution and Waste Water Treatment." In *Water Treatment*, 335–354. InTech.

Senthilkumar, R., et al. 2019. "Green alga-mediated treatment process for removal of zinc from synthetic solution and industrial effluent." *Environmental Technology* 40 (10):1262–1270. doi: https://doi.org/10.1080/09593330.2017.1420696.

Sode, Sidsel, et al. 2013. "Bioremediation of reject water from anaerobically digested waste water sludge with macroalgae (Ulva lactuca, Chlorophyta)." *Bioresource Technology* 146:426–435. doi: https://doi.org/10.1016/j.biortech.2013.06.062.

Torres, M.D., et al. 2019. "Seaweed biorefinery." *Reviews in Environmental Science and Bio/Technology* 18 (2):335–388. doi: https://doi.org/10.1007/s11157-019-09496-y.

Trianti, Cindy Martiana, and Ratih Ida Adharini. 2020. "The Utilization of Gracilaria verrucosa as shrimp ponds wastewater biofilter." 3rd International Symposium on Marine and Fisheries Research, Indonesia, 8–9 July 2019. EDP Sciences.

Venkatesan, Jayachandra, et al. 2017. "Chapter 1 - Introduction to Seaweed Polysaccharides." In *Seaweed Polysaccharides*, edited by Jayachandran Venkatesan, Sukumaran Anil and Se-Kwon Kim, 1–9. Elsevier.

Verstraete, Frans. 2013. "Risk management of undesirable substances in feed following updated risk assessments." *Toxicology and Applied Pharmacology* 270 (3):230–247. doi: https://doi.org/10.1016/j.taap.2010.09.015.

Wang, Shuang, et al. 2020. "A state-of-the-art review on dual purpose seaweeds utilization for wastewater treatment and crude bio-oil production." *Energy Conversion and Management* 222:113253. doi: https://doi.org/10.1016/j.enconman.2020.113253.

Wei, Zhangliang, et al. 2017. "Bioremediation using Gracilaria lemaneiformis to manage the nitrogen and phosphorous balance in an integrated multi-trophic aquaculture system in Yantian Bay, China." *Marine Pollution Bulletin* 121 (1–2):313–319. doi: https://doi.org/10.1016/j.marpolbul.2017.04.034.

Younes, Maged, et al. 2018. "Re-evaluation of carrageenan (E 407) and processed Eucheuma seaweed (E 407a) as food additives." *EFSA Journal* 16 (4):e05238. doi: https://doi.org/10.2903/j.efsa.2018.5238.

10 Recent Insights of Algal-Based Bioremediation and Energy Production for Environmental Sustainability

Sunil Kumar, Nitika Bhardwaj, S. K. Mandotra, A. S. Ahluwalia

CONTENTS

DOI: 10.1201/9781003155713-10

10.1 INTRODUCTION

The exponential growth of human population and their incessant demands have caused serious threats to the environment. Freshwater bodies are being continuously contaminated by industrial effluents containing various harmful and toxic chemicals (Agrawal et al., 2020a). Pollution and contamination are causing serious threats to various aquatic and terrestrial habitats; therefore, sustainable efforts are needed to overcome such an alarming situation (Kumar et al., 2018). The world is suffering from various problems like poverty, hunger, pandemics, unsustainable energy sources, environmental problems like water pollution, and global warming, and all these are of serious concern and affecting life significantly (Agrawal and Verma 2020; Agrawal et al., 2020b). Due to the increasing population and technological advancements, life on Earth is facing the serious threat of water pollution. Polluted water comes into the aquatic bodies from municipal, domestic, agricultural, and various industrial processes (Bhatt et al., 2014; Kumar et al., 2019a). The report by India's Central Pollution Control Board (CPCB, 2011), New Delhi, indicates that approximately 40 billion liters of wastewater is generated every day, and urban areas are the main contributors to such wastewater generation. Only 20–30% of such polluted wastewater is subjected to treatment. All over the globe, mainly in developing nations, wastewater from different sources is released into aquatic bodies without treatment or with inefficient treatment (Kivaisi, 2001). Due to the nutrient load of the wastewater, such bodies are facing the problem of eutrophication and the formation of nuisance toxic algal blooms (Smith et al., 1999).

Pollutants normally enter our environment because of the release of industrial waste, improper agricultural practices, and waste disposal. The effluent discharge from municipalities and industries constitutes a major source of water pollution (Agrawal and Verma 2021a,b). Increasing concentration and long-term presence of these pollutants in the environment cause a chronic threat to environmental sustainability, human health, and the safety of even wildlife. The contamination of water is continuously rising in the absence of seriousness on the part of stakeholders. Various anthropogenic processes and the release of polluted effluents have been resulting in the continuous increase in the concentration of a variety of toxic substances in the water bodies (Singh and Borthakur, 2018).

Several approaches have emerged for the elimination of such polluted substances like ion exchange, evaporation, precipitation, etc. The development of biological treatment systems is considered eco-friendly and economically fruitful (Valderrama et al., 2002; Kumar and Verma 2021). To prevent water-borne diseases and improving the quality of life, the availability of good-quality water is essential (Oluduro and Adewoye, 2007). Algae are one of the important biological agents that are being used by several researchers at laboratory or small pilot-scale experiments for wastewater treatment (Goswami et al., 2021a; Mehariya et al., 2021a,b). In spite of its vast diversity, till now, algal biomass has not been commercially exploited for wastewater treatment in a big way, although many studies have been undertaken (Volesky, 1990; Wase and Forster, 1997, Renuka et al., 2015). The potential of algae in wastewater remediation, however, is much wider in scope than its current role. This chapter highlights the role and effectiveness of algae-based technologies for cleaning water bodies and using the biomass produced in the process for biofuel energy.

10.2 WHAT ARE POLLUTANTS?

Pollutants are substances that enter the environment through anthropogenic activities. The number of toxic chemicals that are produced and used by humans is enormous. Pollutants are defined as contaminants that enter the natural environment by natural/anthropogenic activities, beyond permissible values. They cause detrimental consequences to the inhabitants in an evident way (Panigrahi et al., 2019). The use of different microorganisms plays a crucial role in the remediation of environmental pollutants in a sustainable way; this approach is cost-effective in nature and helps in reinstating a natural sustainable environment for a long time (Dixit et al., 2015).

10.2.1 Dyes and Heavy Metals

A colored material is made up of three integrated systems, i.e., chromophore, auxochrome, and a conjugated double bond. A compound is considered to be a dye that absorbs wavelengths of light in a visible light spectrum ranging from 400 to 700 nm (Christie, 2014). Dyes are obtained from both natural and synthetic sources. One of the major advantages of using natural dyes is that these are free from the harmful effect on life. The primary drawbacks of natural colors include the requirement for several stages in the coloring procedure, the assorted variety of resources, and associated applications of coloring methodology. The quick change in patterns and the interest for better abilities on various substrates would require a total database portraying conceivable application (Bechtold et al., 2003). Synthetic dyes generally have composite coal-tar-oriented components like naphthalene, benzene, anthracene, xylene, and toluene (Nigam et al., 2000). A brief introduction of synthetic dyes has been given in Figure 10.1. The utilization of synthetic dyes of different industrial processes has been continuously increasing every day due to their modest cost of production, durability, and variety of color in comparison to dyes of natural origin. The complex molecular and highly conjugated structures of synthetic dyes make these hard for biological degradation and hence the pollution problem (Ansari and Mosayebzadeh, 2010; Singh et al., 2011; Chaturvedi et al., 2013a; Agrawal and Verma 2019).

Everyone knows that the existence or release of these dyes in freshwater bodies is undesirable. However, the release of colored effluents from textile companies into

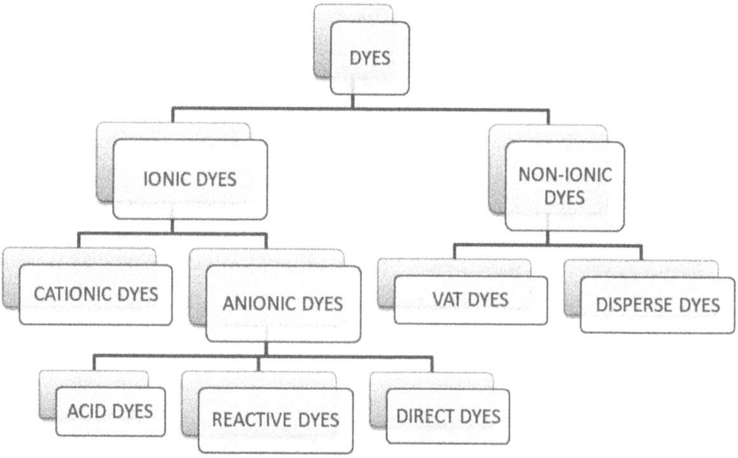

FIGURE 10.1 Classification of dyes (adopted from Tan et al., 2015).

freshwater bodies has been continuing, adversely affecting the photosynthetic activity of phytoplanktons by blocking the penetration of light into aquatic bodies (Kumar et al., 2019b).

Pollution caused by heavy metals, due to its toxic effects, poses a huge risk to human life as well as other living organisms in the environment. These heavy metals are extremely reactive even at very low concentrations and can pile up in the food web, resulting in adverse health concerns (Glombitza and Reichel, 2014). Heavy metals are usually released into the environment via different industrial processes. Details of industries responsible for heavy metal pollution have been compiled in Table 10.1. Biomineralization, biosorption, biotransformation, and bioaccumulation

TABLE 10.1
Heavy Metals Found in Major Industries (Bond and Straub, 1974)

Industries	Cd	Hg	Al	As	Cr	Zn	Cu	Pb	Ni
Chlorine and alkalis	✓	✓	–	✓	✓	✓	–	✓	
Organo chemicals	✓	✓	✓	✓	✓	✓	–	✓	
Agro fertilizers	✓	✓	✓	✓	✓	✓	✓	✓	✓
Paper and pulp	–	✓	–	–	✓	✓	✓	✓	✓
Fossil fuel refining	✓	✓	✓	✓	✓	✓	✓	✓	✓
Cementing and flat glass	–	–	–	–	✓	–	–	–	–
Iron alloy (steel)	✓	✓	–	✓	✓	✓	✓	✓	✓
Hydropower					✓				
Aeronautical (plating and finishing)	✓	✓	✓	–	✓	–	✓	–	✓
Tanning					✓				
Textile	–		–	–	✓				

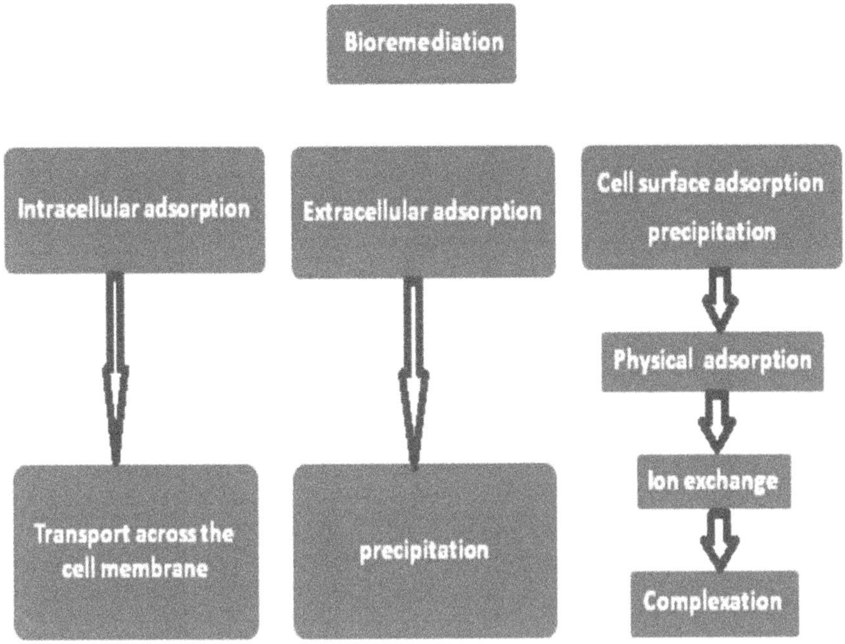

FIGURE 10.2 Possible mechanisms involved in heavy metal bioremediation by microbes.

are the biological phenomena used by a number of microorganisms for their continued existence in an environment polluted by heavy metals. The organisms selected for heavy metal removal are either the inhabitants of the polluted environment or are isolated from another environment and brought to the contaminated site (Sharma et al., 2000). The mechanisms of metal uptake by various biosorbents are based on the cellular surface of the microorganisms, along with the exchange of metal ions and complex formation with the metal ions present on the active sites of the cell surface. The detailed mechanism of heavy metal removal is shown in Figure 10.2.

All the classes of algae have cellulose in their cell wall, whereas red and brown algae have the presence of sulfonated polysaccharides. In addition to these, other binding sites are polysaccharides like xylans, mannan, alginic acid, proteins, and glycan (Lesmana et al., 2009).

10.2.2 Water Pollution

A major portion of water pollution is contributed by the effluents from industrial and residential discharge. Most of the industrial wastes are released into the nearby water bodies/rivers, contributing to toxicity and destabilization of aquatic ecosystems due to increased nutrient loading and higher oxygen demand of such water bodies (Morrison et al., 2001; Kumari et al., 2006; Upadhyay et al., 2021). The presence of good-quality water is essentially required for checking various diseases and for improving quality of life (Oluduro and Adewoye, 2007). Types of impurities and their amount present in the natural water varies as per the source of the water; for instance, various natural (soil leaching, weathering of rocks, and dissolution of atmospheric aerosol particles) and anthropogenic activities (leather processing, use of metals, mining, etc.) lead to the introduction of heavy metals into the aquatic system (Adeyeye, 1994; Asaolu et al., 1997; Ipinmoroti, 1993). Fertilizer industry is one of the major culprits in deteriorating water quality; excess fertilizers used in agricultural fields are being washed and discharged into the surrounding freshwater bodies, causing ecological imbalance (Singh et al., 2006).

10.3 BIOREMEDIATION

Bioremediation is the process of treating contaminated water and materials by the use of particular microorganisms by altering their growth conditions; these conditions of the microorganism are altered according to the target pollutants. In other words, it is the process of biodegradation of organic compounds by living organisms. During this process, organic materials could be broken down in the presence of oxygen (aerobically) or the absence of oxygen (anaerobically). Biomineralization is another biological phenomenon related to biodegradation through which organic compounds are converted to minerals like CH_4, H_2O, and CO_2. Biological entities via the process of bioremediation either degrade or transform the more toxic form of compounds to their less toxic forms. Currently, it is one of the most efficient and cost-effective processes to check environmental pollution. At present, the disposal of sewage is becoming a challenging problem in urban colonies (Moore, 1998), and bioremediation is emerging as an effective and sustainable tool in large-scale setups. Most of the time, metals are accumulated inside the cell through its metabolic activity; the process of bioaccumulation helps in reducing harmful metals from a particular area (Volesky, 1990; Wase and Forster, 1997). In recent times, researchers are focusing on various algal organisms as a potential bioremediation agent which helps in the efficient, sustainable, and cost-effective removal of heavy metals from polluted freshwater bodies.

10.3.1 Factors Affecting Bioremediation

Organism(s) or biomass involved in bioremediation are key factors in the bioremediation process. The efficiency of bioremediation is defined by the intensity and combination of different factors, i.e., environmental, contaminants, and microbes. Figure 10.3. comprises the different factors involved and affecting the mechanism of bioremediation.

10.3.2 Algal Status in Bioremediation

During the past few years, algae have come up with the enormous potential to bioremediate metals and dyes; as a result, they are considered an effective means of treating wastewater at a larger scale (Mehta and Gaur, 2005; Cepoi et al., 2020; Kumar et al., 2019b). Heavy metal removal by different species of algae such as *Cladophora glomerata*, *Oedogonium rivulare* (Dwivedi, 2012), *Oscillatoria tenuis* (Ajavan et al., 2011), *Spirogyra hatillensis* (Dwivedi, 2012), *Oscillatoria quadripunctulata* (Rana et al., 2013), *Chlorella vulgaris* (Aung et al., 2012), *Spirogyra*

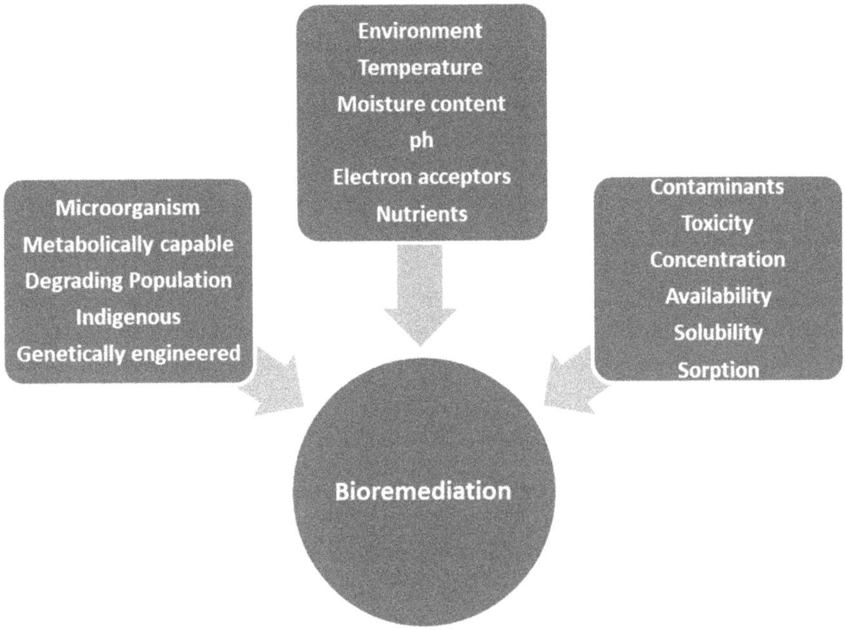

FIGURE 10.3 Factors affecting bioremediation.

hyaline (Kumar and Oommen, 2012) and *Spirulina maxima, Chlorella vulgaris* (Chan et al., 2014) and algal consortia (Renuka et al., 2016) is well known. A number of cyanobacterial species have also been reported for the treatment of municipal wastewater; algal species such as *Nostoc commune* and *Oscillatoria limosa* have efficient removal capabilities of Cl_2, NO_3, SO_4, and PO_4 (Azarpira et al., 2014). Dubey, et al., (2011) reported the biosorption and biodegradation potential of five cyanobacterial species (*Cyanothece* sp., *Oscillatoria* sp., *Nostoc* sp., *Nodularia* sp., and *Synechococcus* sp.). In another 15-day study conducted by Pathak et al. (2014), the green microalga *Chlorella pyrenoidosa* was found efficient for the phycoremediation of textile waste. Similarly, *Kappaphycus alvarezii, Ulva lactuca*, species of *Scenedesmus, Chlorella, Cyanothece, Phormidium, Nostoc, Botryococcus, Nodularia, Chlamydomonas, Arthrospira, Spirulina, Oscillatoria*, etc. are few such genera of algae that have been used in wastewater remediation. Many cyanobacterial species inhabitant of soil and water ecosystem are also known for their biosorption capabilities (Zinicovscaia and Cepoi, 2016).

Being one of the most primitive photosynthetic prokaryotes, cyanobacteria show a unique assemblage of organisms that occupy a vast array of habitats. They are susceptible to sudden chemical and physical alterations in nutrient composition, light, temperature, and salinity (Dubey, et al., 2011); hence they have a vast potential in industrial and wastewater treatment, bioremediation of heavy metals, as biofertilizers, source of various chemicals and also in feed, food, and fuel industry (Cairns and Dickson, 1971). A number of cyanobacterial species (*Synechococcus* sp., *Oscillatoria salina, Aphanocapsa* sp., and *Plectonema terebrans*) have been used as mats in the aquatic environment and have shown potential bioremediation capabilities of oil spills (Raghukumar et al., 2001; Cohen, 2002).

Till now, in most of the studies, different bacterial strains have been used in biological treatment strategies. However, in recent years, single species and consortia of microalgae have shown wide applicability in absorbing excess phosphorus and nitrogen. Besides this, they are also used to lower the chemical oxygen demand of wastewater and removal of heavy metals such as lead (Aksu and Kutsal, 1991). *Sargassum* sp. and *Ascophyllum* sp. are marine algae that curtail pollutants via the process of biosorption (Fourest and Volesky, 1996; Yu et al., 1999). Immobilization of algae using bioreactors is a recent development in the field of bioremediation. A study performed with microalga *Chlorella vulgaris* immobilized in alginate beads showed significant removal of phosphate and ammonia from wastewater (Tam et al., 1994).

10.3.3 Why Algae?

Microalgae are exceptional in bioremediation process as they offer several benefits over other biological materials. In comparison to other organisms/methods, the use of microalgae offers a significant improvement in remediation of heavy metals across the world (Wilde and Benemann, 1993). Microalgal use to get rid of the problem of colored industrial waste effluents has offered huge potential due to their capacity to fix carbon dioxide from surroundings. In addition, the algal biomass also has great capability in the production of biofuels (Huang et al., 2010; Bhardwaj et al., 2020; Mandotra et al., 2021). The phytoremediation capabilities of several microalgal species make them useful for the sustainability of the environment (Lim et al., 2010; Ellis et al., 2012; Renuka et al., 2015; Mandotra et al., 2020).

Integrated algal biomass production with industrial waste leads to the economical and environmental friendly bioremediation process; algal biomass thus recovered can be further utilized for various products, including algal biofuels (Munoz and Guieysse, 2006; Chaturvedi et al., 2013b; Lizzul et al., 2014; Mandotra et al., 2020; Agrawal et al., 2020c; Kumar and Verma 2021; Sharma et al., 2021). Owing to the presence of elevated concentrations of N and P in wastewater, it becomes a favorable feedstock for algal growth. These conditions are in use by many conventional water treatment processes using mix algal consortia (Zhu and Ketola, 2012; Renuka et al., 2015). Algae can also be used in nutraceutical, pharmaceutical industry, cosmetics, fertilizers, and as feed additives (Kumar et al., 2019c; Mehariya et al., 2021c). A few species of marine algae such as *Dunaliella tertiolecta*, *Emiliania huxleyi*, and *Haematococcus pluvialis* have been used for the production of various value-added products and pigments like astaxanthin, b-carotene, omega-3 fatty acids, Vitamin-E, and pigments (Pangestuti and Kim, 2011; Saini et al., 2021; Goswami et al., 2021b).

Selection of microalgae for remediation process should be done on the basis of the following characteristics (Singh et al., 2016):

- Adaptability to wastewater as a nutrient medium
- High nutrient removal ability
- Adaptability to industrial flue gases as carbon dioxide source
- High growth rate and homogenous suspension

- High lipid content and lipid productivity
- Capability to grow in outdoor cultivation

10.3.4 Algal Interaction with Wastewater

Algal interaction and behavior for remediation process differ with algal species and type of the environment in which the alga and the target pollutant exists (Table 10.2). Chemical composition and binding groups present on the algal walls play a key role in the biosorption/bioremediation process.

10.4 LARGE-SCALE PRODUCTION

Large-scale production of algae is carried out by growing microalgae in large water bodies like oceans, ponds, stagnant water bodies, etc. For the bulk cultivation of green cells, this method is preferred. It includes:

10.4.1 Raceway Ponds

These are based on the gyration of microalgae in a racetrack loop system. They are used for the cultivation of algae and then the extraction of biomass. They are further divided into open ponds and covered ponds.

10.4.1.1 Open Ponds

Open raceway ponds are made up of concrete or compressed earth. As microalgal growth requires sunlight for photosynthesis, therefore, the depth of the pond is generally kept shallow (0.3 m approximately). This helps in the proper distribution of light inside the pond which further helps in exposing cells to light and enhances the growth of algal cells (Suparmaniam et al., 2019). The raceway pond system has a special feature of the paddle-wheel, which has a great role in blending and gyration of microalgae during the cultivation stage. This function of the paddle-wheel prevents algae from settling down in the pond bottom and provides a continuous motion for the raising algal cells. A feeding port is located near the paddle-wheel. As the name suggests, the feeding port is used for the introduction of nutrients and suitable culture media to the growing green cells in the pond. Once the circulation of microalgae in the raceway loop is complete, the algae-containing water is separated through the harvesting port. The raceway pond system has various challenges. A few of these are:

- Open raceway ponds are more susceptible to contamination as the risk of predators is quite high in an open atmosphere.
- This is an open type of system; therefore, the system is at the stake of environmental risks such as rainfall, pollution, etc.
- Evaporation from open raceway pond is more and has a direct impact on the CO_2 usage of growing microalgae.
- Other factors like temperature, pH, light intensity, large area for cultivation, etc., are also some such factors that impede the growth of algae in an open water system.

TABLE 10.2
Algal interaction with Different Pollutants Present in Wastewater Outcomes

Target Pollutant	Test Algae	Output	References
Copper (Cu) and Iron (Fe)	*Anabaena doliolum*	Immobilized algal cells are efficiently removed metals from wastewater in comparison to free algal cells	Rai and Mallick, 1992
Ammonia (NH_3) and Phosphorous	*Chlorella vulgaris* and *Scenedesmus dimorphus*	Both algae efficiently remove target pollutants which further depends upon the design of bioreactor	Gonzalez et al., 1997
Lead (Pb)	*Spirulina* sp.	Test alga adsorbed up to 74% metal within 12 min of reaction time and showed adsorption capacity of 0.62 mg Pb per 10^5 algal cells	Chen and Pan, 2005
Mercury (Hg), Cadmium (Cd) and Lead (Pb)	*Dunaliella* sp.	Test alga shows the great ability for the adsorption of heavy metals. Metal sorption found independent of reaction time	Imani et al., 2011
Biological oxygen demand (BOD) and chemical oxygen demand (COD)	*Oscillatoria* sp.	Results from the cyanobacterial treatment of textile effluent show 57.6% and 39.82% decrease in COD and BOD	Abraham and Nanda, 2010
Phosphate, Nitrate, COD and BOD	*Chlorella vulgaris* and *Scenedesmus quadricauda*	*C. vulgaris* shows the removal efficiencies for COD, BOD, Nitrate, and Phosphate from polluted water as 80.64%, 70.91%, 78.08%, and 62.73% while *S. quadricauda shows* 70.96%, 89.21%, 70.32%, and 81.34% removal with 15 days of reaction time	Kshirsagar, 2013
Phosphorous and Nitrate	*Chlorella minutissima*	Test alga shows very high removal efficiencies for phosphorous which is reduced from 4.47ppm to 1.15ppm while reduction of Nitrate is from 3.6ppm to 0.3ppm.	Sharma and Khan, 2013
Tectilon yellow 2G	*Chlorella vulgaris*	88% adsorption	Acuner and Dilek, 2004
Remazol black B	*Chlorella vulgaris*	Up to 419.5 mg dye adsorbed by one gram algal biomass	Aksu and Tezer, 2005
Malachite green	*Cosmarium* sp.	91% adsorption	Khataee et al., 2010
Indigo blue	*Scenedesmus quadricauda*	100% adsorption	Chia et al., 2014
Congo red	*Haematococcus* sp.	98% adsorption	Mahalakshmi et al., 2015
Tracid red BS	*Pseudoanabaena* sp.	78.44% removal after 7 days of reaction time	El-Sheekh, et al., 2017
Methylene blue	*Microspora* sp.	98%	Maurya et al., 2014
Orange-g	*Acutodesmus obliquus*	56.49 mg dye adsorbed per gram algal biomass	Sarwa and Verma, 2013
Fossil fuel	*Scenedesmus obliquus* GH2	Crude oil degradation	Tang et al., 2010

All the above-mentioned disadvantages have a direct impact on the yield of microalgae. Therefore, generally, the biomass obtained from the open raceway is quite low compared to a bioreactor. To overcome the above-mentioned disadvantages, covered ponds were developed.

10.4.1.2 Covered Ponds

These ponds have a greenhouse covering them. The sheet of the greenhouse is made up of plexiglass and it acts as a shield and secures the growing green cells from any sort of contamination, both in terms of predators and some environmental factors. But with the advancement in research and technologies, a special type of algal cultivation vessel called *photobioreactor* was introduced for overcoming these shortcomings.

10.4.2 Enclosed Photobioreactor

A photobioreactor is a closed type of device used for growing microalgae. Satpati and Pal (2018) stated that only partial light enters the photobioreactor, which helps in the growth of microalgae. The designing of a photobioreactor helps in maintaining sterile conditions inside and thus avoids contamination of cultures and contributes to higher biomass yield (Mandotra et al., 2014). Keeping in view the manufacturing and functioning, there are several types of photobioreactors.

10.4.2.1 Tubular Photobioreactor

These are the most common type of photobioreactors and are made up of glass or plastic tubes. The arrangement can be horizontal, vertical, or fence like in spiral or helix design (Tan et al., 2018). It is an enclosed device due to which the chance of any contamination is negligible. The transparency of glass helps sunlight penetrate the reactor. This penetrating light helps in the growth of microalgae and thus has a positive impact on biomass yield. The tubular photobioreactor has a pump or airlift for the interchange of gases and vital nutrients across the culture media. It is comprised of a sparger that is responsible for mixing and mass transport of carbon dioxide. As a result of photosynthesis, as the amount of oxygen increases, the airlift pumps help in the elimination of excess oxygen (Faried et al., 2017).

10.4.2.2 Flat Plate Photobioreactor

It is cuboidal in shape and is composed of thin glass, polycarbonate, or plexiglass plate. Nowadays, it is in demand due to its thin structure allowing immense light to pass through it, leading to an increase in cell density inside this reactor (Faried et al., 2017). This type of bioreactor is commonly used in laboratories, but many scientists like Richmond et al. (2003) and Feng et al. (2011) have reported its use in the outdoor atmosphere too. According to Molina et al. (2001), 1.2–12.3 cm of the light path will generate 20 gL^{-1} of algal biomass and 0.25–3.64 $gL^{-1}d^{-1}$ of biomass yield if cultivated outdoor in semi-continuous conditions. Due to this, a flat plate bioreactor is highly commendable for the mass growth of microalgae. The high surface area to volume ratio, together with cell densities, contributes to pure algal cultivation (Faried et al., 2017). On the basis of exposure to light, the bioreactor is organized into two types:

indoor and outdoor (Ting et al., 2017). In an indoor bioreactor, artificial light is employed, whereas in outdoor, natural sunlight with a nominal path is used. The mixing process is carried out mechanically by a motor or through air bubbling with the help of a perforation tube. This bioreactor is better than the tubular bioreactor as it allows less aggregation of dissolved oxygen as well as shows higher photosynthesis efficiency. The study carried out by Kiran et al. (2014) reviewed some drawbacks of this photobioreactor like reduced control of culture temperature, growth of algae on the walls of the bioreactor, and the need for many compartments for improved algal cultivation. Lee (2001) suggested that steady sterilization by heat was difficult to perform in this photobioreactor, because of the large surface area to volume ratio.

In terms of the commercial purpose of biofuel yield, the real productivity of photobioreactor is much higher than open ponds. The tendency of seizing solar energy in a photobioreactor is also high; these require a minimal amount of energy in mixing gases and liquid. Jorquera et al. (2010) indicated that 55 W/m^3 and 2000–3000 W/m^3 power density is needed for the flat plate and tubular photobioreactor, respectively, whereas in the case of open ponds, only 4 W/m^3 power density is utilized, which is due to lower synthesis cost and requirement of low energy for mixing purpose.

10.4.2.3 Bio-Film Photobioreactor

These photobioreactors came into origin to overcome the segregation of biomass from the treated wastewater. This problem was one of the major obstacles in the treatment of wastewater with the help of microalgae (Ting et al., 2017). Bio-film photobioreactors are designed in such a manner that the microalgae bind with the supporting material and wastewater is dispensed through a bio-film. This leads to lowering the nutrient requirements of the microalgae. Commonly, bio-films have an association of algal and bacterial colonies, during the time when their activity is steady; there is a balance between CO_2 and O_2. This leads to the removal of extra CO_2 (Li et al., 2017). In this photobioreactor, the important factor is to choose the proper substrate material for the algal attachment. Ting et al. (2017) reviewed the texture of substrate materials; the rougher the surface, more will be the chances of microalgal attachment. Moreover, other parameters like type of microalgal strain and quality of wastewater used are also important to select along with substrate material.

10.5 HARVESTING

Once the microalgae achieve their optimum growth, harvesting is done to recover the biomass. It is a series of processes to separate water from microalgae to enhance biomass yield. The main feature is to choose a suitable harvesting method. While performing harvesting of algal biomass, the input cost and energy required must be kept minimal to make it economical at a large scale. The biomass harvesting process is significantly influenced by the specific strain/species of the algal cell, and depending upon the strain, some algae are easy to harvest and vice versa. The main objective of this process is to generate 2–7% dry cell weight of algae through various harvesting techniques. The method of harvesting is chosen based on the type of microalgae, its size, density, etc. (Brennan and Owende., 2010). Harvesting is categorized into two parts, *viz.* bulk harvesting followed by thickening. In bulk harvesting, the suspension of biomass is allowed to settle down. During the thickening process, separation of biomass is done, which then leads to dense slurry matter with the help of filtration and centrifugation.

10.5.1 Centrifugation

This is a widely accepted method with a 95% recovery rate of microalgae (Tan et al., 2018). In this process, centrifugal force is responsible for segregating microalgae from the liquid. The recovery rate of biomass through centrifugation depends on various factors such as:

- Sedimentation rate of biomass
- Size of algal biomass cells
- Duration of centrifugal force
- Biomass sedimentation distance

This method is highly productive but not suitable for commercial scale, due to its cost (Sim et al., 1988). It is mentioned that centrifugation can be used for all types of algal strains, and the device is easy to clean with minimum chances of contamination (Tan et al., 2018). There are many types of centrifugation systems such as hydro cyclone, solid bowl decanter, nozzle type, and solid ejecting disc, etc.

10.5.2 Filtration

Brennan and Owende (2010) defined filtration as a process of filtering solid matter from liquid or gas by placing a fine medium between them through which only liquid/gas can pass. The filters used in this process have specific pore sizes named micros trainers which help in the filtering of desired products. They work in a circular manner with periodic backwash (Grima et al., 2003). Generally, filamentous algae are suitable for undergoing filtration but unicellular/small colonial microalgal species like *Chlorella, Scenedesmus,* etc. are not. Filtration is of many types, like vacuum filtration, ultrafiltration, dead-end filtration, pressure filtration and tangential flow filtration (TFF), etc.

The consumption of energy is low among TFF and pressure types. Depending upon the pore size of filters, they are divided into two parts (Azhar et al., 2019), *viz.* microfiltration and ultrafiltration. In microfiltration, the pore size of the membrane filters is in the range of 100 nm to 10,000 nm, whereas in ultrafiltration, the pore size of the membrane is between 1 nm to 100 nm. One of the main disadvantages associated with this technique is that the process requires regular changes of filters and membranes, making it cost-intensive (Mohn, 1980).

10.5.3 Flocculation

Flocculation is defined as the clumping of algal cells with each other with the help of flocculants. These flocculants are polymers that help in increasing the setting rate of algal cells. During the aggregation of algal cells, two types of forces play a significant role. In the case of long distance, the force of electrostatic repulsion is there, and in the shorter distance, Vander wall forces are responsible for algal flocculation. Uduman et al. (2010) described an auto flocculation method that flocculates cells utilizing pH. The microalgal cells carry a negative charge, as their neutralization cationic polymers are used for their aggregation. Although the flocculation process is suitable for harvesting any type of algal strain, however, at a large scale, this process is not suitable due to its cost-intensive nature and toxicity associated with flocculants (Oh et al., 2001).

10.5.3.1 Chemical Flocculation

Chemical flocculants are categorized into inorganic and organic. Inorganic flocculants are multivalent salts, whereas organic flocculants are polyelectrolytes or polymers. In spite of their wide applicability for almost all types of algal species with low operating cost, the use of these flocculants is considered dangerous to the environment.

10.5.3.2 Auto-Flocculation

This process is carried out consequently at basic pH levels. Carbon dioxide in the form of carbonate precipitates the algal cells, resulting in auto-flocculation. Sodium hydroxide facilitates this process by maintaining the overall medium basic. Normally, carbon dioxide-limited algal medium, placed under sufficient sunlight, auto-flocculates the algal cells. During this process, large algal colonies in the range of 50–200 μm are formed followed by settling under the influence of gravitational force. It is one of the simplest and cost-effective methods for biomass recovery without the addition of any harmful chemical and energy input (Pragya et al., 2013).

10.5.3.3 Bio-Flocculation

Bio-flocculation is an efficient and recent emerging technique of flocculating algal biomass. The process uses microorganisms (fungi, bacteria, algae) and their combined use for the sedimentation of biomass. Extracellular polymeric substance (EPS) is a biopolymer as the main component causing bio-flocculation. EPS are basically bio-molecules such as polysaccharides, sugars, and their derivatives (uronic acid, cellulose, xylose, mannose, glucose, pectins, etc.). Guar gum, one of the plant-based bio-flocculants, has shown a flocculation efficiency of 94% in *Chlamydomonas* sp. (Banerjee et al., 2014). Various bio-flocculants such as inulin, *Strychnos potatorum* (seed powder), and eggshell powder are also used efficiently for algal biomass harvesting (Vasistha et al., 2020).

10.6 LIPID EXTRACTION FROM ALGAE

Microalgae are considered an important source of biofuels. Various types of biofuels such as bio-ethanol, biodiesel, biohydrogen, bio-butanol, bio-methane jet fuels, etc. are produced from microalgae (Brennan and Owende, 2010; Dasgupta et al., 2015; Mandotra et al., 2016; Goswami et al., 2021c). Experiments are being conducted for increasing the yield of lipid content from microalgae. The process of lipid production from microalgae is the most important step for biofuel formation. Some of the methods for the extraction of lipids from algal cells are known (Mandotra et al., 2019).

10.6.1 Extraction by Chemicals and Solvent Cells

10.6.1.1 Folch Method

Axelsson and Gentili (2014) proposed a method for lipid extraction from the algal cells using various types of organic solvents or a mixture of solvents (Kumar et al., 2016; Upadhyay et al., 2016; Mandotra et al., 2019). The most widely used solvent mixture is of chloroform:methanol in the ratio of 2:1 to a final volume that is twenty times (5 g of sample in 100 ml of solvent mixture) the volume of algae cells for the extraction of lipids (Folch et al., 1957). Once the sample and solvent mixture is assorted with each other, a careful agitation is given to the mixture for 20–25 minutes. The mixture is then centrifuged at 2000 rpm and the liquid phase is recovered finally with the help of filter paper or funnel (filtration) (Folch et al., 1957). The main benefit of this method is that it is easy and quick for operating the bulk as well as small amount of algal samples.

10.6.1.2 Bligh and Dryer Method

This method is closely related to the Folch method and varies mainly in the ratio of solvent to the tissues (Bligh and Dryer, 1959). Both the steps of lipid extraction and then lipid partitioning is accomplished at once. The choice of solvents can be made on the basis of the polarity of lipids. Jensen (2008) demonstrated that on using the salt solution, the yield of lipids increases with less separation time. Till now, many methods for lipid extraction have been given by various researchers, but the overall outcome (in terms of lipid content) showed by these two methods is significantly higher than the other methods. Some studies using a high sample/solvent ratio showed no significant improvement in the yield of lipid content. If the content of lipids is high in the microalgal cells, only then there is a need of using the high sample to solvent ratio during the extraction process (Iverson et al., 2001).

10.6.2 Extraction by Mechanical Process

10.6.2.1 Expeller Press

This is the simplest and anciently adopted method for extraction of oil from the biomass. This method is based on the principle of breaking the algal cells and then the oil is extracted from the biomass whether dried or wet by means of high pressure. The pressure should be gentle as improper handling may lead to less quantitative and qualitative lipid amounts (Hara and Radin, 1978). The expeller press is a screw-type apparatus to squeeze the dried biomass. This dry biomass is allowed to enter from one side and its product is collected from the other side of the press. For the squeezing of dried biomass, constant pressure and friction are applied. Once the high pressure is used, the oil will flow from the minute openings. These minute openings prevent other undesired components to pass through. The temperature reaches 60–100°C as the high pressure is applied (Hara and Radin, 1978).

10.6.2.2 Bead Beating

This method is based on a high-speed mechanism using agitating beads and vibrating containers (Lee et al., 2012). Vibrations from the vibrating container lead to the breaking up of algal cells using beads to agitate the whole algal culture (Kumar et al., 2015). During this procedure, a large amount of heat is produced that further heats the sample. Therefore, a cooling jacket is usually used to protect the cells from degradation by high temperatures.

10.6.2.3 Ultra-Sonication Extraction

This method is eco-friendly in every aspect, easy to conduct, timesaving, and with high reproducibility as compared to other methods (Kumar et al., 2015). In ultrasonication, sound waves are used to travel into the liquid to produce pressure. During this process, bubbles are produced to achieve a particular size and rupture abruptly. This phenomenon is termed cavitation (Suslick and Flannigan, 2008). At the onset of the cavitation process, ultrasound is propagated into the medium, which finally allows the cell to break. This method needs no handling with chemicals, no temperature maintainers, and no heating of medium. Thus it is easy to work with for the extraction of lipids.

10.6.3 Enzyme-Assisted Extraction

The use of enzymes, like cellulose and trypsin, to the algal biomass was reviewed by Kumar et al. (2015). The cell wall of microalgae is broken by enzymolysis, i.e., by using enzymes. This method requires low temperature which is expensive to maintain. Species of microalgae such as *Scenedesmus* have been used for extracting lipids by using this method (Kumar et al., 2015). Zuorro et al. (2016) extracted lipid content from the microalga *Nannochloropsis* sp. with the help of two enzymes, namely cellulase and mannanase. For this extraction, the pH was kept at 4, the temperature was maintained at 53°C, pre-treatment was given for about 210 minutes and cellulase dosage was 13.8 mg g^{-1} whereas that of mannanase was 1.5 mg g^{-1}.

10.7 CONCLUSIONS

In the era of high population, pollution of freshwater bodies and energy crises are posing serious threats to human civilization. Advancement in science research and technology suggests the utilization of environmentally friendly feedstocks. Industrial, agricultural, and municipal waste is a major concern and is being treated using numerous conventional technologies that are energy-intensive and costly in nature. Microalgae, being cosmopolitan in their distribution in nature, are available worldwide. Algae have been proved to be potential candidates for bioremediation and subsequent utilization for biofuel production due to their high growth rate and flexible growth conditions. A combination of open raceway ponds and airlift tubular photobioreactors can be used for large-scale operations in algal cultivation, leading to low chances of contamination and high biomass yield. However, more research should be carried in-depth to isolate potential algal strains with improved remediation capabilities and high lipid and biomass content. Overall, microalgae have emerged as a sustainable replacement for conventional wastewater treatment technologies and as an alternative to fossil fuels.

ABBREVIATIONS

CPCB, Central Pollution Control Board; BOD, biological oxygen demand; COD, chemical oxygen demand; TFF, tangential flow filtration

REFERENCES

Abraham, J., and Sonil Nanda. "Evaluation of textile effluents before and after treatment with cyanobacteria." *Journal of Industrial Pollution Control* 26 (2010): 149–152.

Acuner, E., and F. B. Dilek. "Treatment of tectilon yellow 2G by *Chlorella vulgaris*." *Process Biochemistry* 39, no. 5 (2004): 623–631.

Adeyeye, Emmanuel I. "Determination of trace heavy metals in Illisha africana fish and in associated water and soil sediments from some fish ponds." *International Journal of Environmental Studies* 45, no. 3–4 (1994): 231–238.

Agrawal, K. and Verma, P. "Biodegradation of synthetic dye Alizarin Cyanine Green by yellow laccase producing strain *Stropharia* sp. ITCC-8422." *Biocatalysis and Agricultural Biotechnology* 21 (2019): 101291.

Agrawal, K. and Verma, P. "Degradation and detoxification of waste via bioremediation: a step toward sustainable environment". In *Emerging Technologies in Environmental Bioremediation*, (2020). (pp. 67–83). Elsevier.

Agrawal, K. and Verma, P. ""Omics"—a step toward understanding of complex diversity of the microbial community". In *Wastewater Treatment Cutting Edge Molecular Tools, Techniques and Applied Aspects*, (2021a). (pp. 471–487). Elsevier.

Agrawal, K. and Verma, P. "Metagenomics: a possible solution for uncovering the "Mystery Box" of microbial communities involved in the treatment of wastewater". In *Wastewater Treatment Cutting Edge Molecular Tools, Techniques and Applied Aspects*, (2021b). (pp. 41–53). Elsevier.

Agrawal, K., Bhatt, A., Bhardwaj, N., Kumar, B. and Verma, P. "Integrated approach for the treatment of industrial effluent by physico-chemical and microbiological process for sustainable environment". In *Combined Application of Physico-Chemical & Microbiological Processes for Industrial Effluent Treatment Plant*, (2020a). (pp. 119–143). Springer, Singapore.

Agrawal, K., Bhatt, A., Chaturvedi, V. and Verma, P. "Bioremediation: an effective technology toward a sustainable environment via the remediation of emerging environmental pollutants". In *Emerging Technologies in Environmental Bioremediation*, (2020b). (pp. 165–196). Elsevier.

Agrawal, K., Bhatt, A., Bhardwaj, N., Kumar, B., and Verma, P. "Algal biomass: potential renewable feedstock for biofuels production–part I". In *Biofuel Production Technologies: Critical Analysis for Sustainability*, (2020c). (pp. 203–237). Springer, Singapore.

Aksu, Zümriye, and Sevilay Tezer. "Biosorption of reactive dyes on the green alga Chlorella vulgaris." *Process Biochemistry* 40, no. 3–4 (2005): 1347–1361.

Aksu, Zümriye, and Tülin Kutsal. "A bioseparation process for removing lead (II) ions from waste water by using C. vulgaris." *Journal of Chemical Technology & Biotechnology* 52, no. 1 (1991): 109–118.

Ansari, R., and Z. Mosayebzadeh. "Removal of basic dye methylene blue from aqueous solutions using sawdust and sawdust coated with polypyrrole." *Journal of the Iranian Chemical Society* 7, no. 2 (2010): 339–350.

Asaolu, S. S., K. O. Ipinmoroti, C. E. Adeyinowo, and O. Olaofe. "Interrelationship of heavy metals concentration in water, sediment as fish samples from Ondo State coastal Area, Nig. Afr." *Journal of Science* 1 (1997): 55–61.

Aung W. L., Aye K. N., Hlain, N. N. *Biosorption of Lead (Pb21) by Using Chlorella Vulgaris.* International Conference on Chemical Engineering and its Applications. (2012). Bangkok, Thailand.

Axelsson, Martin, and Francesco Gentili. "A single-step method for rapid extraction of total lipids from green microalgae." *PLOS ONE* 9, no. 2 (2014): e89643.

Azarpira, Hossein, Pajman Behdarvand, Kondiram Dhumal, and Gorakh Pondhe. "Wastewater remediation by using Azolla and Lemna for selective removal of mineral nutrients." *International Journal of Biological Sciences* 4 (2014): 66–73.

Azhar, Afriyanti, Abdi Dharma, Nasril Nasir, and Zulkarnain Chaidir. "Two stage feeding evaluation of cattle waste medium use in *Chlorella vulgaris* culture and *Chlorella pyrenoidosa* culture for simultaneous production of biomass." *Journal of Physics: Conference Series* 1175, no. 1 (2019): 012011.

Banerjee, Chiranjib, Sandipta Ghosh, Gautam Sen, Sumit Mishra, Pratyoosh Shukla, and Rajib Bandopadhyay. "Study of algal biomass harvesting through cationic cassia gum, a natural plant based biopolymer." *Bioresource Technology* 151 (2014): 6–11.

Bechtold, Thomas, A. Turcanu, Erika Ganglberger, and Susanne Geissler. "Natural dyes in modern textile dyehouses—how to combine experiences of two centuries to meet the demands of the future?." *Journal of Cleaner Production* 11, no. 5 (2003): 499–509.

Bhardwaj, N., Agrawal, K., and Verma, P. "Algal biofuels: an economic and effective alternative of fossil fuels". In *Microbial Strategies for Techno-Economic Biofuel Production*, (2020). (pp. 59–83). Springer.

Bhatt, Neha Chamoli, Amit Panwar, Tara Singh Bisht, and Sushma Tamta. "Coupling of algal biofuel production with wastewater." *The Scientific World Journal* 2014 (2014): 210504.

Bligh, E. Graham, and W. Justin Dyer. "A rapid method of total lipid extraction and purification." *Canadian Journal of Biochemistry and Physiology* 37, no. 8 (1959): 911–917.

Bond, Richard G., and Conrad P. Straub, eds. *Handbook of Environmental Control: Waste Water: Treatment and Disposal.* (1974).

Brennan, Liam, and Philip Owende. "Biofuels from microalgae—a review of technologies for production, processing, and extractions of biofuels and co-products." *Renewable and Sustainable Energy Reviews* 14, no. 2 (2010): 557–577.

Cairns Jr, John, and Kenneth L. Dickson. "A simple method for the biological assessment of the effects of waste discharges on aquatic bottom-dwelling organisms." *Journal (Water Pollution Control Federation)* 43 (1971): 755–772.

Central Pollution Control Board (CPCB), Government of India, 2011. <http://cpcb.nic.in/GeneralStandards.pdf>.

Cepoi, Liliana, Inga Zinicovscaia, Ludmila Rudi, Tatiana Chiriac, Vera Miscu, Svetlana Djur, Ludmila Strelkova, and Dmitrii Grozdov. "S pirulina platensis as renewable accumulator for heavy metals accumulation from multi-element synthetic effluents." *Environmental Science and Pollution Research* 27 (2020): 1–19.

Chan, Alison, Hamidreza Salsali, and Ed McBean. "Heavy metal removal (copper and zinc) in secondary effluent from wastewater treatment plants by microalgae." *ACS Sustainable Chemistry & Engineering* 2, no. 2 (2014): 130–137.

Chaturvedi, V., Bhange, K., Bhatt, R., and Verma, P. "Biodetoxification of high amounts of malachite green by a multifunctional strain of *Pseudomonas mendocina* and its ability to metabolize dye adsorbed chicken feathers". *Journal of Environmental Chemical Engineering* 1, no. 4 (2013a): 1205–1213.

Chaturvedi, V., Chandravanshi, M., Rahangdale, M., and Verma, P. An integrated approach of using polystyrene foam as an attachment system for growth of mixed culture of cyanobacteria with concomitant treatment of copper mine waste water. *Journal of Waste Management* 2013 (2013b): 282798.

Chen, H., & Pan, S. S. (2005). Bioremediation potential of spirulina: toxicity and biosorption studies of lead. *Journal of Zhejiang University. Science. B*, *6*(3), 171.

Chia, Mathias Ahii, Ojone Anne Odoh, and Zakari Ladan. "The indigo blue dye decolorization potential of immobilized Scenedesmus quadricauda." *Water, Air, & Soil Pollution* 225, no. 4 (2014): 1–9.

Christie, Robert. *Colour Chemistry*. (2014). Royal Society of Chemistry.

Cohen, Yehuda. "Bioremediation of oil by marine microbial mats." *International Microbiology* 5, no. 4 (2002): 189–193.

Dasgupta, Chitralekha Nag, M. R. Suseela, S. K. Mandotra, Pankaj Kumar, Manish K. Pandey, Kiran Toppo, and J. A. Lone. "Dual uses of microalgal biomass: an integrative approach for biohydrogen and biodiesel production." *Applied Energy* 146 (2015): 202–208.

Dixit, Ruchita, Deepti Malaviya, Kuppusamy Pandiyan, Udai B. Singh, Asha Sahu, Renu Shukla, Bhanu P. Singh et al. "Bioremediation of heavy metals from soil and aquatic environment: an overview of principles and criteria of fundamental processes." *Sustainability* 7, no. 2 (2015): 2189–2212.

Dubey, Sanjay Kumar, Jaishree Dubey, Sandeep Mehra, Pradeep Tiwari, and A. J. Bishwas. "Potential use of cyanobacterial species in bioremediation of industrial effluents." *African Journal of Biotechnology* 10, no. 7 (2011): 1125–1132.

Dwivedi, Seema. "Bioremediation of heavy metal by algae: current and future perspective." *Journal of Advanced Laboratory Research in Biology* 3, no. 3 (2012): 195–199.

Ellis, Joshua T., Neal N. Hengge, Ronald C. Sims, and Charles D. Miller. "Acetone, butanol, and ethanol production from wastewater algae." *Bioresource Technology* 111 (2012): 491–495.

El-Sheekh, Mostafa M., Ghada W. Abou-El-Souod, and Hayam A. El Asrag. "Biodegradation of some dyes by the cyanobacteria species Pseudoanabaena sp. and Microcystis aeruginosa Kützing." *Egyptian Journal of Experimental Biology (Botany)* 13, no. 2 (2017): 233–243.

Faried, M., M. Samer, E. Abdelsalam, R. S. Yousef, Y. A. Attia, and A. S. Ali. "Biodiesel production from microalgae: processes, technologies and recent advancements." *Renewable and Sustainable Energy Reviews* 79 (2017): 893–913.

Feng, Pingzhong, Zhongyang Deng, Zhengyu Hu, and Lu Fan. "Lipid accumulation and growth of Chlorella zofingiensis in flat plate photobioreactors outdoors." *Bioresource Technology* 102, no. 22 (2011): 10577–10584.

Folch, Jordi, Mark Lees, and G. H. Sloane Stanley. "A simple method for the isolation and purification of total lipides from animal tissues." *Journal of Biological Chemistry* 226, no. 1 (1957): 497–509.

Fourest, Eric, and Bohumil Volesky. "Contribution of sulfonate groups and alginate to heavy metal biosorption by the dry biomass of Sargassum fluitans." *Environmental Science & Technology* 30, no. 1 (1996): 277–282.

Glombitza, Franz, and Susan Reichel. "Metal-containing residues from industry and in the environment: Geobiotechnological urban mining." *Advances in Biochemical Engineering/Biotechnology* 141 (2014): 49–107.

González, L. E., Cañizares, R. O., & Baena, S. (1997). Efficiency of ammonia and phosphorus removal from a Colombian agroindustrial wastewater by the microalgae Chlorella vulgaris and Scenedesmus dimorphus. *Bioresource technology*, *60*(3), 259–262.

Goswami, R. K., Mehariya, S., Verma, P., Lavecchia, R., and Zuorro, A. "Microalgae-based biorefineries for sustainable resource recovery from wastewater." *Journal of Water Process Engineering* 40 (2021a): 101747.

Goswami, R. K., Agrawal, K., and Verma, P. "An overview of microalgal carotenoids: advances in the production and its impact on sustainable development." In *Bioenergy Research: Evaluating Strategies for Commercialization and Sustainability*, (2021b). (pp. 105–128). John Wiley & Sons.

Goswami, R. K., Mehariya, S., Karthikeyan, O. P. K., and Verma, P. "Advanced microalgae-based renewable biohydrogen production systems: a review." *Bioresource Technology* 320, A (2021c): 124301.

Grima, E. Molina, E.-H. Belarbi, F.G. AciénFernández, A. Robles Medina, and Yusuf Chisti. "Recovery of microalgal biomass and metabolites: process options and economics." *Biotechnology Advances* 20, no. 7–8 (2003): 491–515.

Hara, Atsushi, and Norman S. Radin. "Lipid extraction of tissues with a low-toxicity solvent." *Analytical Biochemistry* 90, no. 1 (1978): 420–426.

Huang, GuanHua, Feng Chen, Dong Wei, XueWu Zhang, and Gu Chen. "Biodiesel production by microalgal biotechnology." *Applied Energy* 87, no. 1 (2010): 38–46.

Imani, Saber, Saeed Rezaei-Zarchi, Mehrdad Hashemi, Hojjat Borna, Amaneh Javid, and Hossein Bari Abarghouei. "Hg, Cd and Pb heavy metal bioremediation by Dunaliella alga." *Journal of Medicinal Plants Research* 5, no. 13 (2011): 2775–2780.

Ipinmoroti, K. O. "Determination of trace metals in fish, associated waters and soil sediments from fish ponds." *Discovery and Innovation* 5, no. 2 (1993): 135–138.

Iverson, Sara J., Shelley L. C. Lang, and Margaret H. Cooper. "Comparison of the Bligh and Dyer and Folch methods for total lipid determination in a broad range of marine tissue." *Lipids* 36, no. 11 (2001): 1283–1287.

Jensen, Søren K. "Improved Bligh and Dyer extraction procedure." *Lipid Technology* 20, no. 12 (2008): 280–281.

Jorquera, Orlando, Asher Kiperstok, Emerson A. Sales, Marcelo Embiruçu, and Maria L. Ghirardi. "Comparative energy life-cycle analyses of microalgal biomass production in open ponds and photobioreactors." *Bioresource Technology* 101, no. 4 (2010): 1406–1413.

Khataee, Ali R., Mahmoud Zarei, and Minoo Pourhassan. "Bioremediation of malachite green from contaminated water by three microalgae: neural network modeling." *Clean–Soil, Air, Water* 38, no. 1 (2010): 96–103.

Kiran, Bala, Ritunesh Kumar, and Devendra Deshmukh. "Perspectives of microalgal biofuels as a renewable source of energy." *Energy Conversion and Management* 88 (2014): 1228–1244.

Kivaisi, Amelia K. "The potential for constructed wetlands for wastewater treatment and reuse in developing countries: a review." *Ecological Engineering* 16, no. 4 (2001): 545–560.

Kshirsagar, Ayodhya D. "Bioremediation of wastewater by using microalgae: an experimental study." *International Journal of Life Science Biotechnology and Pharma Research* 2, no. 3 (2013): 339–346.

Kumar, B., Agrawal, K., Bhardwaj, N., Chaturvedi, V., and Verma, P. Advances in concurrent bioelectricity generation and bioremediation through microbial fuel cells. In *Microbial Fuel Cell Technology for Bioelectricity*, (2018). (pp. 211–239). Springer, Cham.

Kumar, B., Agrawal, K., Bhardwaj, N., Chaturvedi, V., and Verma, P. "Tech-no-economic assessment of microbe-assisted wastewater treatment strategies for energy and value-added product recovery". In *Microbial Technology for the Welfare of Society*, (2019a). (pp. 147–181). Springer.

Kumar, B., and Verma, P. "Techno-economic assessment of biomass-based integrated biorefinery for energy and value-added product". In *Biorefineries: A Step Towards Renewable and Clean Energy*, (2021). (pp. 581–616). Springer.

Kumar, Sunil, Amrik Singh Ahluwalia, and Mayank Uday Charaya. "Adsorption of Orange-G dye by the dried powdered biomass of Chlorella vulgaris Beijerinck." *Current Science* 116, no. 4 (2019b): 604.

Kumar, Jatinder, Shahanshah Khan, S. K. Mandotra, Priyanka Dhar, Amol B. Tayade, Sheetal Verma, Kiran Toppo, Rajesh Arora, Dalip K. Upreti, and Om P. Chaurasia. "Nutraceutical profile and evidence of alleviation of oxidative stress by Spirogyra porticalis (Muell.) Cleve inhabiting the high altitude Trans-Himalayan Region." *Scientific Reports* 9, no. 1 (2019c): 1–13.

Kumar, J. I. Nirmal, and Cini Oommen. "Removal of heavy metals by biosorption using freshwater alga Spirogyra hyalina." *Journal of Environmental Biology* 33, no. 1 (2012): 27.

Kumar, Pankaj, Sachin Kumar Mandotra, M. R. Suseela, Kiran Toppo, and Pushpa Joshi. "Characterization and transesterification of fresh water microalgal oil." *Energy Sources, Part A: Recovery, Utilization, and Environmental Effects* 38, no. 6 (2016): 857–864.

Kumar, Ranjith, Ramanathan Polur, Hanumantha Rao, and Muthu Arumugam. "Lipid extraction methods from microalgae: a comprehensive review." *Frontiers in Energy Research* 2 (2015): 61.

Kumari, S. Binu, A. Kavitha Kirubavathy, and Rajammal Thirumalnesan. "Suitability and water quality criteria of an open drainage municipal sewage water at Coimbatore, used for irrigation." *Journal of Environmental Biology* 27, no. 4 (2006): 709–712.

Lee, Andrew K., David M. Lewis, and Peter J. Ashman. "Disruption of microalgal cells for the extraction of lipids for biofuels: processes and specific energy requirements." *Biomass and Bioenergy* 46 (2012): 89–101.

Lee, Yuan-Kun. "Microalgal mass culture systems and methods: their limitation and potential." *Journal of Applied Phycology* 13, no. 4 (2001): 307–315.

Lesmana, Sisca O., Novie Febriana, Felycia E. Soetaredjo, Jaka Sunarso, and Suryadi Ismadji. "Studies on potential applications of biomass for the separation of heavy metals from water and wastewater." *Biochemical Engineering Journal* 44, no. 1 (2009): 19–41.

Li, Tong, Marc Strous, and Michael Melkonian. "Biofilm-based photobioreactors: their design and improving productivity through efficient supply of dissolved inorganic carbon." *FEMS Microbiology Letters* 364, no. 24 (2017): fnx218.

Lim, Sing-Lai, Wan-Loy Chu, and Siew-Moi Phang. "Use of Chlorella vulgaris for bioremediation of textile wastewater." *Bioresource Technology* 101, no. 19 (2010): 7314–7322.

Lizzul, A. M., P. Hellier, S. Purton, F. Baganz, N. Ladommatos, and L. Campos. "Combined remediation and lipid production using Chlorella sorokiniana grown on wastewater and exhaust gases." *Bioresource Technology* 151 (2014): 12–18.

Mahalakshmi, S., D. Lakshmi, and U. Menaga. "Biodegradation of different concentration of dye (Congo red dye) by using green and blue green algae." *International Journal of Environmental Research* 9, no. 2 (2015): 735–744.

Mandotra, S. K., A. S. Ahluwalia, and P. W. Ramteke. "Production of high-quality biodiesel by Scenedesmus abundans". In *The Role of Microalgae in Wastewater Treatment*, (2019). (pp. 189–198). Springer, Singapore.

Mandotra, S. K., Afreen J. Lolu, Sunil Kumar, P. W. Ramteke, and Amrik S. Ahluwalia. "Integrated approach for bioremediation and biofuel production using algae". In *Restoration of Wetland Ecosystem: A Trajectory Towards a Sustainable Environment*, (2020). (pp. 145–160). Springer, Singapore.

Mandotra, S. K., Chitra Sharma, N. Srivastava, A. S. Ahluwalia, and P. W. Ramteke. "Current prospects and future developments in algal bio-hydrogen production: a review." *Biomass Conversion and Biorefinery* (2021): 1–18.

Mandotra, S. K., Pankaj Kumar, M. R. Suseela, and P. W. Ramteke. "Fresh water green microalga Scenedesmus abundans: a potential feedstock for high quality biodiesel production." *Bioresource Technology* 156 (2014): 42–47.

Mandotra, S. K., Pankaj Kumar, M. R. Suseela, S. Nayaka, and P. W. Ramteke. "Evaluation of fatty acid profile and biodiesel properties of microalga Scenedesmus abundans under the influence of phosphorus, pH and light intensities." *Bioresource Technology* 201 (2016): 222–229.

Maurya, Rahulkumar, Tonmoy Ghosh, Chetan Paliwal, Anupama Shrivastav, Kaumeel Chokshi, Imran Pancha, Arup Ghosh, and Sandhya Mishra. "Biosorption of methylene blue by de-oiled algal biomass: equilibrium, kinetics and artificial neural network modelling." *PLOS ONE* 9, no. 10 (2014): e109545.

Mehariya, S., Kumar, P., Marino, T., Casella, P., Lovine, A., Verma, P., Musuma, and Molino, Antonio. "Aquatic weeds: a potential pollutant removing agent from wastewater and polluted soil and valuable biofuel feedstock". In *Bioremediation Using Weeds*, (2021a). (pp 59–77). Springer.

Mehariya, S., Goswami, R., Verma, P., Lavecchia, R., and Zuorro, A. "Integrated approach for wastewater treatment and biofuel production in microalgae biorefineries." *Energies* 14, no. 8 (2021b): 2282.

Mehariya, S., Goswami, R. K., Karthikeysan, O. P., and Verma, P. Microalgae for high-value products: a way towards green nutraceutical and pharmaceutical compounds. *Chemosphere* (2021c). 130553.

Mehta, S. K., and J. P. Gaur. "Use of algae for removing heavy metal ions from wastewater: progress and prospects." *Critical Reviews in Biotechnology* 25, no. 3 (2005): 113–152.

Mohn, F. H. "Experiences and strategies in the recovery of biomass from mass cultures of microalgae." In *Algae Biomass: Production and Use/[Sponsored by the National Council for Research and Development, Israel and the Gesellschaft fur Strahlen-und Umweltforschung (GSF)*, (1980). Munich, Germany]; editors, Gedaliah Shelef, Carl J. Soeder.

Molina, Emilio, J. Fernández, F. G. Acién, and Yusuf Chisti. "Tubular photobioreactor design for algal cultures." *Journal of Biotechnology* 92, no. 2 (2001): 113–131.

Moore, P. D. "Essential elements from waste." *Nature* 333 (1998): 706.

Morrison, G., O. S. Fatoki, L. Persson, and A. Ekberg. "Assessment of the impact of point source pollution from the Keiskammahoek sewage treatment plant on the Keiskamma River-pH, electrical conductivity, oxygen-demanding substance (COD) and nutrients." *Water SA* 27, no. 4 (2001): 475–480.

Munoz, Raul, and Benoit Guieysse. "Algal–bacterial processes for the treatment of hazardous contaminants: a review." *Water Research* 40, no. 15 (2006): 2799–2815.

Nigam, Poonam, G. Armour, Ibrahim M. Banat, Dave Singh, and Roger Marchant. "Physical removal of textile dyes from effluents and solid-state fermentation of dye-adsorbed agricultural residues." *Bioresource technology* 72, no. 3 (2000): 219–226.

Oh, Hee-Mock, Seog June Lee, Myung-Hwan Park, Hee-Sik Kim, Hyoung-Chin Kim, Jung-Hoon Yoon, Gi-Seok Kwon, and Byung-Dae Yoon. "Harvesting of Chlorella vulgaris using a bioflocculant from Paenibacillus sp. AM49." *Biotechnology Letters* 23, no. 15 (2001): 1229–1234.

Oluduro, A. O., and B. I. Adewoye. "Efficiency of *Moringaoleifera* seed extract on the microflora of surface and ground water." *Journal of Plant Sciences* 6 (2007): 453–438.

Pangestuti, Ratih, and Se-Kwon Kim. "Biological activities and health benefit effects of natural pigments derived from marine algae." *Journal of Functional Foods* 3, no. 4 (2011): 255–266.

Panigrahi, Satyanarayan, Parthiban Velraj, and Toleti Subba Rao. "Functional microbial diversity in contaminated environment and application in bioremediation." In *Microbial Diversity in the Genomic Era*, (2019). (pp. 359–385). Academic Press.

Pathak, V. V., D. P. Singh, Richa Kothari, and A. K. Chopra. "Phycoremediation of textile wastewater by unicellular microalga Chlorella pyrenoidosa." *Cellular and Molecular Biology* 60, no. 5 (2014): 35–40.

Pragya, Namita, Krishan K. Pandey, and P. K. Sahoo. "A review on harvesting, oil extraction and biofuels production technologies from microalgae." *Renewable and Sustainable Energy Reviews* 24 (2013): 159–171.

Raghukumar, C., V. Vipparty, J. David, and D. Chandramohan. "Degradation of crude oil by marine cyanobacteria." *Applied Microbiology and Biotechnology* 57, no. 3 (2001): 433–436.

Rai, L. C., and N. Mallick. "Removal and assessment of toxicity of Cu and Fe to Anabaena doliolum and Chlorella vulgaris using free and immobilized cells." *World Journal of Microbiology and Biotechnology* 8, no. 2 (1992): 110–114.

Rana, L., S. Chhikara, and R. Dhankar. "Assessment of growth rate of indigenous cyanobacteria in metal enriched culture medium." *Asian Journal of Experimental Biology* 4, no. 3 (2013): 465–471.

Renuka, N., A. Sood, R. Prasanna, and A. S. Ahluwalia. "Phycoremediation of wastewaters: a synergistic approach using microalgae for bioremediation and biomass generation." *International Journal of Environmental Science and Technology* 12 (2015): 1443–1460.

Renuka, Nirmal, Radha Prasanna, Anjuli Sood, Amrik S. Ahluwalia, Radhika Bansal, Santosh Babu, Rajendra Singh, Yashbir S. Shivay, and Lata Nain. "Exploring the efficacy of wastewater-grown microalgal biomass as a biofertilizer for wheat." *Environmental Science and Pollution Research* 23, no. 7 (2016): 6608–6620.

Richmond, Amos, Zhang Cheng-Wu, and Yair Zarmi. "Efficient use of strong light for high photosynthetic productivity: interrelationships between the optical path, the optimal population density and cell-growth inhibition." *Biomolecular Engineering* 20, no. 4–6 (2003): 229–236.

Saini, K. M., Yadav, D. S., Mehariya, S., Rathore, P., Kumar, B., Marino, T., Leone, G. P., Verma, P., Musmarra, D., Molino, A. "Overview of extraction of astaxanthin from Haematococcus pluvialis using CO_2 supercritical fluid extraction technology vis-a-vis quality demands". In *Global Perspectives on Astaxanthin from Industrial Production to Food, Health, and Pharmaceutical Applications*, (2021). (pp: 341–354). Academic Press. Elsevier.

Sarwa, Prakash, and Sanjay Kumar Verma. "Decolourization of Orange G Dye by microalgae Acutodesmusobliquues strain PSV2 isolated from textile industrial site." *International Journal of Applied Sciences and Biotechnology* 1, no. 4 (2013): 247–252.

Satpati, Gour Gopal, and Ruma Pal. "Microalgae-biomass to biodiesel: a review." *Journal of Algal Biomass Utilization* 9, no. 4 (2018): 11–37.

Sharma, Chitra, Sunil Kumar, Nitika Bhardwaj, S. K. Mandotra, and A. S. Ahluwalia. "Mitigation of heavy metals utilizing algae and its subsequent utilization for sustainable fuels". In *Algae*, (2021). (pp. 41–62). Springer, Singapore.

Sharma, Gulshan Kumar, and Shakeel Ahmad Khan. "Bioremediation of sewage wastewater using selective algae for manure production." *International Journal of Environmental Engineering and Management* 4, no. 6 (2013): 573–580.

Sharma, Pramod K., David L. Balkwill, Anatoly Frenkel, and Murthy A. Vairavamurthy. "A new Klebsiellaplanticola strain (Cd-1) grows anaerobically at high cadmium concentrations and precipitates cadmium sulfide." *Applied and Environmental Microbiology* 66, no. 7 (2000): 3083–3087.

Sim, T-S., A. Goh, and E. W. Becker. "Comparison of centrifugation, dissolved air flotation and drum filtration techniques for harvesting sewage-grown algae." *Biomass* 16, no. 1 (1988): 51–62.

Singh, Jivan, N. S. Mishra, Sushmita Banerjee, and Yogesh C. Sharma. "Comparative studies of physical characteristics of raw and modified sawdust for their use as adsorbents for removal of acid dye." *Bioresources* 6, no. 3 (2011): 2732–2743.

Singh, Pardeep, and Anwesha Borthakur. "A review on biodegradation and photocatalytic degradation of organic pollutants: a bibliometric and comparative analysis." *Journal of Cleaner Production* 196 (2018): 1669–1680.

Singh, Prabhakar Pratap, Manisha Mall, and Jaswant Singh. "Impact of fertilizer factory effluent on seed germination, seedling growth and chlorophyll content of gram (Cicer aeritenum)." *Journal of Environmental Biology* 27, no. 1 (2006): 153–156.

Singh, Vaishali, Archana Tiwari, and Moumita Das. "Phyco-remediation of industrial wastewater and flue gases with algal-diesel engenderment from micro-algae: a review." *Fuel* 173 (2016): 90–97.

Smith, Val H., G. David Tilman, and Jeffery C. Nekola. "Eutrophication: impacts of excess nutrient inputs on freshwater, marine, and terrestrial ecosystems." *Environmental Pollution* 100, no. 1–3 (1999): 179–196.

Suparmaniam, Uganeeswary, Man Kee Lam, Yoshimitsu Uemura, Jun Wei Lim, Keat Teong Lee, and Siew Hoong Shuit. "Insights into the microalgae cultivation technology and harvesting process for biofuel production: A review." *Renewable and Sustainable Energy Reviews* 115 (2019): 109361.

Suslick, Kenneth S., and David J. Flannigan. "Inside a collapsing bubble: sonoluminescence and the conditions during cavitation." *Annual Review of Physical Chemistry* 59 (2008): 659–683.

Tam, N. F. Y., Pui Sang Lau, and Yuk Shan Wong. "Wastewater inorganic N and P removal by immobilized Chlorella vulgaris." *Water Science and Technology* 30, no. 6 (1994): 369.

Tan, Kok Bing, Mohammadtaghi Vakili, Bahman Amini Horri, Phaik Eong Poh, Ahmad Zuhairi Abdullah, and Babak Salamatinia. "Adsorption of dyes by nanomaterials: recent developments and adsorption mechanisms." *Separation and Purification Technology* 150 (2015): 229–242.

Tan, Xin Bei, Man Kee Lam, Yoshimitsu Uemura, Jun Wei Lim, Chung Yiin Wong, and Keat Teong Lee. "Cultivation of microalgae for biodiesel production: a review on upstream and downstream processing." *Chinese Journal of Chemical Engineering* 26, no. 1 (2018): 17–30.

Tang, X., L. Y. He, X. Q. Tao, Z. Dang, C. L. Guo, G. N. Lu, and X. Y. Yi. "Construction of an artificial microalgal-bacterial consortium that efficiently degrades crude oil." *Journal of Hazardous Materials* 181, no. 1–3 (2010): 1158–1162.

Ting, Han, Lu Haifeng, Ma Shanshan, Yuanhui Zhang, Liu Zhidan, and Duan Na. "Progress in microalgae cultivation photobioreactors and applications in wastewater treatment: A review." *International Journal of Agricultural and Biological Engineering* 10, no. 1 (2017): 1–29.

Uduman, Nyomi, Ying Qi, Michael K. Danquah, Gareth M. Forde, and Andrew Hoadley. "Dewatering of microalgal cultures: a major bottleneck to algae-based fuels." *Journal of Renewable and Sustainable Energy* 2, no. 1 (2010): 012701.

Upadhyay, A. K., S. K. Mandotra, N. Kumar, N. K. Singh, Lav Singh, and U. N. Rai. "Augmentation of arsenic enhances lipid yield and defense responses in alga Nannochloropsis sp." *Bioresource Technology* 221 (2016): 430–437.

Upadhyay, Atul Kumar, and S. K. Mandotra. "Constructed wetland and microalgae: a revolutionary approach of bioremediation and sustainable energy production". In *Algae*, (2021). (pp. 27–40). Springer, Singapore.

Valderrama, Luz T., Claudia M. Del Campo, Claudia M. Rodriguez, Luz E. de-Bashan, and Yoav Bashan. "Treatment of recalcitrant wastewater from ethanol and citric acid production using the microalga Chlorella vulgaris and the macrophyte Lemnaminuscula." *Water Research* 36, no. 17 (2002): 4185–4192.

Vasistha, S., A. Khanra, M. Clifford, and M. P. Rai. "Current advances in microalgae harvesting and lipid extraction processes for improved biodiesel production: A review." *Renewable and Sustainable Energy Reviews* (2020): 110498.

Volesky, B. (1990). *Removal and recovery of heavy metals by biosorption* (pp. 3–43). Boca Ratón: CRC press.

Wase J., and Forster C. F. *Biosorbents for Metals Ions*. (1997). Taylor& Francis.

Wilde, Edward W., and John R. Benemann. "Bioremoval of heavy metals by the use of microalgae." *Biotechnology Advances* 11, no. 4 (1993): 781–812.

Yu, Qiming, Jose T. Matheickal, Pinghe Yin, and Pairat Kaewsarn. "Heavy metal uptake capacities of common marine macro algal biomass." *Water Research* 33, no. 6 (1999): 1534–1537.

Zhu, Liandong, and Tarja Ketola. "Microalgae production as a biofuel feedstock: risks and challenges." *International Journal of Sustainable Development & World Ecology* 19, no. 3 (2012): 268–274.

Zinicovscaia, I., & Cepoi, L. (Eds.). (2016). *Cyanobacteria for bioremediation of wastewaters*. Springer International Publishing.

Zuorro, Antonio, Gianluca Maffei, and Roberto Lavecchia. "Optimization of enzyme-assisted lipid extraction from *Nannochloropsis* microalgae." *Journal of the Taiwan Institute of Chemical Engineers* 67 (2016): 106–114.

Index

D

E

F

G

H

S

T

W